TURBULENT JETS AND PLUMES
- A Lagrangian Approach

TURBULENT JETS AND PLUMES - A LAGRANGIAN APPROACH

JOSEPH H. W. LEE
Department of Civil Engineering
The University of Hong Kong
Hong Kong, China

VINCENT H. CHU
Department of Civil Engineering and Applied Mechanics
McGill University
Montreal, Canada

Kluwer Academic Publishers
Boston/Dordrecht/London

Distributors for North, Central and South America:
Kluwer Academic Publishers
101 Philip Drive
Assinippi Park
Norwell, Massachusetts 02061 USA
Telephone (781) 871-6600
Fax (781) 681-9045
E-Mail: kluwer@wkap.com

Distributors for all other countries:
Kluwer Academic Publishers Group
Post Office Box 322
3300 AH Dordrecht, THE NETHERLANDS
Telephone 31 786 576 000
Fax 31 786 576 254
E-Mail: services@wkap.nl

Electronic Services <http://www.wkap.nl>

Library of Congress Cataloging-in-Publication Data

Lee, J. H. W. (Joseph H. W.)
Turbulent jets and plumes : a Lagrangian approach / by Joseph H.W. Lee, Vincent Chu.
p. cm.
Includes bibliographical references and index.
ISBN 1-4020-7520-0
1. Turbulence. 2. Jets--Fluid dynamics. 3. Plumes (Fluid dynamics) 4. Lagrange equations. I. Chu, Vincent, 1942- II. Title.
TA357.5.T87L44 2003
620.1'064--dc21 2003047599

Printed on acid-free paper.
Printed in the United States of America.

To Winnie and Alice

Contents

Preface

Jets and plumes are turbulent flows produced by momentum and buoyant forces. Examples include cooling tower and smokestack emissions, fires and volcano eruptions, deep sea vents, atmospheric thermals, marine outfall sewage discharges, thermal effluents from power stations and ocean dumping of sludge. The prediction of the path and mixing of jets and plumes is important for sound environmental control and impact assessment. The study of this important turbulent shear flow is a key to understanding a variety of environmental and industrial mixing problems.

Many technical papers have been written on this subject. For a variety of reasons, the newcomer often finds it difficult to sieve through the enormous literature on the subject. Most of the published literature is concerned with laboratory measurements near the source. Without proper interpretation, the laboratory results are not suitable for engineering application. The user manuals of mathematical models used in industrial practice also do not cover adequately the fundamentals of the basic principles. A great variety of turbulent closures, ranging from simple entrainment hypothesis to fully three-dimensional Reynolds-stress-transport models, have been introduced. Often the student fails to see the correct underlying physical concepts, and is sometimes misled by the common mistakes found in the literature.

This book is an outgrowth of two decades of research and teaching by the authors on the subject. The objective of this simple text is to introduce the fundamental concepts of the mixing of turbulent buoyant flows as well as incorporate recent advances. Jets and plumes are introduced as slender turbulent shear flows. We give a consistent and unified development of a Lagrangian approach for prediction of free jets and plumes, including effects of crossflow and ambient density stratification. The connection of Lagrangian and Eulerian formulations is discussed. Regardless of the detailed formulation, it is our view that the conceptual understanding of the process is greatly facilitated by a Lagrangian thinking or interpretation. This theme persists throughout the text, starting from simple cases and building towards the practically important case of a turbulent buoyant jet in a density-stratified crossflow. Basic ideas are illustrated by ample use of flow visualization using the laser-induced fluorescence technique. An extensively validated Lagrangian computer code for the computation of buoyant jets is also described. The treatment reflects the authors' belief that notwithstanding the rapid advances in numerical methods and power of modern computers, the integral approach plays a pivotal role in explaining important and practically useful concepts to the engineer, and that the subject can best be understood by observation conducted from simple experiments.

Besides developing the theoretical framework, the book will also address many recent experimental measurements and computational fluid dynamic (CFD) results. In particular, the measurement of the scalar concentration field has been greatly facilitated by the Laser-induced fluorescence and video imaging techniques. The robustness of the unified integral method is illustrated in a variety of complex engineering applications. A practical Lagrangian modelling tool in the form of an interactive virtual reality software, VISJET, is offered to the reader to apply the techniques to solve his or her own problems.

The text includes many worked examples, comparisons between model predictions with laboratory and field data, and classroom-tested problems. It is suitable for senior undergraduate or first year postgraduate students as part of a curriculum in Water and Environmental Engineering. The book also complements standard fluid mechanics textbooks in that it covers the fundamental environmental fluid mechanics concepts not normally covered in these texts - such as turbulent mixing, buoyancy and density stratification. The book can serve as an introduction to turbulent shear flow in a course on environmental fluid mechanics, and may be used as a reference by students in physics, mechanical and chemical engineering, earth and environmental sciences.

We would like to acknowledge the notable contributions and continual support of our past students and associates, in particular Reverend Dr Valiant Cheung, Dr Paul Chu, and Prof Chen Guoqian, who have worked productively and maintained an interest in this subject. We would also like to express our special appreciation for the stimulating exchanges and discussions with our colleagues Prof Gerhard Jirka and Prof Ian Wood. We thank Dr Wenping Wang for the development of advanced computer graphics and visualization tools in VISJET. The assistance of Dr C.P.

Kuang, Dr H.C. Chan, and Dr Daeyoung Yu on the preparation of the manuscript is well-appreciated.

The research embodied in this work has been supported at various times by the National Research Council of Canada, the Croucher Foundation, the Alexander von Humboldt Foundation, and the Hong Kong Research Grants Council. The VISJET development is supported by the Hong Kong Innovation and Technology (ITF) Fund. The writing of this book has been facilitated by a Universitaes 21 Fellowship from the University of Hong Kong. This book has been typeset using LaTeX 2ε.

JOSEPH HUN-WEI LEE AND VINCENT H. CHU

Chapter 1

INTRODUCTION

Turbulent jets and plumes are a class of turbulent flows produced by momentum and buoyancy sources. A jet is the flow produced by a continuous source of momentum (Figure 1.1) while a plume by a continuous source of buoyancy (Figure 1.2). The appearance of the jet is similar to that of the plume although the turbulent motions in the two flows are produced by different mechanisms. Mixing in the jet is directly related to the inertia of the turbulent eddies. In the plume, the buoyant force produces the inertia, which leads to mixing.

Besides a problem of basic interest in fluid mechanics, the study of jets and plumes has applications in many practical problems. Examples of jets and plumes can be readily found. A cigarette smoker observes (maybe unknowingly!) a plume break up into eddies many times a day. Smoke plumes from factories and fossil-fueled electric power stations carry undesirable products of combustion such as sulphur dioxide and particulates which can cause severe pollution under unfavourable meteorological conditions. The supercritical high-velocity flow in many hydraulic structures and flood protection works can be considered a jet. Fire plumes have also been extensively studied to improve the design of smoke detectors and sprinkler systems in buildings and atriums of large shopping plazas. In many coastal cities, pre-treated sewage are discharged as buoyant jets through submarine outfalls located on the sea bed. Similarly, large flows of condenser cooling water from steam-electric power generation are often discharged at an elevated temperature of $5 - 10^o$ C above the natural background value (thermal effluent) through this type of submerged outlets.

A good grasp of jets and plumes greatly facilitates the understanding of water quality control and quantitative environmental impact and risk

Figure 1.1. High Reynolds number turbulent jet produced by the test of a rocket (three million pound thrust) by Lockheed in the hills of Redland, California (from *Los Angeles*, Sunset-Lane, Menlo Park, California 1968. Courtesy of Lockheed Martin.)

Figure 1.2. Thermal plume generated by the spectacular fire that destroyed Barcelona's Gran Teatro del Liceo (from *Time*, February 14, 1994. Courtesy of Gamma-Imaginechina)

assessment. Environmental regulations and water quality objectives are mostly set in terms of pollutant concentrations in receiving waters. It is important to be able to predict the concentration distribution in the vicinity (the near field) of a discharge for a given discharge design and location, waste load, and environmental conditions. This is required for defining mixing zones and to minimize the impact of discharges on sensitive receivers (e.g. nearby beaches for amenities, wetland reserve, or fisheries). As we will see, the proper design of a submarine outfall — a hydraulic structure — can have a profound effect on the water quality actually observed near the discharge.

Jets and plumes are also related to many natural environmental processes. Hot "black smokers" released from deep sea hydrothermal vents, volcanic eruptions and forest fires, and sinking brine plumes due to freezing of openings on polar ice cap, are some examples. The local natural environment can also be altered by jets and plumes for the benefit of mankind – dissolved oxygen level in the hypolimnion of a lake can be raised through jet mixing with the oxygen rich upper layer waters. On the other hand, nutrient-rich bottom waters in some fjords can be effectively brought up to the surface layer by plume action to increase the productivity of fisheries.

Environmental discharges as turbulent jets and plumes form a class of 'active' designs for which the engineer can control the amount of mixing, and hence the degree of dilution, near the discharge point. Within this near field, the discharge itself generates the flow field and mixing. The velocity and concentration fields are hence coupled and have to be solved simultaneously. An important engineering parameter is the dilution. The dilution at a point is defined as the ratio of the discharge concentration (or concentration excess above the ambient) to the concentration (excess) at the point. This is a measure of the mixing capacity of the discharge.

Jets and plumes are often analysed as momentum and buoyancy sources. Since velocity is momentum per unit mass, any means of producing velocity in a fluid is a momentum source. Under the earth's gravitational attraction, the buoyancy force is associated with the density variation in the fluid (Archimedes' Principle). Many physical processes can lead to changes in fluid density and hence sources of buoyancy, e.g. heating by radiation and conduction, diffusion of solute, or simply the discharging of fluids and mixtures into a receiving environment of different density. With buoyancy, momentum would be generated at a rate equal to the buoyancy force (Newton's second law of motion). The Euler equation governing this momentum generation process is

$$\rho \frac{D\mathbf{u}}{dt} = -\nabla p + \rho g' \hat{\mathbf{k}} \tag{1.1}$$

where ρ = fluid density, $\mathbf{u}$ = velocity = momentum per unit mass of the fluid, ∇p = pressure gradient, $\rho g'$ = buoyancy force per unit volume of the fluid, $g' = g(\rho_a - \rho)/\rho$ = reduced gravity = buoyancy force per unit mass of the fluid (subscript a denotes reference ambient values), and $\hat{\mathbf{k}}$ = unit vector in the +z-direction (opposite to direction of gravity). The effect of viscosity is often ignored in the formulation of the turbulent jets and plumes at high Reynolds numbers.

The hot gas discharging from a rocket engine as shown in Figure 1.1 is a continuous momentum source. In this example, the momentum is generated by the rocket engine at a rate equal to the thrust of the rocket engine, which is three million pounds! Since the initial momentum flux is huge, the role of buoyancy in the momentum generation is insignificant. Figure 1.2 shows the turbulent motion of a plume of smoke generated by the fire that destroyed most of Barcelona's Gran Teatro del Liceo. The source of the buoyancy in this plume is the heat produced by the fire. The initial momentum flux of the fire on the ground is negligible. The subsequent generation of momentum in the plume is determined by the buoyancy flux of the plume. If both momentum and buoyancy are present at the source, momentum would be the dominant effect in the region near the source. Buoyancy becomes significant only when the momentum produced by the buoyancy force has exceeded the initial momentum at the source. Figure 1.3 is another example of turbulent motion produced by momentum and buoyancy. The steam, smoke and ash are generated by the eruption of Mount St. Helens. The sudden release of momentum has produced essentially a *puff*. The flow may eventually become like a *thermal* as buoyancy become the dominant effect far from the source. Puffs and thermals are produced by instantaneous sources of momentum and buoyancy respectively, while jets and plumes by continuous sources (see Table 1.1 and Example 1.2).

Motions of enormous magnitude can be generated by minute density differences if the buoyancy force acts on the fluid for a long period of time. Examples can be found in the atmosphere, lakes, and oceans where buoyancy forces act over a column of fluid of large vertical extent. The updraft produced by the action of thermals in the atmosphere is an example. The initial density difference due to uneven heating of the ground by solar radiation is quite small. However, the motion of the thermals produced by this density difference is massive; glider pilots

Table 1.1. Turbulent shear flows generated by continuous and instantaneous sources of momentum and buoyancy in a stationary environment of uniform density. M_o, F_o are the source momentum and buoyancy fluxes for a jet and plume respectively. M_{po}, B_o are the source momentum and buoyancy force for the puff and thermal respectively. D, d_o are the dimensions of the point and line sources, and w_o is the source velocity. z=vertical co-ordinate above source; t=time after release. The expressions for volume, V, or volume flux, Q, are deduced by dimensional analysis

Flows	Description	Source strength and dimensions	Volume flux or Volume
Round jet	Point source of momentum flux	$\frac{M_o}{\rho_o} = \frac{\pi}{4} D^2 w_o^2 \sim \frac{L^4}{T^2}$	$Q \sim z\sqrt{\frac{M_o}{\rho_o}}$
Plane jet	Line source of momentum flux	$\frac{M_o}{\rho_o} = d_o w_o^2 \sim \frac{L^3}{T^2}$	$Q \sim \sqrt{\frac{M_o z}{\rho_o}}$
Round puff	Instantaneous point source of momentum	$\frac{M_{po}}{\rho_o} = w_o \frac{\pi}{6} D^3 \sim \frac{L^4}{T}$	$V \sim [\frac{M_{po}}{\rho_o}]^{\frac{3}{4}} t^{\frac{3}{4}}$
Line puff	Instantaneous line source of momentum	$\frac{M_{po}}{\rho_o} = w_o \frac{\pi}{4} d_o^2 \sim \frac{L^3}{T}$	$V \sim [\frac{M_{po}}{\rho_o}]^{\frac{2}{3}} t^{\frac{2}{3}}$
Round plume	Point source of buoyancy flux	$\frac{F_o}{\rho_o} = g_o' \frac{\pi}{4} D^2 w_o \sim \frac{L^4}{T^3}$	$Q \sim [\frac{F_o}{\rho_o}]^{\frac{1}{3}} z^{\frac{5}{3}}$
Plane plume	Line source of buoyancy flux	$\frac{F_o}{\rho_o} = g_o' d_o w_o \sim \frac{L^3}{T^3}$	$Q \sim [\frac{F_o}{\rho_o}]^{\frac{1}{3}} z$
Round thermal	Instantaneous point source of buoyancy force	$\frac{B_o}{\rho_o} = g_o' \frac{\pi}{6} D^3 \sim \frac{L^4}{T^2}$	$V \sim [\frac{B_o}{\rho_o}]^{\frac{3}{4}} t^{\frac{3}{2}}$
Line thermal	Instantaneous line source of buoyancy force	$\frac{B_o}{\rho_o} = g_o' \frac{\pi}{4} d_o^2 \sim \frac{L^3}{T^2}$	$V \sim [\frac{B_o}{\rho_o}]^{\frac{2}{3}} t^{\frac{4}{3}}$

Figure 1.3. Mammoth fists of steam, smoke and ash - some thrusting up to 6000 m - rose from the peak of Mount St.Helens near Portland, Oregon in a volcanic eruption (from *Life*, May 1980. Courtesy of Associated Press)

depend on these to keep the aircraft afloat. Another example of utilizing the buoyancy force can be found in the offshore disposal of wastewater through a diffuser located at the floor of the ocean. The mixing of the effluent close to the diffuser is controlled by the action of the jet. With a 2-3 percent initial density difference between the wastewater (essentially

Figure 1.4. Thermal plume in a stratified cross wind. The plume from the Lamma Power Station, Hong Kong, is observed to be bent over by the crossflow and level off at an elevation where the density of the plume is the same as the cross wind (photo taken in August 1982)

freshwater) and the sea water, the dilution of the wastewater is chiefly determined by the action of the plume. The dilution of a wastewater effluent in large depth is, therefore, primarily a function of the buoyancy flux of the discharge.

Figure 1.4 shows a plume rise in a stably stratified atmosphere. In this more complex example, the plume is affected by the velocity of the cross wind and the density variation in the atmosphere. The plume is bent over by the crossflow as the momentum and buoyancy of the surrounding fluid is entrained into the plume. The entrainment or mixing process causes the momentum and buoyancy in the plume to approach ambient values; the plume eventually levels off as the relative buoyancy is reduced to zero. The development of the plume in this example is therefore a problem of a concentrated source of buoyancy being modified by the momentum and buoyancy of its surrounding environment.

Many predictive methods have been developed over past decades to study the turbulent mixing processes in jets and plumes. Experimental data have been accumulated for flows with different source and ambient conditions. It is the purpose of this book to provide a systematic review of these findings and the various aspects of the phenomena as momentum and buoyancy interaction problems. The book is divided into ten chapters. The basic concepts of turbulent mixing and entrainment and techniques of analysis are treated at length in Chapters 2 and 3 for straight jets and plumes. Inclined buoyant jets in stagnant environment are considered in Chapter 4 as a prelude to the general integral formulation, which includes the treatment of the initial potential core. The effect of ambient density stratification is examined in Chapter 5. In Chapters 6 to 8 we look at several important basic flows necessary for understanding the complicated and practically important problem of a turbulent buoyant jet in a stratified crossflow: coflowing jets, puffs and thermals. In Chapter 9, a general Lagrangian formulation of the jet in crossflow problem is discussed. In Chapter 10, an extensively validated Lagrangian model based on the projected area entrainment hypothesis is illustrated by a number of practical example applications.

Throughout this text we aim at developing a solid 1D analytical framework in analysing jet and plume problems, and demonstrate the advantages of using a Lagrangian approach. A general Lagrangian spreading hypothesis can be used in the formulation of turbulent entrainment, and this can be interpreted using the concept of dominant eddies. The Lagrangian method is also simpler and more convenient in its implementation because the important results can be obtained without actually specifying the velocity and concentration profiles, which are not generally known for complex flow situations. In many instances, more meaningful physical interpretation of the model predictions can be obtained using this approach.

Numerical examples and problems are provided in each chapter for the reader to acquire a quantitative appreciation of the material. Results of computer simulations are often included to illustrate the need for both mathematical models and order of magnitude estimates based on a physical understanding of the problem. An interactive virtual reality Lagrangian model, VISJET , is provided to assist the reader to obtain a better physical understanding of jets and plumes. We end this chapter with three examples to show how turbulent motions are generated by momentum and buoyancy forces, and how useful engineering estimates can be obtained using simple models without the need to consider the details of the turbulent motion.

EXAMPLE 1.1 <u>*Sand dumping from a dredger*</u>
A 100 kg of sand is dumped overboard from a dredging vessel into a lake. A sand-water mixture is formed as the cloud of sand mixes with its surrounding. The mixture is a heavy fluid element which descends down the water body by virtue of its buoyancy. Find the buoyancy force

associated with the thermal. Assuming that the radius of the sand cloud, R, spreads at a rate proportional to the vertical velocity, $w = dz/dt$,

$$\frac{dR}{dt} = \beta \frac{dz}{dt}, \tag{1.2}$$

find the velocity and radius of the sand cloud after it has reached a depth of 5 m below the water surface. Calculate the depth when the sand particles begin to separate from the cloud. Assume a spreading rate of $\beta = 0.34$ and an average fall velocity of w_s=5 cm/s for the sand particles (relative density 2.6).

SOLUTION: This is an example of a round thermal (an instantaneous source of buoyancy) in a stagnant environment. The heavier sand particles are the source of buoyancy which generates the turbulence and causes the sand to mix with the surrounding water.

The downward buoyancy force associated with the 100 kg of sand in water is the weight of the sand minus the weight of the displaced water (Archimedes Principle). The weight of the sand is $(\rho_s \mathcal{V}_o)g$ = 100kg $\times$ 9.81m/s^2. The weight of the displaced water is 2.6 times lighter. Hence, the buoyancy force

$$B = \rho_s g \mathcal{V}_o - \rho_w g \mathcal{V}_o = 100 \times 9.81 \times (1 - \frac{1}{2.6}) = \underline{603.7} \text{ N}$$

where ρ_s = density of sand, ρ_w = density of water, ρ_s/ρ_w = specific gravity of the sand $\simeq 2.6$, $\mathcal{V}_o$ = volume of water displaced by the sand, $\rho_s \mathcal{V}_o$ = mass of sand = 100 kg. As the thermal (cloud of sand-water mixture) sinks, it mixes with the surrounding water by turbulent entrainment; the radius of the thermal R increases with the distance z from the source. According to Eq. 1.2,

$$R = R_o + \beta z$$

(see Chapter 9 for the hypothesis leading to Eq. 1.2). The volume $\mathcal{V}$ and mass $\mathcal{M}$ of the thermal also increase correspondingly as (approximating the cloud as a sphere with characteristic radius R):

$$\mathcal{V} = \frac{4\pi}{3} R^3$$

$$\mathcal{M} = \mathcal{M}_o + \rho_w(\mathcal{V} - \mathcal{V}_o)$$

where R_o and $\mathcal{M}_o$ are the initial radius and initial mass of the thermal at the source, respectively. If the density of the surrounding water is uniform, the downward buoyancy force associated with the thermal (603.7 N in this example) would stay constant. The sand-water cloud

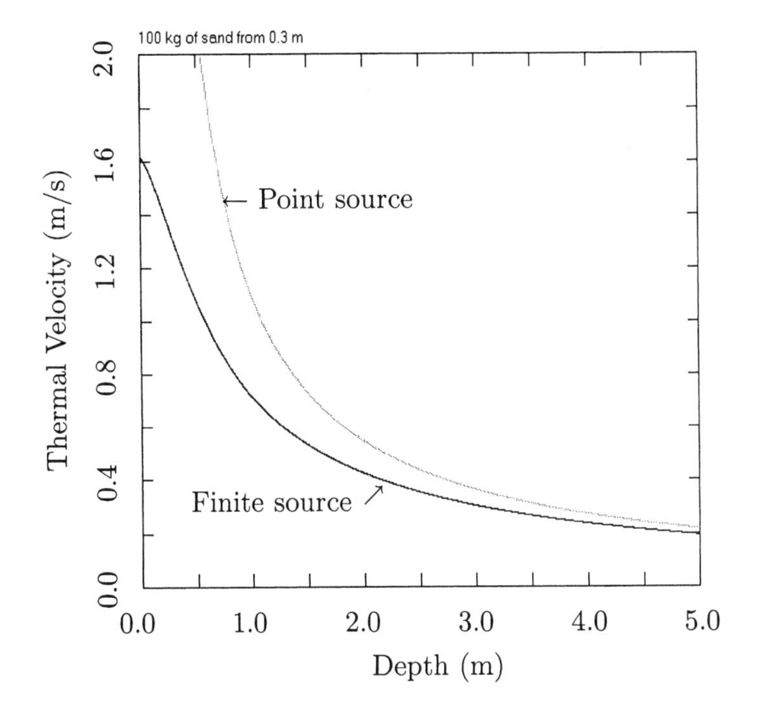

Figure 1.5. The thermal velocity produced by the dumping of 100 kg of sand at an elevation of 0.3 m above the water surface. Dashed line is the point source approximation.

(thermal) moves collectively with a velocity that is much greater than the sand particle fall velocity.

According to Newton's second law of motion, the momentum of the thermal, $(1+k)\mathcal{M}w$, increases with time in proportional to the buoyancy force, B, as follows (see also Eq. 5.23):

$$(1+k)\mathcal{M}w = Bt$$

where $w = dz/dt$ is downward velocity and k an *added mass* coefficient associated with the motion outside of the thermal. For a sphere accelerating in an irrotational fluid, $k = 0.5$. For large z, the initial mass $\mathcal{M}_o$ and initial volume $\mathcal{V}_o$ may be ignored as a first approximation. Therefore,

$$\mathcal{M} \simeq 4\rho_w \pi R^3/3 = 4\rho_w \pi \beta^3 z^3/3$$

and the momentum equation becomes

$$\frac{4\pi}{3}(1+k)\rho_w \beta^3 z^3 \frac{dz}{dt} = Bt.$$

Integrating this equation gives the trajectory of the thermal with respect to time, $z(t)$:

$$z = [\frac{3B}{2\pi\beta^3\rho_w(1+k)}]^{\frac{1}{4}} t^{\frac{1}{2}}.$$

The velocity (momentum per unit mass) of the thermal is then:

$$w = \frac{dz}{dt} = [\frac{3B}{32\pi\beta^3\rho_w(1+k)}]^{\frac{1}{4}} t^{-\frac{1}{2}} = [\frac{3B}{8\pi\beta^3\rho_w(1+k)}]^{\frac{1}{2}} z^{-1} \qquad (1.3)$$

At a depth of 5 m below the water surface, the sand-cloud mixture has a velocity of

$$w_{5\mathrm{m}} = \sqrt{\frac{3 \times 603.7}{8\pi \times 0.34^3 \times 1.5 \times 1000}} \frac{1}{5} = \underline{0.22} \ \mathrm{m/s}.$$

which is still four times larger than the fall velocity of the constituent sand particles. The corresponding radius of the cloud is $R = 0.34 \times 5 = \underline{1.7}\ m$, with a Reynolds number (based on the diameter) of $Re = Vd/\nu = 0.22 \times 3.4/10^{-6} = 7.5 \times 10^5$. Rather significant turbulent motions have been generated by the buoyancy. At this depth of 5 m, the sand and water stay together in the thermal as a mixture since the fall velocity $w_s \ll w$. The sand particles do not begin to separate from the thermal until the thermal velocity is reduced to a velocity comparable to the fall velocity at a depth of about 22 m below the water surface.

The above result is based on the 'point source' assumption that $\mathcal{V}_o$ and $\mathcal{M}_o$ are negligible. This assumption has led to the unrealistic infinite velocity at the point source (see Equation 1.3). The realistic solution as shown in Figure 1.5 is obtained by numerical integration of the equation of motion including the finite-source effects of $\mathcal{V}_o$ and $\mathcal{M}_o$ (Prob. 1.4).

EXAMPLE 1.2 <u>*Sand slurry plume in crossflow*</u>
A sand-water slurry is pumped continuously from a 0.4 m diameter jet nozzle at the bottom of a barge into a lake. The jet velocity is 3 m/s and the waste material has a relative density of 1.03. The discharge forms a plume in crossflow. A line thermal of negatively buoyant material is formed as the vessel is moving with a velocity of 0.5 m/s relative to the lake water. Find the buoyancy force per unit length of the line thermal. Assuming that the radius of thermal, R, spreads at a rate proportional to the vertical velocity, $w = dz/dt$,

$$\frac{dR}{dt} = \beta\frac{dz}{dt}, \qquad (1.4)$$

find the velocity and radius of the cloud after the waste has reached a depth of 20 m below the water surface. Assume a spreading rate of $\beta = 0.4$ and an average fall velocity of 5 cm/s for the sand particles.

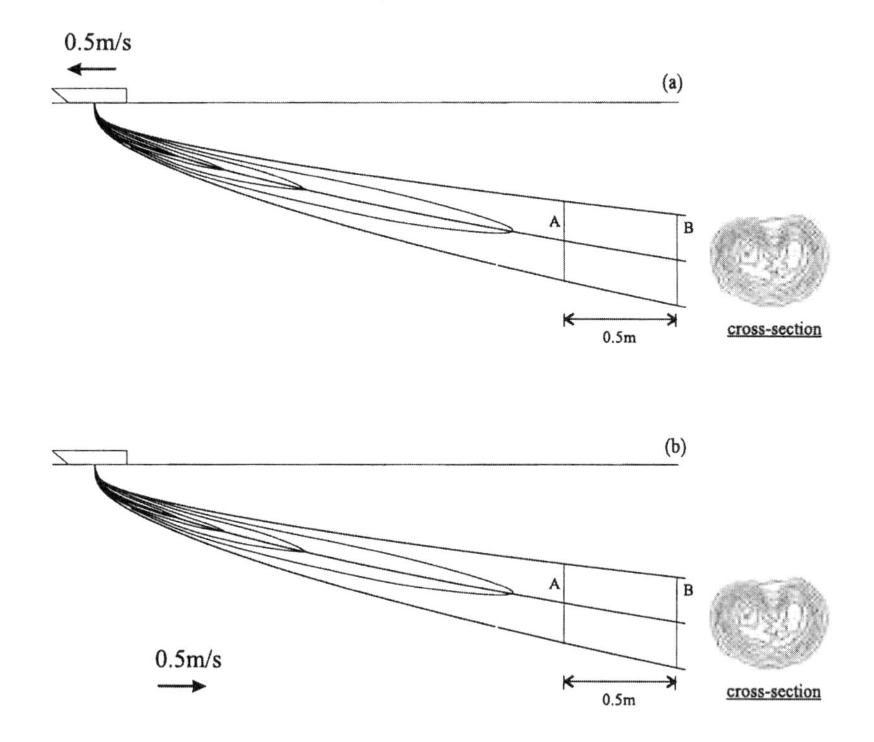

Figure 1.6. (a) Line thermal produced by discharging a sand-water slurry from a dredging vessel moving at a velocity of 0.5 m/s. (b) The equivalent plume-in-crossflow problem. The elemental volume AB is an element along the path of the buoyant jet produced by the buoyant flux at the source for a period of one second. A buoyant force of 222 N is associated with the heavy slurry in the elemental volume AB. The length of the volume is 0.5 m.

SOLUTION: The path of the waste plume from the moving vessel forms a line thermal as shown in figure 1.6a. Calculation of the thermal velocity is made on an elemental volume AB produced by the release of the heavy material over a period of one second. Over a period of one second, the vessel moves forward a distance of 0.5 m. The heavy material is distributed over this same distance to form the elemental volume. Therefore the length of the elemental volume is 0.5 m. The jet discharge flow is $Q_o = 3 \times \pi \times 0.4^2/4 = 0.378\ m^3/s$. The buoyancy flux is $F_o = (\rho_o - \rho_a)gQ_o = (1030 - 1000) \times 9.81 \times 0.38 = 111$ N/s. Therefore, the volume per unit length of this element is $\mathcal{V}_o = 0.38/0.5 = 0.76$ m^3/m and the buoyancy force per unit length of this element is

$$B = (\rho_s - \rho_a)g\mathcal{V}_o = (1030 - 1000) \times 9.81 \times 0.76 = \underline{222}\ \mathrm{N/m}$$

The calculations for the thermal velocity follows the same procedure as the previous example. As before, the momentum of the thermal, $(1+k)\mathcal{M}w$, increases with time in proportional to the buoyancy force, B, as follows:

$$(1+k)\mathcal{M}w = Bt$$

where $w = dz/dt$ is downward velocity and k the *added mass* coefficient associated with the irrotational motion outside of the thermal. The radius of the thermal R increases linearly with distance from the source. Ignoring the source size,

$$R \simeq \beta z,$$

We have assumed the distance to be z as a first approximation by ignoring the horizontal movement. The justification of this assumption will become clear as we examine the problem of advection line thermal in chapter 8. The volume $\mathcal{V}$ per unit length of the cylinderical line thermal is $\mathcal{V} = \pi R^2$. The mass per unit length of the same is $\mathcal{M} = \rho_w \pi R^2 = \rho_w \pi \beta^2 z^2$. The momentum equation now become

$$\pi(1+k)\rho_w \beta^2 z^2 \frac{dz}{dt} = Bt.$$

The velocity of the thermal can then be obtained as (Eq. 8.23):

$$w = \frac{dz}{dt} = \frac{2}{3}[\frac{3B}{2\pi\beta^2\rho_w(1+k)}]^{\frac{1}{3}} t^{-\frac{1}{3}} = [\frac{2B}{3\pi\beta^2\rho_w(1+k)}]^{\frac{1}{2}} z^{-\frac{1}{2}} \qquad (1.5)$$

For a circular cylinder accelerating in an irrotational fluid, $k = 1.0$ (see Chapter 7 and 8 for this added mass coefficient). At a depth of 20 m below the water surface, this velocity is

$$w_{20\mathrm{m}} = \sqrt{\frac{2 \times 222}{3\pi \times 0.4^2 \times 2.0 \times 1000 \times 20.0}} = \underline{0.086}\ \mathrm{m/s}.$$

At this depth, the sand and water may not stay together in the thermal as a mixture since the fall velocity of the sand (5 cm/s) is now comparable to the velocity of the thermal (8.5 cm/s).

The thermal velocities in the previous two examples are determined by two parameters: the added-mass coefficient k and the spreading coefficient β. The added mass coefficient is to account for the irrotational motion surrounding the thermal. The spreading coefficient on the other hand is the parameter that characterizes the turbulent motion within the thermal. Similar parameterization is employed to obtain one-dimensional solutions for a variety of jet-and-plume problems examined in this monograph. The value of the added mass coefficient has been determined using two- and three-dimensional turbulence models for line puffs and thermals in Chapter 7 and 8.

EXAMPLE 1.3 Horizontal buoyant jet for sewage disposal
In a small coastal town, an innovative sewage disposal strategy calls for discharging the fine-screened wastewater through a bored hole of 5 cm diameter (bored through a rocky cliff) into the sea in a horizontal offshore direction at the rate of 0.04 m^3/s. The density of sewage is 2.5 % smaller than the density of ambient sea water. Calculate the momentum flux and buoyancy flux of the discharge at the source. Define a length scale based on these source fluxes. Assuming that buoyancy effect is negligible, find the jet dilution at the distance of x=5 m from the source. Estimate the ammonia nitrogen concentration at this distance if the effluent concentration is 10 mg/L.

SOLUTION: This is another example of a continuous source since the wastewater (with momentum and buoyancy) is discharged from the pipe continuously at a rate of Q_o=0.04 m^3/s. The rate of supply of momentum by the source is the momentum *flux* (momentum per unit time). The rate of supply of buoyancy is the buoyancy flux. First, the mass *flux* (mass per unit time) is calculated:

$$\frac{\text{mass}}{\text{time}} = \rho Q_o$$

As momentum = mass × velocity, the momentum *flux* (momentum per unit time), M_o, is given by:

$$M_o = \frac{\text{momentum}}{\text{time}} = \frac{\text{mass}}{\text{time}} \times \text{velocity} = \rho Q_o V_o \tag{1.6}$$

For this example, $\rho = 1000$ kg/m^3, $A_o = \pi D^2/4 = \pi(0.05)^2/4 = 1.964 \times 10^{-3}\text{m}^2$, and $Q_o = 0.04$ m^3/s. The discharge velocity is $V_o = Q_o/A_o = 20.37$ m/s. The sewage is directed at this high velocity away from the shore. Hence, the specific momentum flux (momentum flux per unit mass) is

$$\frac{M_o}{\rho} = Q_o V_o = \underline{0.815}\ \text{m}^4/\text{s}^2$$

As the buoyancy force per unit mass is $\frac{\Delta\rho_o g \mathcal{V}}{\rho \mathcal{V}} = g'_o$, the buoyancy flux, F_o, is given by:

$$\frac{\text{buoyancy force}}{\text{time}} = \frac{\text{mass}}{\text{time}} \times \frac{\text{buoyancy force}}{\text{mass}} = \rho Q_o g'_o$$

The specific buoyancy flux (buoyancy flux per unit mass) is:

$$\frac{F_o}{\rho} = Q_o g_o' = 0.04 \times 0.025 \times 9.81 = \underline{9.81 \times 10^{-3}} \text{ m}^4/\text{s}^3$$

We see from the above calculations that the dimensions of the specific momentum flux and buoyancy flux are

$$[\frac{M_o}{\rho}] = \frac{L^4}{T^2}, \quad [\frac{F_o}{\rho}] = \frac{L^4}{T^3}.$$

Based on these fluxes, a length scale can be defined by eliminating the dependence on time:

$$l_s = \frac{(M_o/\rho)^{\frac{3}{4}}}{(F_o/\rho)^{\frac{1}{2}}}$$

For $M_o/\rho = 0.815 \text{ m}^4/\text{s}^2$ and $F_o/\rho = 9.81 \times 10^{-3} \text{m}^4/\text{s}^3$,

$$l_s = \underline{8.7}\ m.$$

This length scale of around 9 m defines the relative effect of momentum and buoyancy. Momentum is significant close to the discharge. Buoyancy does not become significant until the distance from the source has exceeded this length scale (see Chapters 3 and 4).

At $x = 5$ m, since $x < l_s$, we may neglect the effect of buoyancy as a first approximation, and assume that the jet is straight and that the momentum flux in the jet is constant within the first 5 m. By virtue of the jet momentum, the wastewater is mixed with the ambient fluid. The jet volume flux at a distance x downstream is given by (Chapter 2):

$$Q = \sqrt{\pi}\beta\sqrt{\frac{M_o}{\rho}}\ x,$$

where $\beta = 0.17$ is the jet spread rate. Therefore, at $x = 5$ m,

$$Q = \sqrt{\pi} \times 0.17 \times \sqrt{0.815} \times 5 = \underline{1.36} \text{ m}^3/s.$$

The jet is diluted by the fluid entrained from its surrounding environment. The dilution ratio is

$$S = \frac{Q}{Q_o} = \frac{1.36}{0.04} = \underline{34}.$$

Due to mixing of the jet with the entrained fluid, the tracer concentration is reduced on average by a factor equal to this dilution ratio. To calculate the concentration of ammonia nitrogen (which in its un-ionised form can be toxic to marine life), a simple mass balance with the source nitrogen flux gives:

$$c_o Q_o = CQ$$

where c_o and C are the ammonia nitrogen concentration of the discharge and at x=5 m respectively. If the source concentration is c_o = 10 mg/L, the ammonia nitrogen concentration at x = 5 m would be

$$C = 10/34 = \underline{0.29} \text{ mg/L}$$

The calculations made in this example are based on the assumption that the jet is straight and that the buoyancy effect is negligible; average jet properties have been adopted. The ambient tracer concentration has also been neglected (Prob. 1.2). Further details concerning velocity and concentration distribution and the important role of buoyancy in turbulent mixing are given in Chapters 2 and 3.

PROBLEMS

1.1 Consider a round submerged outlet (diameter D) in depth H and discharging buoyant effluent at an angle ϕ_o to the horizontal in otherwise stagnant fluid. The discharge velocity, density, and tracer (or pollutant) concentration are u_o, ρ_o, c_o; the ambient current u_a is assumed to be uniform and horizontal, and of uniform density ρ_a. The kinematic viscosity of the discharge and ambient fluid are both ν. g is the acceleration due to gravity (Fig. 1.7). Of interest is the dilution $S = \dfrac{c_o}{c(x,y,z)}$, where $c(x,y,z)$ = concentration.

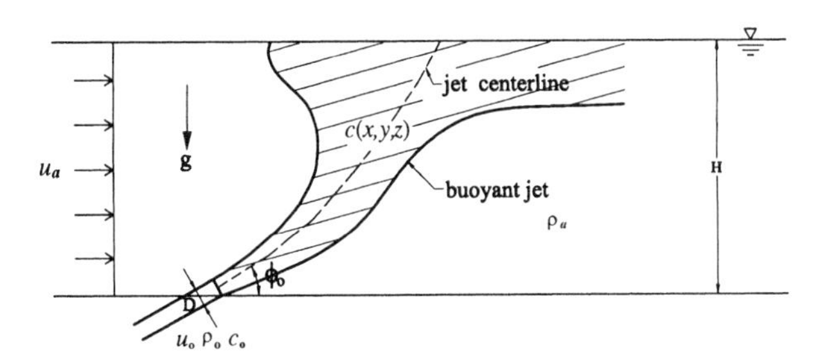

Figure 1.7. A submerged buoyant jet

Based on physical considerations, it is reasonable to postulate a dependence of $c(x,y,z)$ on the discharge and ambient parameters, and the vertical location z, in the following manner:

$$fn(c, c_o, u_o, \frac{\Delta\rho_o}{\rho_a} g, D, z, \phi_o, \nu, u_a) = 0$$

Show that the following dimensionless relation can be obtained by dimensional analysis.

$$\frac{c_o}{c} = fn(\frac{u_o}{\sqrt{g\frac{\Delta\rho_o}{\rho_a}D}}, \frac{z}{D}, \frac{u_o D}{\nu}, \phi_o, \frac{u_o}{u_a})$$

The dilution $S = c_o/c_c$ is governed by a jet densimetric Froude number $Fr = \frac{u_o}{\sqrt{g\frac{\Delta\rho_o}{\rho_a}D}}$, the discharge angle ϕ_o, the relative depth $\frac{z}{D}$, the jet Reynolds number $Re = \frac{u_o D}{\nu}$, and $\frac{u_o}{u_a}$, the ratio of discharge to ambient velocity. For a discharge in stagnant fluid, $u_a = 0$, the dilution is determined by Fr, z/D, and ϕ_o. The densimetric Froude number is a measure of the ratio of the inertia force to buoyancy force per unit mass; it is an important buoyant jet parameter, especially in cases when the source dimension cannot be neglected. $Fr \to \infty$ corresponds to the limiting case of a pure momentum jet ($\Delta\rho \to 0$); high values of Fr can be achieved also by high jet velocity, $u_o \to \infty$, or small source dimensions, $D \to 0$. Similarly, the limiting case of a plume is given by $Fr \to 0$. Plume behaviour is enhanced by low velocities, and relatively large source buoyancy and dimension.

1.2 For the buoyant jet in Example 1.3, determine the dissolved oxygen concentration at 5 m from the discharge if the dissolved oxygen concentration in the source and ambient fluid are 1 mg/L and 7 mg/L respectively.

1.3 The round puff is produced by a 'point instananeous source' of momentum. The eruption of the Mount St. Helens as shown in Figure 1.3 is an example. The impulse produced by the eruption is a product of the initial mass of the eruption and the corresponding initial velocity; i.e., M_o = mass $\times$ velocity $= \rho\frac{\pi}{6}D^3 w_o$. As fluid from the surrounding is drawn into the round puff through a process known as turbulent entrainment, the volume of the eruption, V, increases with time t following a functional relation that is dependent on the initial impulse and time, $V = \text{fn}(M_o, t, \rho)$.

Using dimensional analysis, derive the following relation (see Table 1.1):

$$V \sim (\frac{M_o}{\rho})^{\frac{3}{4}} t^{\frac{3}{4}} \tag{1.7}$$

A similar analysis can be made for the line puff (an instantaneous line source of momentum), with $V \sim (M_o/\rho)^{\frac{2}{3}} t^{\frac{2}{3}}$, where V and M_o now represent the volume and momentum per unit length of the line puff respectively (see Table 1.1 and Chapter 7).

1.4 For the sand dumping problem in Example 1.2, if the lump of sand is released at 0.3 m above the water surface, determine a) the free fall velocity of the sand particles just before impinging the free surface; and b) the initial velocity and momentum of the sand-cloud mixture in the water. State any assumptions made (see also chapter 5 and Probs. 5.5 and 8.3).

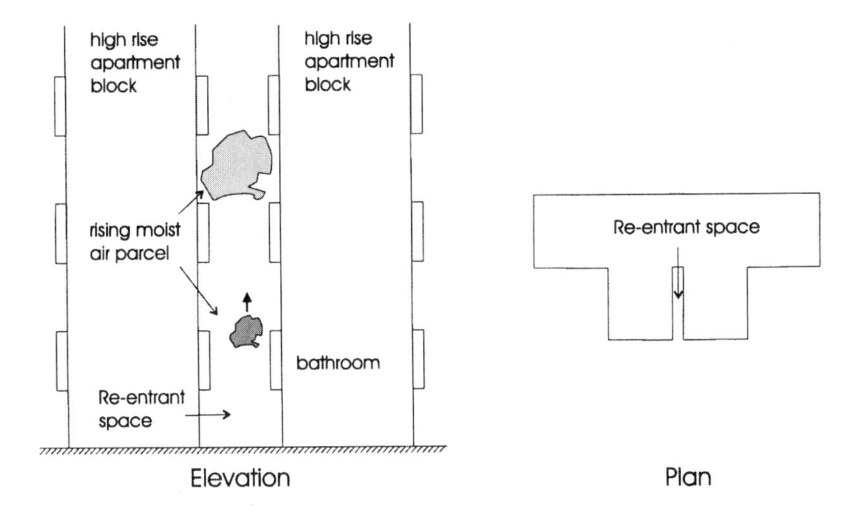

Figure 1.8.

1.5 Spread of infectious disease by moist air in housing estate

In a densely populated housing estate in a city, residential flats are vertically arranged in 30 storey-buildings (Fig. 1.8). The high rise block is separated from an adjacent block by a narrow re-entrant space (light well). The air movement in such confined space (with plan dimensions of only 1.5 m wide by 6 m) is very limited. A possible way to spread an infectious disease (such as a deadly virus) in these poorly-ventilated spaces is through the transport by moist air released from the bathroom of flats (which typically face the re-entrant space) after a hot shower of an infected patient(s).

Assume the virus is carried in a volume of moist air of initial volume $V_o = 5$ m^3, relative density difference $\Delta\rho_o/\rho_a \approx 0.001$. Assume a linear spreading hypothesis for the round thermal (moist air parcel), $R = \beta z$, where R=radius of the thermal, and $\beta = 0.34$. Neglecting lateral boundaries, use Eq. 1.3 to find the dilution (inverse of relative concentration) and velocity of the virus-laden moist air after rising a vertical distance of $z = 3$ m (one floor), and $z = 30$ m (10 floors). Comment on your results.

Chapter 2

TURBULENT JETS

A jet is the flow generated by a continuous source of momentum. The Reynolds number of a jet can be conveniently defined as $Re = \frac{u_o D}{\nu}$ where u_o = jet velocity; D = jet diameter; ν = kinematic viscosity of the fluid. Experiments have shown that if Re exceeds about 2000, the jet flow will be turbulent. In most problems of concern to the engineer, the Reynolds number would be large and the jet flow will be turbulent. The round jet shown in Figure 1.1 is an example. Turbulent eddies of many sizes and shapes are observed due to the presence of the smoke as a tracer of the turbulent motion. There is a general tendency for the length scale of the turbulent motion to increase as the eddies move along the jet. Induced by the motion of the eddies, fluid from the surrounding environment is drawn into the jet through a process called *turbulent entrainment.* The extent of the turbulent zone increases and the concentration of the tracer decreases with distance from the source as the source fluid is diluted by the fluid entrained from the surroundings. The jet in this example produces a thrust (three million pounds) for the rocket engine. In many engineering applications, the jet is a device for mixing of source fluid with the ambient fluid. Figure 2.1 is a turbulent water jet produced in the laboratory from a high velocity discharge into a weak coflowing current. Rhodamine dye is introduced into the jet as a tracer. The turbulent structure in the jet is revealed by the fluorescence image, which is induced by a sheet of laser light through the center plane of the jet. Despite the enormous difference in the size, and the absence of the fine-scale turbulent structure, the laboratory jet is observed to spread at the same rate as the jet produced by the rocket engine. Apparently, the entrainment process and the spreading rate of a turbulent jet is determined by the large and dominant eddies extending across the full width of the jet. The ultimate mixing of the entrained fluid with the source fluid is carried out by the small eddies that circulate around the dominant eddies. It can also be seen that the initial development of the jet is confined to a mixing layer at the edge of the jet. There is a core of essentially irrotational fluid close to the source, about $6D$ in length, that is not affected by the jet diffusion. Beyond this potential core, all the source fluid is mixed with the ambient fluid, and the mean flow is fully established.

EXAMPLE 2.1 *Waste water is discharged into a lake at a rate of 1.6 m^3/s through a 1 m diameter pipe at the outfall. Assuming momentum flux conservation, find the volume flux of the jet at a distance of 10 m from the source. Assume that the jet radius, $\tilde{b}$, increases with the distance from the source, x, at a rate $d\tilde{b}/dx = 0.17$.*

Assuming tracer mass flux conservation, find the tracer concentration at the same location if the concentration at the source is $c_o = 100$ mg/L, and the ambient concentration is zero (Fig. 2.2).

SOLUTION: The cross sectional area of the jet at the source is

$$A_o = \frac{\pi}{4}D^2 = \frac{\pi}{4} \times 1^2 = 0.785\,\mathrm{m}^2$$

and the initial jet volume flux (discharge) is $Q_o = 1.6$ m^3/s. The jet velocity at the source (i.e. the momentum per unit mass of fluid) is

$$V_o = \frac{Q_o}{A_o} = \frac{1.6}{0.7853} = 2.038\,\mathrm{m/s}$$

With this velocity, the jet is discharging its momentum at a rate equal to

$$M_o = V_o \rho Q_o = \text{momentum flux} = \frac{\text{momentum}}{\text{mass}} \times \frac{\text{mass}}{\text{volume}} \times \frac{\text{volume}}{\text{time}}$$

Dividing this momentum flux of the jet by the fluid density gives

$$\text{specific momentum flux} = \frac{M_o}{\rho} = V_o Q_o = 2.038 \times 1.6 = 3.261\,\mathrm{m}^4/\mathrm{s}^2$$

Due to its momentum, the jet entrains ambient fluid outside the jet boundary into the main turbulent stream thus increasing the volume flux of the jet. The mixing results in a change in both the jet velocity and

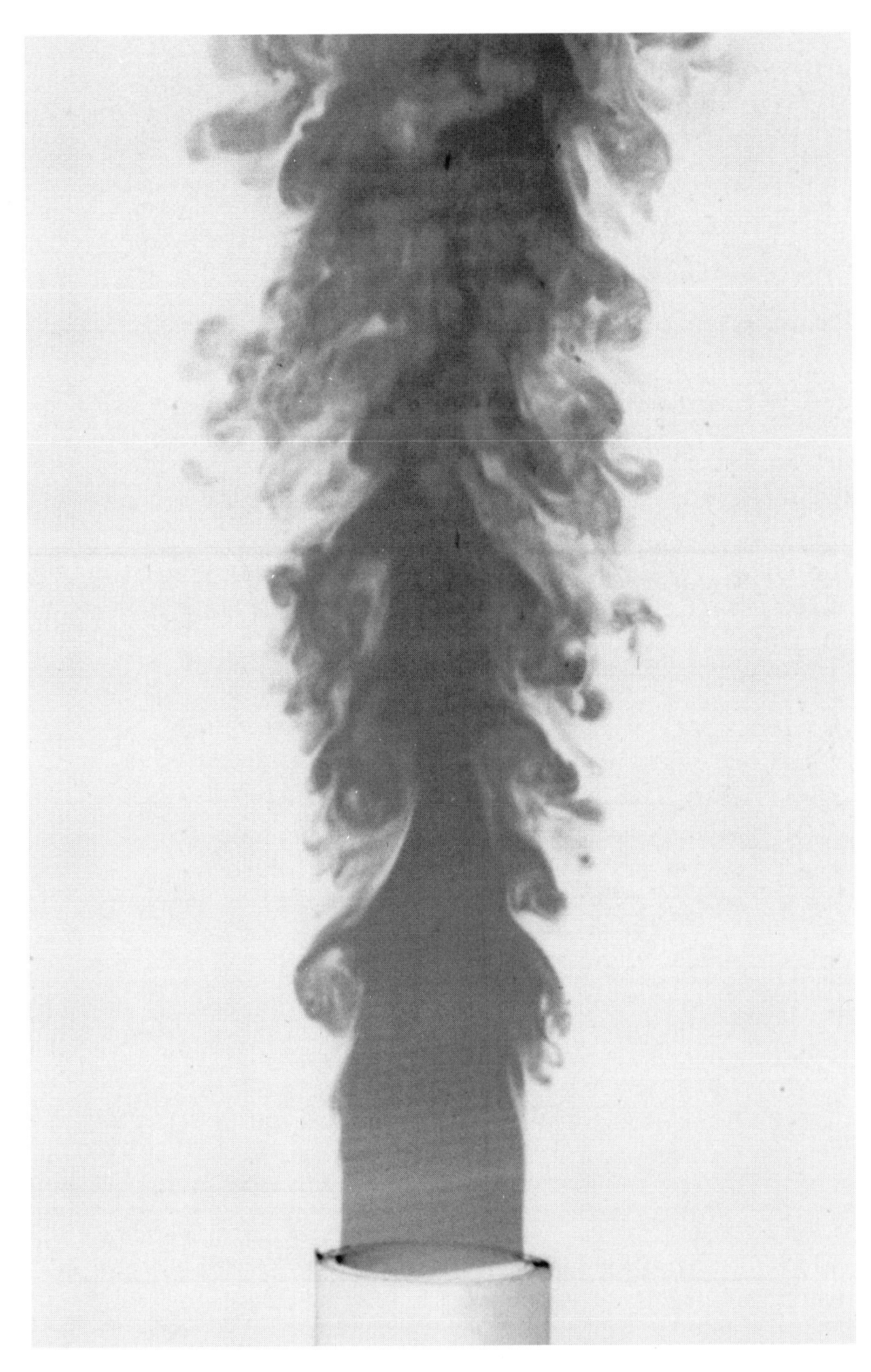

Figure 2.1. Potential core of a turbulent jet in a coflow; $u_o = 22.1$ cm/s, $D = 2$ cm; coflow velocity $u_a = 1.37$ cm/s.

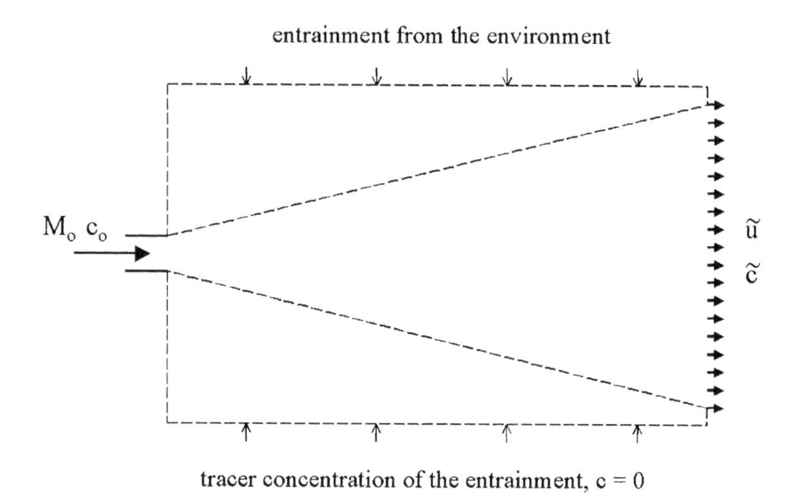

Figure 2.2. Control volume and top-hat profile for mass and momentum fluxes in a round jet. Tracer concentration of the entrained ambient fluid is zero.

width. The radius of the jet $\tilde{b}$ is expected to depend on the momentum flux of the jet M_o, the fluid density ρ and distance from the source x; i.e.,

$$\tilde{b} = \text{fn}(M_o, \rho, x)$$

The only dimensionless parameter in this functional relation is $\tilde{b}/x$, which must be a constant. According to this dimensional consideration, the width of the jet increases linearly with distance from the source. The rate is $d\tilde{b}/dx = 0.17$, which is obtained from experimental observation. The momentum flux of the jet is constant. Everything else is not: the volume flux increases and the energy flux decreases with distance from the source. Turbulence is generated as energy is dissipated in the jet. At a distance of $x = 10$ m, the jet properties are:

$$\text{jet radius} = \tilde{b} = 0.5 + 0.17 \times 10 = 2.2\,\text{m}$$

$$\text{cross sectional area} = \tilde{A} = \pi \times 2.2^2 = 15.21\,\text{m}^2$$

Momentum flux: In an environment of negligible pressure gradient, the momentum flux of the turbulent jet is constant. Equating the momentum flux at $x = 10$ m to the momentum flux at the source gives

$$\frac{M}{\rho} = \tilde{u}^2 \tilde{A} = 3.261\,\text{m}^4/\text{s}^2$$

The average velocity of the jet at 10 m distance from the source is

$$\tilde{u} = [\frac{3.261}{15.21}]^{\frac{1}{2}} = 0.463\,\mathrm{m/s}.$$

The volume flux at the same location is

$$Q = \tilde{u}\tilde{A} = 0.463 \times 15.21 = \underline{7.04}\,\mathrm{m}^3/\mathrm{s}.$$

The jet is diluted. The dilution ratio is

$$S = \frac{Q}{Q_o} = \frac{7.04}{1.6} = 4.4.$$

Over a distance of 10 m, the volume flux and jet radius are both increased by 4.4 times through jet mixing.

<u>Tracer mass flux</u>: The mass flux of a conservative tracer (e.g. dye) is constant:

$$\text{tracer mass flux} = \frac{\text{tracer mass}}{\text{volume}} \times \frac{\text{volume}}{\text{time}} = c_o Q_o = \tilde{c}Q$$

Therefore, the tracer concentration in the jet is

$$\tilde{c} = c_o \frac{Q_o}{Q} = \frac{100}{4.4} = \underline{22.7}\ mg/L$$

<u>Energy Flux</u>: At 10 m, the specific kinetic energy flux is

$$\frac{E}{\rho} = \frac{1}{2}\tilde{u}^2 Q = \frac{1}{2} \times (0.463)^2 \times 7.042 = 0.755\,\mathrm{m}^5/\mathrm{s}^3$$

This is compared with the flux at the source, which is

$$\frac{E_o}{\rho} = \frac{1}{2}V_o^2 Q_o = \frac{1}{2} \times (2.038)^2 \times 1.6 = 3.323\,\mathrm{m}^5/\mathrm{s}^3.$$

The kinetic energy flux of the jet decreases with distance from the source. Nearly 77% of the initial kinetic energy is dissipated within 10 m. The dissipated energy is utilized by the jet to generate the turbulence, which induces the mixing and entrainment of surrounding fluid into the turbulent jet. The fluid power needed to produce the turbulent jet is

$$E_o = \frac{1}{2}\rho V_o^2 Q_o = 1000 \times 3.323 = 3323\ \mathrm{W}$$

<u>Top-hat Profile</u>: For convenience, we have carried out the calculations in this introductory example using a 'top-hat' profile (i.e. assuming

constant velocity distribution across the jet). The velocity $\tilde{u}$ and concentration $\tilde{c}$ of the top-hat profile are assumed to be the average velocity and concentration of the dominant eddies - the large eddies occupying the jet cross section that are responsible for the overall transport of mass and momentum. Small eddies circulating within the dominant eddies are asssumed to have no net effect on the overall transport, so that the mass and momentum fluxes are calculated based on the top-hat profiles.

Gaussian Profile: The top-hat profile is convenient in a Lagrangian formulation of the jet using a reference frame following the motion of the dominant eddies. In a fixed reference frame, and from an Eulerian point of view, the velocity and concentration profiles across the jet are Gaussian. Mass and momentum in the turbulent jet move back and forth, and left and right, by the random action of the turbulent motion. The result of this random advection is an essentially Gaussian probability density function for the tracer concentration. The profile for the velocity is Gaussian as well, since momentum transport is analogous to mass transport, and velocity is momentum per unit mass of the fluid. Typical mean velocity profile in the fully developed region of the jet is Gaussian as shown in Figure 2.3.

The Lagrangian formulation of jets and plumes using the top-hat profile is convenient. A justification for using the top-hat profile is the dominant-eddy hypothesis. The hypothesis, and the connection between the top-hat and the Gaussian profiles, will be further discussed in this and later chapters.

1. PLANE JET

We begin the study of turbulent jets using the conventional procedure. First the analytical framework for the Eulerian formulation is developed for the plane jet, which sometimes is referred to as the slot jet. Figure 2.3 shows the mean flow of the slot jet on the x- and y-plane. At the source the velocity and concentration is u_o and c_o, respectively. The turbulent flow in the jet is produced by discharging the source fluid from a slender slot. The development of the jet is divided into two zones: Zone of Flow Establishment (ZFE) and Zone of Established Flow (ZEF). The velocity and concentration in the potential core of the ZFE are constant. Surrounding the potential core is the mixing layer. The exchange of momentum between the core and the quiescent fluid across the mixing layer leads to the profiles in the mixing layer as follows:

$$u(x,y) = u_o \exp[-\frac{(y-r)^2}{b^2}],\ c(x,y) = c_o \exp[-\frac{(y-r)^2}{(\lambda b)^2}]; y > r$$

$$u(x, y) = u_o, \qquad c(x, y) = c_o; \qquad y \leq r \quad (2.1)$$

where $r(x)$ = half-width of the potential core, $b(x)$ = width of the mixing layer, $u(x, y)$ = velocity, $c(x, y)$ = concentration, and y = lateral distance from the center line. The parameter λ is introduced to account for the difference between the diffusion of mass and diffusion of momentum. The width of the concentration profile $\lambda b(x)$ is generally wider than the width of the velocity profiles $b(x)$. At the end of the potential core, $r \to 0$ by definition.

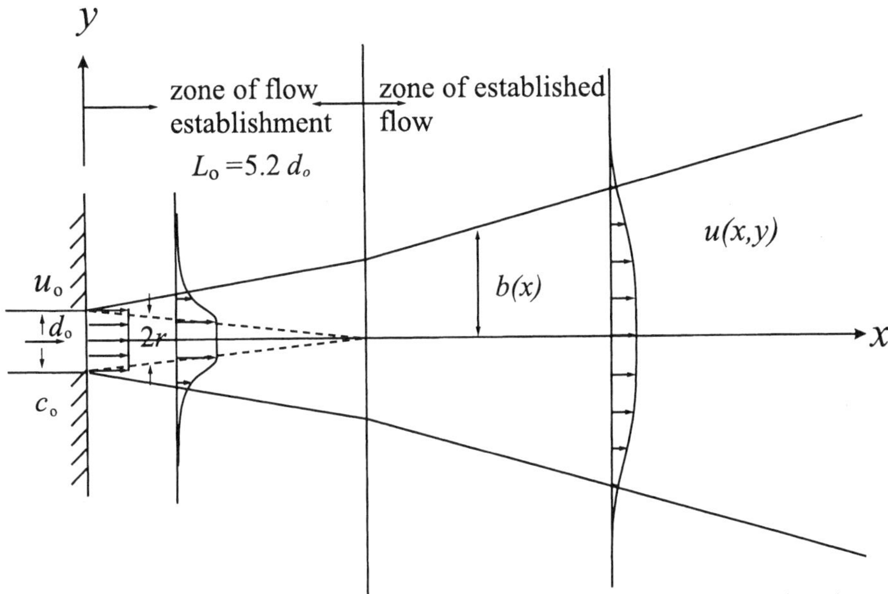

Figure 2.3. Turbulent slot jet showing the Zone of Flow Establishment (ZFE) and Zone of Established Flow (ZEF).

In the Zone of Established Flow (ZEF), $x > 5.2d_o$, when the turbulence has penetrated to the centerline, the velocity and concentration distributions are *self-similar* (this means the profiles at different x all look similar in shape, so that if the profiles at different x are scaled properly, all these should collapse onto one curve), and can be well-approximated by Gaussian distributions:

$$u(x, y) = u_m \exp[-\frac{y^2}{b^2}], \; c(x, y) = c_m \exp[-\frac{y^2}{(\lambda b)^2}]; \; x > 5.2d_o \qquad (2.2)$$

where $u_m(x)$ and $c_m(x)$ are the velocity and concentration maxima along the centerline. The width of the jet, $b(x)$, is defined at a lateral location where the x-component of the velocity is equal to $1/e$ of the center-line value. Experimental measurements of plane jet by Albertson *et al.* (1950), Miller and Comings (1957), Bradbury (1965) have found the jet spreads linearly with a growth rate

$$\frac{db}{dx} \simeq 0.1 \tag{2.3}$$

Since the width of the jet is small compared with the longitudinal length scale, the turbulent jet is a slender shear flow and boundary-layer type of approximation is applicable.

1.1 GOVERNING EQUATIONS

The governing equations for the steady incompressible turbulent-mean flow of the jet are the continuity equation

$$\frac{\partial u}{\partial x} + \frac{\partial v}{\partial y} = 0, \tag{2.4}$$

the momentum equations

$$\rho u \frac{\partial u}{\partial x} + \rho v \frac{\partial u}{\partial y} = -\frac{\partial p}{\partial x} - \frac{\partial \rho \overline{u'^2}}{\partial x} - \frac{\partial \rho \overline{u'v'}}{\partial y}, \tag{2.5}$$

$$\rho u \frac{\partial v}{\partial x} + \rho v \frac{\partial v}{\partial y} = -\frac{\partial p}{\partial y} - \frac{\partial \rho \overline{u'v'}}{\partial x} - \frac{\partial \rho \overline{v'^2}}{\partial y}, \tag{2.6}$$

and the tracer-mass conservation equation

$$u \frac{\partial c}{\partial x} + v \frac{\partial c}{\partial y} = -\frac{\partial \overline{u'c'}}{\partial x} - \frac{\partial \overline{v'c'}}{\partial y}. \tag{2.7}$$

where (u, v) = mean velocity, c = mean concentration (tracer-mass per unit volume), p = pressure, and ρ = fluid density; the prime denote the turbulent fluctuations and overbar the time average of the fluctuations. The boundary conditions for a free jet are: $u \to 0$, $\overline{u'v'} \to 0$, and $\overline{u'c'} \to 0$ as $y \to \pm\infty$. With these boundary conditions the jet is assumed to be free from solid boundary and ambient current effects.

Furthermore, since $b \ll x$, by continuity $v \ll u$, and $\frac{\partial}{\partial x} \ll \frac{\partial}{\partial y}$, the pressure p is approximately constant and $\frac{\partial p}{\partial x} \approx \frac{\partial p}{\partial y} \approx 0$. Hence,

$$p + \rho \overline{v'^2} \simeq p_\infty \tag{2.8}$$

from the boundary-layer approximation of the y-momentum equation (Eq. 2.6). Other equations, according to the boundary-layer approximation, are the continuity equation

$$\frac{\partial u}{\partial x} + \frac{\partial v}{\partial y} = 0, \tag{2.9}$$

the x-momentum equation

$$\rho u \frac{\partial u}{\partial x} + \rho v \frac{\partial u}{\partial y} + \frac{\partial \rho \overline{u'^2}}{\partial x} - \frac{\partial \rho \overline{v'^2}}{\partial x} = -\frac{\partial \rho \overline{u'v'}}{\partial y}, \tag{2.10}$$

and the tracer-mass conservation equation

$$u \frac{\partial c}{\partial x} + v \frac{\partial c}{\partial y} + \frac{\partial \overline{u'c'}}{\partial x} = -\frac{\partial \overline{v'c'}}{\partial y}, \tag{2.11}$$

subjected to the same boundary conditions as specified before. Thus we have 3 variables of the mean flow, u, v, c, and three governing equations. But the turbulent covariances $\overline{u'v'}, \overline{v'c'}$ are the additional unknowns due to the separation of the flow into mean and fluctuation parts. This is the problem of the *closure*. A *turbulence model* must be introduced to relate these covariances with the mean flow.

1.2 INTEGRAL EQUATIONS

A one-dimensional procedure to achieve the turbulent closure of the problem is to integrate across the turbulent jet. First, the x-momentum equation is re-written to a conservative form. Since,

$$\rho u \frac{\partial u}{\partial x} + \rho v \frac{\partial u}{\partial y} = \frac{\partial \rho u^2}{\partial x} + \frac{\partial \rho u v}{\partial y} - \rho u (\frac{\partial u}{\partial x} + \frac{\partial v}{\partial y}) \tag{2.12}$$

the x-momentum equation (Eq.2.5) becomes (on invoking Eq.2.8; see Problem 2.5)

$$\frac{\partial \rho u^2}{\partial x} + \frac{\partial \rho u v}{\partial y} + \frac{\partial \rho \overline{u'^2}}{\partial x} - \frac{\partial \rho \overline{v'^2}}{\partial x} = -\frac{\partial \rho \overline{u'v'}}{\partial y} \tag{2.13}$$

Using the boundary condition and integrating across the jet, from $y = -\infty$ to $y = +\infty$, we have for the left hand side of the equation

$$\frac{d}{dx} \int_{-\infty}^{\infty} \rho u^2 + \rho(\overline{u'^2} - \overline{v'^2}) dy + [\rho u v]_{-\infty}^{\infty}$$

and for the right hand side

$$\int_{-\infty}^{\infty} (-\frac{\partial \rho \overline{u'v'}}{\partial y}) dy = \left[-\rho \overline{u'v'}\right]_{-\infty}^{\infty} = 0.$$

Therefore,

$$\frac{d}{dx} \int_{-\infty}^{\infty} [\rho u^2 + \rho(\overline{u'^2} - \overline{v'^2})] dy = 0. \tag{2.14}$$

The momentum flux is preserved. Thus, the slot jet is a line source of momentum flux. If M_o is this momentum flux per unit length of the slot at the source,

$$M = \int_{-\infty}^{\infty} [\rho u^2 + \rho(\overline{u'^2} - \overline{v'^2})] \, dy = M_o \tag{2.15}$$

Similarly, we have for tracer-mass conservation,

$$u\frac{\partial c}{\partial x} + v\frac{\partial c}{\partial y} = \frac{\partial uc}{\partial x} + \frac{\partial vc}{\partial y} = -\frac{\partial \overline{u'c'}}{\partial x} - \frac{\partial \overline{v'c'}}{\partial y} \tag{2.16}$$

Integrating across the jet as before,

$$\frac{d}{dx}\int_{-\infty}^{\infty} [uc + \overline{u'c'}] \, dy = \left[-\overline{v'c'}\right]_{-\infty}^{\infty}. \tag{2.17}$$

The right hand side of this equation would be zero if the concentration c is the concentration excess above its value in the ambient. Hence, the excess-mass flux is constant. If Γ_o is this excess-mass flux per unit length of the slot,

$$\Gamma = \int_{-\infty}^{\infty} [uc + \overline{u'c'}] \, dy = \Gamma_o. \tag{2.18}$$

Measurements show that the two turbulence quantities in Eq.2.14 are of the same order; a detailed evaluation using the data of Miller and Comings (1957) shows that the contribution of the turbulence terms to the momentum flux integral is only around 4 percent (Prob. 2.7). If the longitudinal fluxes due to the turbulent advection is ignored,

$$M \simeq \int_{-\infty}^{\infty} \rho u^2 \, dy \simeq M_o, \tag{2.19}$$

$$\Gamma \simeq \int_{-\infty}^{\infty} uc \, dy \simeq \Gamma_o. \tag{2.20}$$

These are the *approximated* form of the equations since the parts of fluxes due to turbulent advection have been ignored. The Eulerian integral model of the jet to be described in the next section is based on these approximate equations.

1.3 EULERIAN INTEGRAL MODEL

The conventional integral model of the turbulent jet is based on the assumption of the Gaussian profiles (Equation 2.1 and 2.2). To demonstrate the idea, the calculations for the ZEF is considered in this section. With the velocity and concentration profiles given by Equations 2.2, the specific or kinematic momentum flux M/ρ (consisting of only kinematic dimensions of length and time) and the tracer-mass flux are functions of the width of the jet, b, the maximum velocity, u_m, and the maximum concentration, c_m, as follows

$$\frac{M}{\rho} = \int_{-\infty}^{\infty} u^2 \, dy = \sqrt{\frac{\pi}{2}} u_m^2 b \tag{2.21}$$

$$\Gamma = \int_{-\infty}^{\infty} uc \, dy = \sqrt{\frac{\pi \lambda^2}{1+\lambda^2}} u_m c_m b \tag{2.22}$$

In most cases we will be working with the kinematic fluxes; henceforth the specific momentum flux will be denoted by M unless otherwise stated.

Equating the expressions in the above equations to the flux at the source, $M_o = {u_o}^2 d_o$ and $\Gamma_o = u_o c_o d_o$, and assuming that $b = \beta_G x$,

$$\frac{u_m}{u_o} = [\sqrt{\frac{2}{\pi}} \frac{1}{\beta_G}]^{\frac{1}{2}} (\frac{x}{d_o})^{-\frac{1}{2}}, \tag{2.23}$$

$$\frac{c_m}{c_o} = [\frac{1+\lambda^2}{\lambda^2 \beta_G} \frac{1}{\sqrt{2\pi}}]^{\frac{1}{2}} (\frac{x}{d_o})^{-\frac{1}{2}}. \tag{2.24}$$

The consequence of momentum conservation is the jet dilution. The volume flux (per unit length of the slot) Q at a distance x from the source is greater than the volume flux, Q_o, at the source, since

$$Q = \int_{-\infty}^{\infty} u \, dy = \sqrt{\pi} u_m b = [\sqrt{2\pi}\beta_G]^{\frac{1}{2}} (\frac{x}{d_o})^{\frac{1}{2}} u_o d_o \tag{2.25}$$

The average dilution ratio,

$$S = \frac{Q}{Q_o} = (\sqrt{2\pi}\beta_G)^{\frac{1}{2}} (\frac{x}{d_o})^{\frac{1}{2}} \tag{2.26}$$

The jet spread rate β_G, defined based on the Gaussian velocity profile, has been determined experimentally by Albertson *et al.* (1950) to be $\beta_G = 0.154$. This is somewhat larger than values of 0.13 from measurments of Bradbury (1965), 0.116 from the data of Kotsovinos (1976), and 0.119 from Miller and Comings (1957). Kotsovinos's experiments also give a λ value of 1.35. Adopting a value of $\beta_G = 0.12$ and $\lambda = 1.35$ we have the solution summarized in Table 2.1.

Since $u_m \sim x^{-\frac{1}{2}}, b \sim x$, then the kinetic energy flux, which can be defined as $E = \int_{-\infty}^{\infty} \frac{1}{2} u^3 dy$, varies as $\sim {u_m}^3 b \sim x^{-\frac{1}{2}}$ and $\rightarrow 0$ as $x \rightarrow \infty$. The kinetic energy in the jet is continually converted into the energy

Table 2.1. Plane jet formulae based on slot openning d_o and source velocity u_o for $x \geq 5.2 d_o$.

Jet width	$b = 0.12\ x$
Centerline velocity	$u_m/u_o = 2.58\ (x/d_o)^{-\frac{1}{2}}$
Centerline concentration	$c_m/c_o = 2.27\ (x/d_o)^{-\frac{1}{2}}$
Centerline dilution	$S = 0.44\ (x/d_o)^{\frac{1}{2}}$
Average dilution ratio	$\overline{S} = 0.55\ (x/d_o)^{\frac{1}{2}}$

of turbulence, and the latter steadily decaying through viscous shear. Eventually, far from the source, all the initial kinetic energy is dissipated.

Since

$$d_o = \frac{{u_o}^2 d_o^2}{{u_o}^2 d_o} = \frac{{Q_o}^2}{M_o}, \tag{2.27}$$

the solution can alternatively be expressed in terms of the initial volume and momentum fluxes only. e.g.

$$\text{average dilution} = \overline{S} = 0.55\ \frac{{M_o}^{\frac{1}{2}} x^{\frac{1}{2}}}{Q_o} \tag{2.28}$$

or

$$Q = 0.55\ M_o^{\frac{1}{2}} x^{\frac{1}{2}} \tag{2.29}$$

At a given x, the larger the initial momentum M_o, the greater the entrained flow in the jet up to x.

Despite the approximation made to the fluxes, the expression obtained from the integral prediction of volume flux, $Q = 0.55\ M_o^{1/2} x^{1/2}$, is in excellent agreement with the velocity measurements by Kotsovinos and List (1977), which gives a coefficient of $Q/(\ M_o^{1/2} x^{1/2}) = 0.54$. This type of engineering analysis, in which the functional form of the velocity and concentration profiles are specified, has proven to give rather accurate predictions of jet properties. The functional dependence of the jet properties is independent of the exact form of the profile assumed. The value of the experimental constant (the spreading rate β) is however a function of the chosen profile.

EXAMPLE 2.2 *A slot turbulent jet discharges liquid at a mean velocity 3 m/s into a liquid of the same density. Find the maximum time-averaged velocity, tracer concentration, and the average dilution at a distance of 20 m from the jet slot. Slot width = 0.1 m, and the initial concentration of the tracer = 1000 mg/L.*

Solution: The dimensionless parameter $x/d_o = 20/0.1 = 200$. Substituting x/d_o and $c_o = 1000$ mg/L into the formulae in Table 2.1, the maximum velocity(u_m), maximum concentration(c_m), and the average dilution($\overline{S}$) are found to be 0.55 m/s, 161 mg/L and 7.8 respectively.

The solution given above are valid only in the zone of established flow (ZEF). The solution for the zone of flow establishment (ZFE) is actually implied in the stated formulation; the thickness of the potential core $r(x)$ varies linearly as $2r/d_o = 1 - x/L_o$ (Prob. 2.4). In most practical applications, $x \gg d_o$, and the zone of flow establishment can be ignored altogether.

1.4 ENTRAINMENT HYPOTHESIS

By examining the continuity equation, we can study the closure problem again in a different way. Integrating the continuity equation across the jet, we have

$$\frac{dQ}{dx} = \frac{d}{dx}\int_{-\infty}^{\infty} u dy = [-v]_{-\infty}^{\infty} = 2v_e \tag{2.30}$$

where $v_e = |v|_{y=\pm\infty}$ = entrainment velocity. Thus the closure problem amounts to saying something about the entrainment velocity and its relationship with the local jet characteristics. We can get some insight (or hindsight!) about this relationship by using the solution for $Q(x)$, Eq.(2.25), to back calculate dQ/dx

$$\frac{dQ}{dx} = \frac{(\sqrt{2\pi}\beta_G)^{\frac{1}{2}}}{2} u_o \left(\frac{x}{d_o}\right)^{-\frac{1}{2}} = 2\left(\frac{\sqrt{\pi}\beta_G}{4}\right) u_m \tag{2.31}$$

Comparing Eq.(2.30) and (2.31), we have $v_e \propto u_m$: the entrainment velocity $v_e(x)$ is proportional to the local jet centerline velocity. This means that, alternatively, we could have solved the same problem by beginning from an assumption that $v_e = \alpha u_m$ (where α is an 'entrainment coefficient'), and solving the continuity, momentum, and mass conservation equations. In many buoyant jet problems, turbulent closure can be circumvented by this *entrainment hypothesis* proposed by Morton, Taylor and Turner (1956) to describe the broad mechanics of the turbulent flow rather than the detailed mechanisms. The entrainment velocity is assumed to be proportional to a local characteristic velocity (the centerline velocity) via the entrainment coefficient which can be determined experimentally. The continuity and momentum equations then become:

$$\frac{dQ}{dx} = \frac{d}{dx}(\sqrt{\pi}u_m b) = 2\alpha u_m \tag{2.32}$$

$$\frac{dM}{dx} = \frac{d}{dx}(\sqrt{\pi/2}\; u_m^2 b) = 0 \tag{2.33}$$

Noting that $M = M_o$ and $u_m = \sqrt{2}M/Q$, $b = Q^2/(\sqrt{2\pi}M)$, the continuity equation can be solved to give (neglecting source flows for high dilutions, $Q/Q_o \gg 1$) $Q^2 = 4\sqrt{2}\alpha M_o\, x$. A linear jet spread of $b = (\frac{4}{\sqrt{\pi}}\alpha_G)x$ is hence predicted by this entrainment assumption. The experimental constants in the spreading and entrainment hypotheses are hence related by

$$\beta_G = \frac{4}{\sqrt{\pi}}\,\alpha_G \qquad (2D\ jet) \tag{2.34}$$

where the subscript 'G' denotes the Gaussian profile.

2. ROUND JET

2.1 MEAN FLOW STRUCTURE

The general experimental features described above for a slot jet also apply for the case of the flow from a circular orifice (diameter D)(Fig. 2.4). The diffusion thickness spreads linearly; static pressure is approximately constant; the same type of boundary layer approximations can be made. For this axisymmetric case, the length of the potential core is 6.2D; the mean axial velocity and concentration profiles are found to attain self-similarity beyond the potential core:

1. In the zone of flow establishment (ZFE), $x \leq 6.2D$;

$$u = u_o, c = c_o; \qquad r \leq R \tag{2.35}$$

$$u = u_o \exp[-\frac{(r-R)^2}{b^2}],\ \ c = c_o \exp[-\frac{(r-R)^2}{\lambda^2 b^2}];\ r \geq R \tag{2.36}$$

2. In the zone of established flow (ZEF), $x \geq 6.2D$, the axial velocity and tracer concentration profiles are self-similar and Gaussian:

$$u = u_m \exp[-(\frac{r}{b})^2], c = c_m \exp[-(\frac{r}{\lambda b})^2] \tag{2.37}$$

where (x, r) = streamwise and radial co-ordinates, and $u_m(x)$ and $c_m(x)$ are the centerline maximum velocity and concentration respectively.

3. The turbulent round jet spreads linearly, with

$$b = \beta x \tag{2.38}$$

The above jet structure, qualitatively similar to those of the slot jet, has been extensively confirmed by experiments (Albertson *et al.* 1950; Ricou and Spalding 1961; Papanicolaou and List 1988; Chu 1996). However, the experimental constants (β, λ, α) are in general different from those of the slot jet.

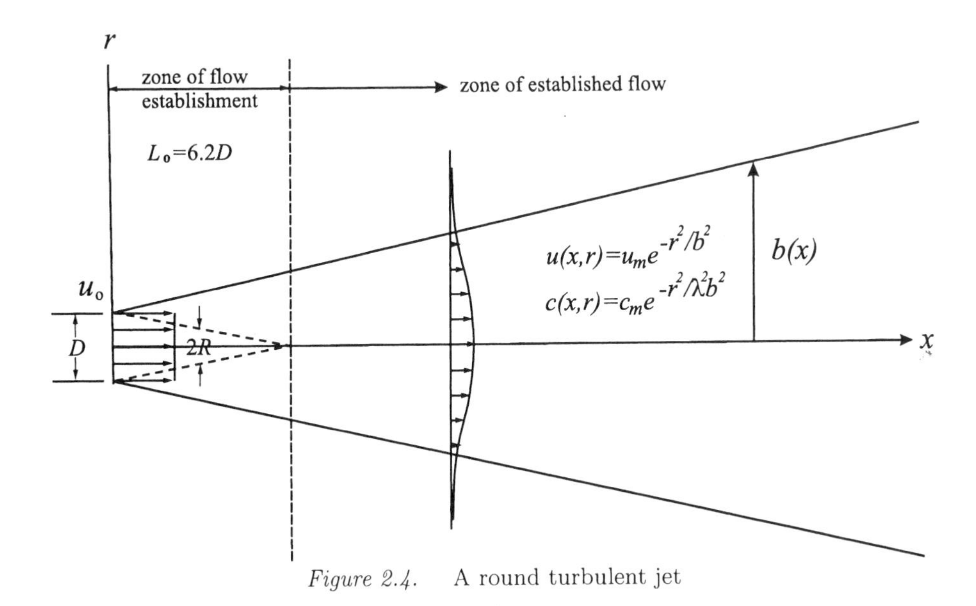

Figure 2.4. A round turbulent jet

The governing equations for the steady flow of a round jet in the (x, r) co-ordinate system are the continuity equation

$$\rho \frac{\partial u}{\partial x} + \rho \frac{1}{r} \frac{\partial}{\partial r}(rv) = 0 \tag{2.39}$$

the x-momentum equation

$$\rho u \frac{\partial u}{\partial x} + \rho v \frac{\partial u}{\partial r} = -\frac{1}{r} \frac{\partial}{\partial r}(r \rho \overline{u'v'}) \tag{2.40}$$

the tracer-mass conservation equation

$$u\frac{\partial c}{\partial x} + v\frac{\partial c}{\partial r} = -\frac{1}{r}\frac{\partial}{\partial r}(r\overline{v'c'}) \tag{2.41}$$

The x-momentum equation can be written as :

$$\rho u\frac{\partial u}{\partial x} + \rho v\frac{\partial u}{\partial r} = \rho u\frac{\partial u}{\partial x} + \rho v\frac{\partial u}{\partial r} + \rho u(\frac{\partial u}{\partial x} + \rho\frac{1}{r}\frac{\partial}{\partial r}(rv))$$

$$= -\frac{1}{r}\frac{\partial}{\partial r}(r\rho\overline{u'v'}) \tag{2.42}$$

Multiplying this equation by $2\pi r$ and then integrating from $r = 0$ to $r = \infty$ we have

$$\frac{d}{dx}\int_0^\infty \rho u^2 2\pi r dr = 0 \tag{2.43}$$

The momentum flux in a turbulent round jet is again preserved. Similarly, the mass flux of tracer across the jet is constant.

$$\frac{d}{dx}\int_0^\infty uc2\pi r dr = 0 \tag{2.44}$$

Substituting the assumed profiles into Eq. 2.43 and Eq. 2.44 we get the kinematic momentum flux

$$M(x) = \int_0^\infty u^2 2\pi r dr = \frac{\pi}{2}({u_m}^2 b^2) = u_o^2\frac{\pi D^2}{4} \tag{2.45}$$

and the mass flux

$$\int_0^\infty uc2\pi r dr = \frac{\pi\lambda^2}{1+\lambda^2}(u_m c_m b^2) = u_o c_o\frac{\pi D^2}{4} \tag{2.46}$$

Assuming a linear spreading

$$b(x) = \beta\ x, \tag{2.47}$$

the solution is then given by:

$$\frac{u_m}{u_o} = \frac{1}{\sqrt{2}\beta}(\frac{x}{D})^{-1} \tag{2.48}$$

$$\frac{c_m}{c_o} = \frac{1+\lambda^2}{2\sqrt{2}\lambda^2\beta}(\frac{x}{D})^{-1} \tag{2.49}$$

The extensive measurements by Albertson *et al.* (1950) and Wygnanski and Fiedler (1969) in round jets have suggested a jet spread rate of $\beta_G = \frac{db}{dx} = 0.114$. This is close to the mean value of 0.107 from different investigators as given by Fischer *et al.* (1979). The more recent measurements of Papanicolaou and List (1988) give $\beta_G = 0.108$ and $\lambda = 1.2$. Adopting $\beta_G = 0.114, \lambda = 1.2$, the jet properties is summarised in Table 2.2.

Table 2.2. Round jet formulae based on nozzle diameter D and source velocity u_o, and the formulae based on jet momentum flux M_o, for $x \geq 6.2D$.

Jet width	$b = 0.114\ x$	
Centerline velocity	$u_m/u_o = 6.2\ (x/D)^{-1}$	$u_m = 7.0 M_o^{1/2} x^{-1}$
Centerline concentration	$c_m/c_o = 5.26/(x/D)^{-1}$	
Centerline dilution	$S = 0.19\ x/D$	
Average dilution ratio	$\overline{S} = 0.32\ x/D$	$Q = 0.286\ M_o^{\frac{1}{2}} x$

2.2 ADDITIONAL REMARKS ON 3D JET:

1. In a turbulent round jet, it is seen that the velocity is inversely proportional to the distances from source, while the volume flux increases linearly with distance. We have $Q \sim x$, $u_m \sim x^{-1}$, $M =$ constant, and the kinetic energy flux $E \sim u_m^3 b^2 \sim x^{-1}$; all the energy (velocity head) is ultimately dissipated (this is the discharge from an orifice into an infinite reservoir – see also Prob. 2.10). Note that the total amount of entrained flow depends only on the momentum flux M_o and x. It is interesting to note that the local Reynolds number $R = \frac{u_m b}{\nu} \sim u_m b$ is equal to a constant in a round jet.

2. <u>The exact momentum conservation</u>
Similar to the 2D case, in the above we have made some bold assumptions and have dropped all the turbulence terms $\overline{u'^2}, \overline{v'^2}$ other than turbulent shear. A more precise analysis retaining the turbulence terms would have resulted in the following global x-momentum conservation (Pratte and Keffer 1972):

$$\frac{d}{dx}\int (u^2 + \overline{u'^2} - \frac{(\overline{v'^2} + \overline{w'^2})}{2})dr = 0 \qquad (2.50)$$

where $\overline{w'}$ is the turbulent fluctuation in the azimuthal direction. Measurements by Wygnanski and Fiedler (1969) show that the magnitude of $v' \approx w'$, and hence Eq. 2.50 is often written as:

$$\frac{d}{dx}\int (u^2 + \overline{u'^2} - \overline{v'^2})dr = 0 \qquad (2.51)$$

As the two turbulence quantities are of the same order and their distributions similar in shape, their net contribution to the momentum flux may be assumed negligible; the final statement of global momentum conservation reverts to Eq.2.43.

3. The Entrainment Hypothesis

By examining the continuity equation, we can study the closure problem again in a different way. Integrating the continuity equation across the jet, it can be shown that:

$$\frac{dQ}{dx} = \frac{d}{dx}(\pi u_m b^2) = 2\pi(rv) = Q_e \tag{2.52}$$

where the right hand side represents the local entrainment flux into the jet. Defining the inflowing entrainment velocity at $r = b$ as the entrainment velocity and invoking the entrainment hypothesis, we have:

$$v_e = |v|_{r=b} = \alpha u_m \tag{2.53}$$

where the entrainment velocity is assumed to be proportional to the centerline velocity via the entrainment coefficient α. The problem can then be closed as:

$$\frac{d}{dx}(\pi u_m b^2) = 2\pi(\alpha u_m)b \tag{2.54}$$

Similar to the 2D case, it can be shown that (Prob. 2.2)

$$\beta_G = 2\alpha_G \qquad (3D\ jet) \tag{2.55}$$

for the Gaussian profile.

EXAMPLE 2.3 *In some cases the turbulent free jet solution can be used to estimate the initial dilution for wastewater discharges with negligible density differences, $Fr \to \infty$ (e.g sewage disposal in fresh water). Consider a submerged vertical round port sewage discharge (diameter 0.15 m) at the bottom of a river of depth 14 m. The density difference is practically zero and the initial velocity and coliform bacteria concentration are 1 m/s and 10^8 counts/mL respectively. Estimate the centerline velocity, average dilution, and bacteria concentration near the surface.*

Solution: The Reynolds number of this jet discharge is $Re = \frac{1\times 0.15}{10^{-6}} = 1.5 \times 10^6$. Using the formulae in Table 2.2 the average dilution, the centerline velocity, and the centerline bacteria concentration are $\overline{S} = 0.32(14/0.15) = \underline{30}$; $u_m = 1.0(6.2)(14/0.15)^{-1} = \underline{0.066}\ m/s$; and $c_m = 10^8 \times 5.26(14/0.15)^{-1} = \underline{5.64 \times 10^6}$ counts/mL.

For these typical depth and port dimensions it is seen that rather modest dilutions are achieved; note that the action of jet diffusion results in a rather rapid velocity decay to 1/15 of the initial value at discharge point. The region of practical interest is well into the zone of established flow (ZEF), with $x/D \sim 90$.

3. THEORY VS EXPERIMENT

We have developed a semi-empirical theory for turbulent jets, which enables us to estimate the important jet characteristics to a degree of accuracy sufficient for many environmental hydraulics or engineering calculations. A summary of the jet properties is given in Table 2.3. These properties have been extensively validated by experiments.

3.1 MEAN PROPERTIES

Fig. 2.5 shows the measured radial profile of the normalized time-mean axial ($x-$) velocity at various downstream locations for a turbulent round jet in stagnant fluid. The measurements were made by laser doppler anemometry (LDA) for a water jet with diameter $D = 1$cm and discharge velocity $u_o = 1.0, 0.9, 0.8, 0.7$ m/s (corresponding to Runs 1 to 4 respectively). It is seen that the streamwise velocity profile is self-similar and can be well-approximated by a Gaussian distribution. In Fig. 2.6, the derived centerline velocity variation for a round turbulent jet (Eq. 2.48) is plotted against the measured values; the classical data of a round air jet by Albertson *et al.*(1950) is also shown. The experimental results show clearly the existence of a potential core for about six diameters from the source, and the predicted variation in the ZEF is well-confirmed; the same result has been obtained by many different investigators (e.g. see Fischer *et al.* 1979). Similar statements can be made for the two-dimensional slot jet.

Fig. 2.7 shows some recent measurements of the tracer concentration (scalar) field in the jet cross-section using the Laser-induced Fluorescence (LIF) technique (Chu 1996). The contours of the time-averaged concentration in the jet cross-section and the radial profile are shown. The measured concentration c normalized by the centerline maximum c_m is plotted against the dimensionless radial distance r/b_{gc}, where $b_{gc} = \lambda b_g$ is the concentration half-width defined by the radial location at which $C = e^{-1}C_m$. It can be seen the radial concentration variation is self-similar; the data at different sections collapse nicely onto one curve and can be well-approximated by the Gaussian distribution. The radial symmetry and self-similar Gaussian concentration distribu-

Table 2.3. Summary of Jet properties (Based on Gaussian Profiles)

Parameter	Round jet	Plane Jet
Jet Volume Flux $Q = \int udA$	Dimensions L^3T^{-1} $Q = \pi u_m b^2$	L^2T^{-1} $Q = \sqrt{\pi} u_m b$
Specific momentum Flux $M = \int u^2 dA = M_o$	L^4T^{-2} $M = \frac{\pi}{2} u_m^2 b^2$	L^3T^{-2} $M = \sqrt{\frac{\pi}{2}} u_m^2 b$
Maximum time-averaged velocity u_m	$u_m = 7.0 M_o^{\frac{1}{2}} x^{-1}$	$u_m = 2.58 M_o^{\frac{1}{2}} x^{-\frac{1}{2}}$
Jet width b	$b = 0.114x$	$b = 0.12x$
Maximum time-averaged concentration excess c_m	$c_m = 5.94 Q_o C_o M_o^{-\frac{1}{2}} x^{-1}$	$c_m = 2.27 Q_o C_o M_o^{-\frac{1}{2}} x^{-\frac{1}{2}}$
Average Dilution $\overline{S} = \frac{Q}{Q_o}$	$\overline{S} = 0.29 M_o^{\frac{1}{2}} x Q_o{}^{-1}$	$\overline{S} = 0.55 M_o^{\frac{1}{2}} x^{\frac{1}{2}} Q_o{}^{-1}$
Jet spreading angle β	$\beta_G = 0.114$ $(\beta = 2\alpha)$	$\beta_G = 0.12$ $(\beta = \frac{4}{\sqrt{\pi}}\alpha)$
Entrainment coefficient α	$\alpha_G = 0.057$	$\alpha_G = 0.053$
Ratio of concentration to velocity width, λ	$\lambda = 1.2$	$\lambda = 1.35$

tions are demonstrated. The predicted linear spread of concentration width ($b_{gc} = \lambda b_g$) and linear increase of centerline dilution with distance downstream are also well-supported by data (Fig. 2.8). The width of the Gaussian concentration profile, b_{gc}, is generally wider than that of the velocity profile, b_g. Measured jet spread rate by various investigators are quite close: 0.114 (Albertson *et al.*1950), and 0.104 by Papanicolaou and List (1988) and Chu (1996) whose data also revealed that $db_{gc}/dx = 0.125$, and $\lambda = b_{gc}/b_g = 1.2$. The predicted linear increase of volume flux (or average dilution), $Q = 0.286 M_o^{1/2} x$ (Table 2.2), is also in excellent agreement with the formula $Q = 0.282 M_o^{1/2} x$ obtained by Ricou and Spalding (1961) from direct volume flux measurement in a

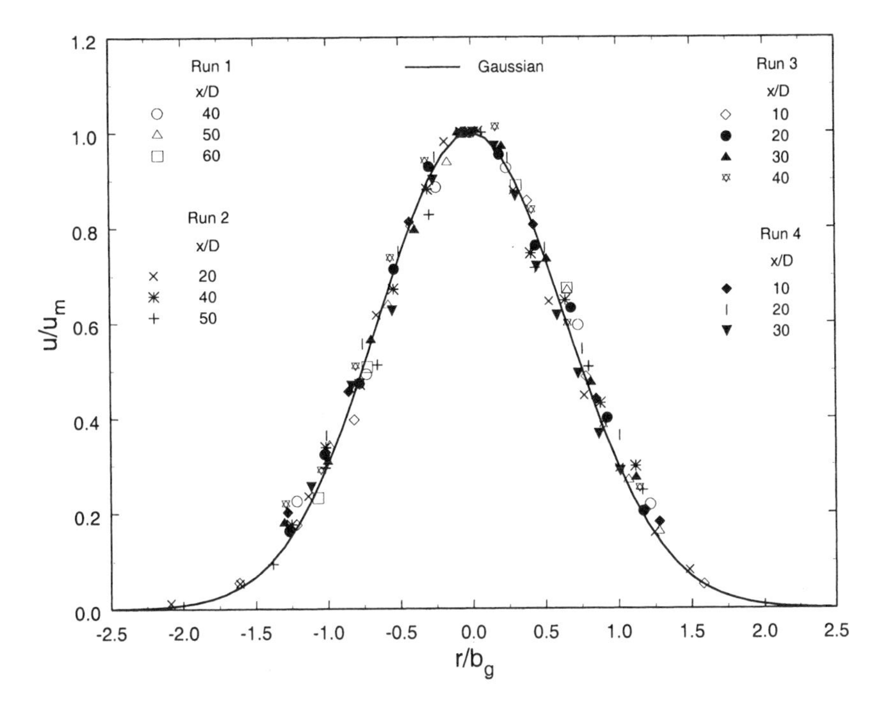

Figure 2.5. Measured radial profile of time-averaged velocity in a round jet

round jet and the more recent value of 0.284 given by Papanicolaou and List (1988).

3.2 TURBULENCE PROPERTIES

The foregoing results can be used to predict time-averaged quantities only. The instantaneous velocity or concentration at any point, however, can differ substantially from the time-mean value. As the eddies move around in a random fashion, a given position may be occupied by either turbulent jet flow or non-turbulent irrotaional fluid. The intermittency factor, γ, is the probability (percentage of time) when turbulent motion is detected at a fixed position. Instrumentation such as the hot-film anemometer can be trained to distinguish turbulent motion from the motion of its irrotational surrounding. Vorticity is measured by the anemometer and the part of the fluid motion with a vorticity level exceeding certain threshold is considered turbulent. The part of the motion with a vorticity below the threshold is irrotational. Figure 2.9 show the γ-profile for the velocity of a turbulent round jet obtained by Corrsin and Kistler (1954) and Wygnanski and Fiedler (1969). Note

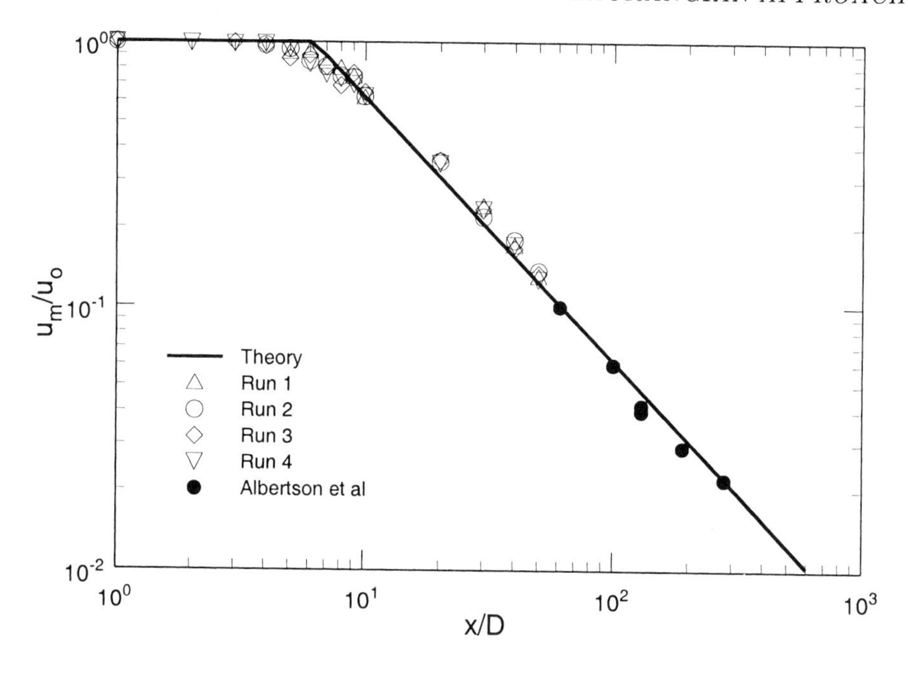

Figure 2.6. Centerline velocity variation of a turbulent round jet. In the zone of established flow, $x \geq 6.2D$, the experimental data follow closely the integral model prediction (solid line)

that if $\gamma = 0.5$ is taken to be the edge of the jet, then the jet boundary is located when $r/x \approx 0.162$, or when $r \approx \sqrt{2}b_g$.

The large turbulent fluctuations are illustrated in Fig. 2.10, which shows measurements of the instantaneous maximum and minimum concentration at different sections of the ZEF in a round turbulent jet (Chu 1996). The ratio of the instantaneous to the mean centerline concentration is plotted against $\frac{r}{b_{gc}}$. It is seen that the peak concentration can exceed the mean by as much as 50%. Fig. 2.11 shows the radial profile of concentration intermittency (γ = fraction of time a given position is occupied by the jet fluid). It can be seen there is a coherent central core of turbulent fluid ($\gamma = 1$). The jet edge can be defined by the radial location at which $\gamma = 0.5$; it is located at around $r \approx 1.2b_{gc} = \sqrt{2}b_g$. In general, although self-similarity of the mean flow is observed beyond the potential core, the turbulence properties (turbulent stresses, intermittency) reach a self-similar state after a longer period of jet development. For example, for a slot jet, the flow only becomes self-preserving (self-similarity of both the mean flow and turbulence quantities) beyond $x/B \approx 30$ (Bradbury 1965). For a round jet, self-preservation is reached

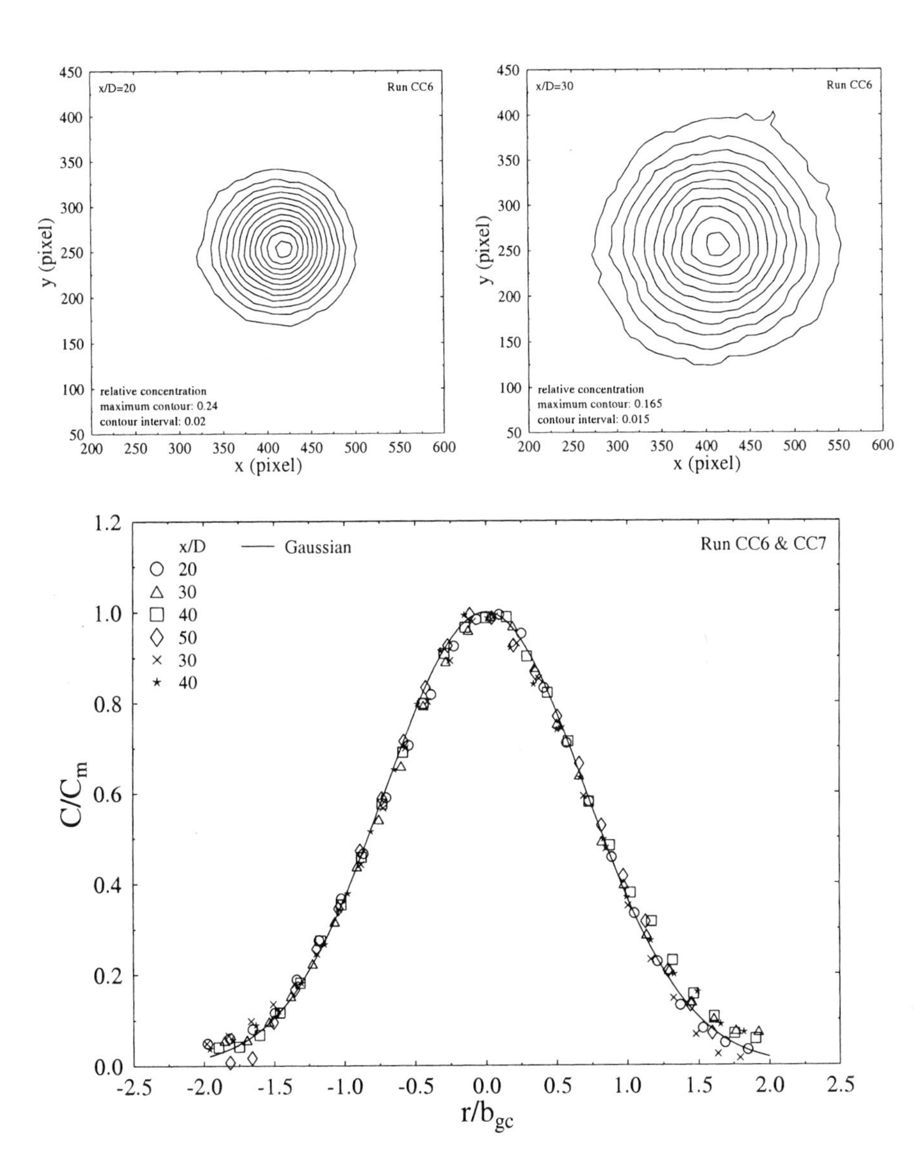

Figure 2.7. Measured cross-section tracer concentration field in a round jet (Chu 1996)

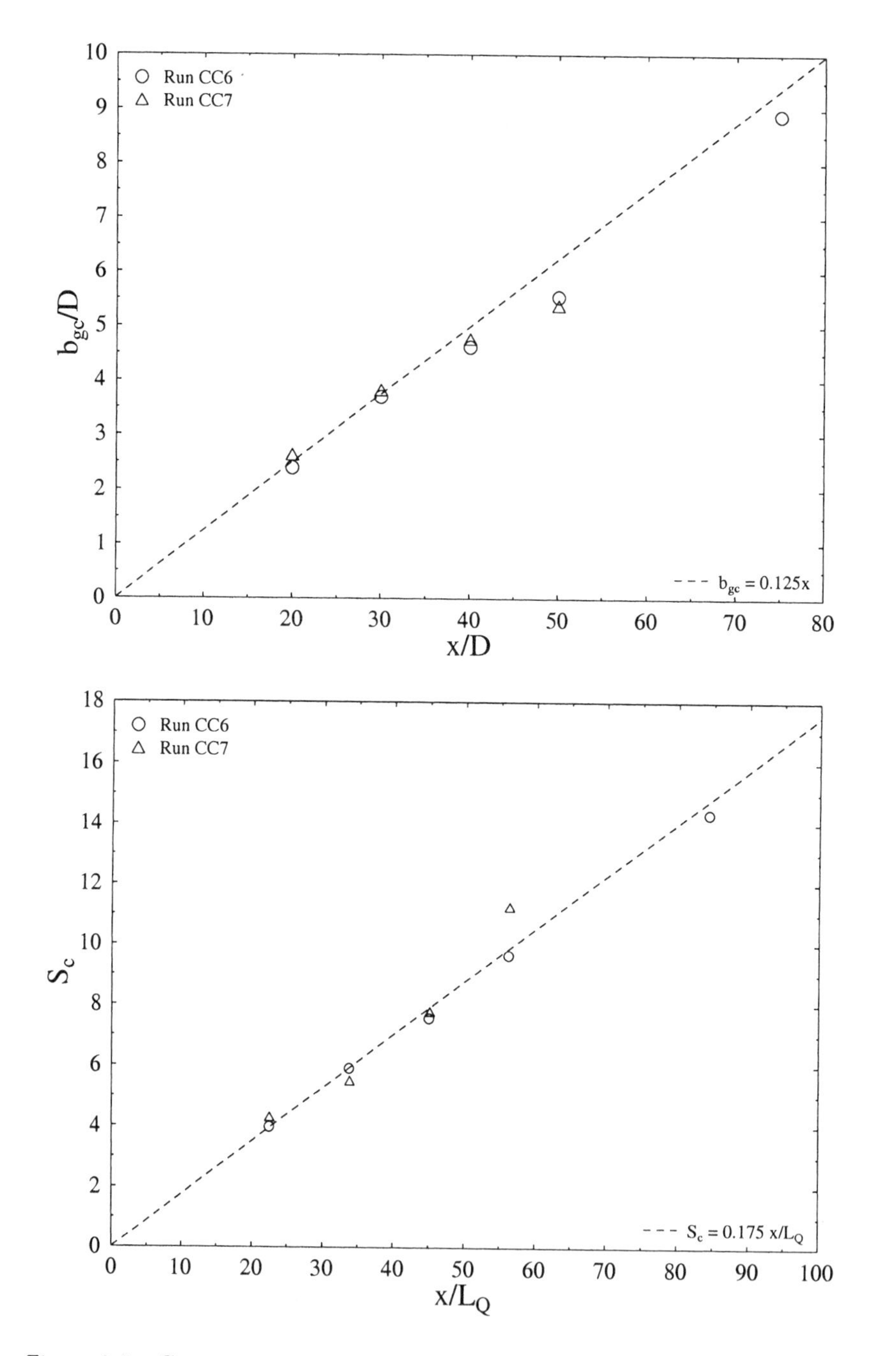

Figure 2.8. Concentration half-width and centerline dilution in a round jet (Chu 1996); $L_Q = (\pi/4)^{1/2} D$

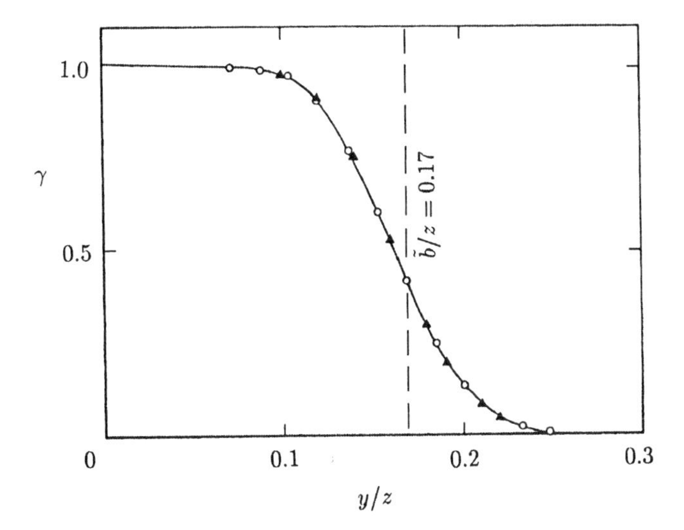

Figure 2.9. The intermittency-factor profile of a round jet. Data are reproduced from Corrsin and Kistler (1954) and Wygnanski and Fiedler (1969).

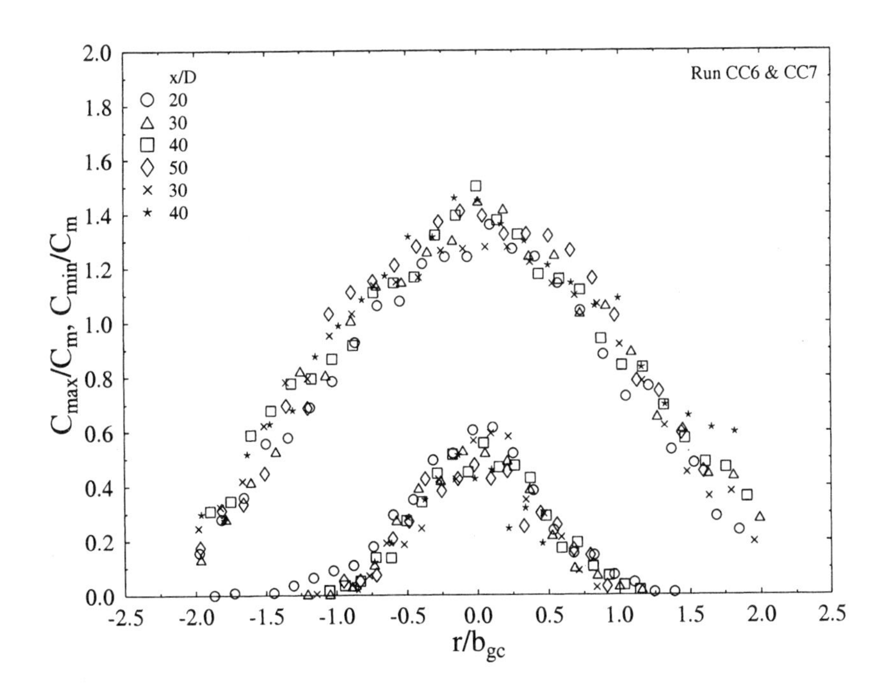

Figure 2.10. Radial profile of instantaneous maximum and minimum concentration in a round jet (Chu 1996)

only after $x/D \approx 50$ (Wygnanski and Fiedler 1969, Papanicolaou and List 1989, Chu 1996).

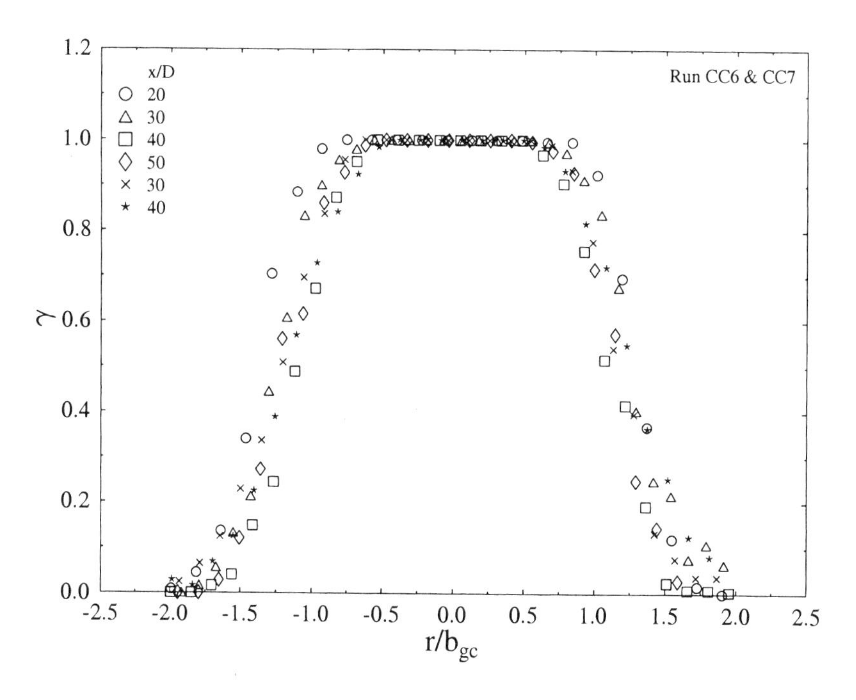

Figure 2.11. Radial profile of concentration intermittency in a round jet

4. THE TOP-HAT PROFILE

It is now apparent that the key physical variable controlling jet mixing is the **jet momentum flux**, and the governing equations can be written solely in terms of the integral fluxes. In fact, based on this physical insight (or hindsight), all the characteristic properties could have been deduced by dimensional reasoning alone to within a constant (Prob. 2.3). It is sometimes useful to represent the mass (volume) and momentum fluxes by a *'top-hat'* profile. The momentum and mass flux integrals can be evaluated by assuming a simple velocity profile:

$$u = \begin{cases} U & \text{if } r \leq B \\ 0 & \text{otherwise} \end{cases} \tag{2.56}$$

where U, and B are the velocity and half-width of an equivalent jet with a sharp boundary and uniform velocity, U, carrying the same mass flow and momentum flux as the actual jet. By equivalence of mass and momentum fluxes, the following relations between the two profiles can be obtained (Fig.2.12):

$$\pi U B^2 = \pi u_m b_g^2 \tag{2.57}$$

$$\pi U^2 B^2 = \frac{\pi}{2} u_m^2 \, b_g^2 \tag{2.58}$$

Thus, we have,

$$U = \frac{u_m}{2} \tag{2.59}$$

$$B = \sqrt{2}\, b_g \tag{2.60}$$

After U and B are obtained, the centerline or average concentration (dilution) can be deduced from tracer mass conservation (Chu *et al.* 1999). Similarly, the top-hat profile variables for a 2D slot jet is uniquely related to the Gaussian equivalents by $U = u_m/\sqrt{2}$, $B = \sqrt{\frac{\pi}{2}} b_g$.

The use of top-hat profiles not only offers analytical simplicity. As discussed above, the definition of top-hat width matches with the point of 50 percent velocity intermittency. The concentration defined by $r = B$ also seems to correspond to the visual edge of the jet. Fig. 2.13 shows a LIF image of the instantaneous concentration field of the jet cross-section; the visual edge of the jet cross-section corresponds to the concentration contour defined by $c(r = \sqrt{2} b_g) \approx 0.25 c_m$. Physically, the use of top-hat profile can be justified using the concept of dominant eddy (see later discussion). For many practical problems, a simple Gaussian profile simply does not exist. The use of top-hat profile avoids the ambiguity in profile assumption in transition between flow regimes while allowing simplicity in mathematical derivations. It has tremendous advantages when the challenging problem of a crossflow is considered - in fact two of the much-tested models that we will describe both use top-hat profiles.

5. PREDICTION OF POTENTIAL CORE LENGTH

We illustrate the ease with which the top-hat profile can be used with the linear jet spread assumption to yield a prediction of the potential core length. As an example, consider the round jet shown in Fig. 2.4. In the potential core, when turbulent diffusion has not penetrated into the centerline axis, we denote the half-width of the potential core by R, and the half-width up to the jet boundary B. In general $R(x)$ is not necessarily linear. The width of the shear/mixing layer is $(B - R)$. At

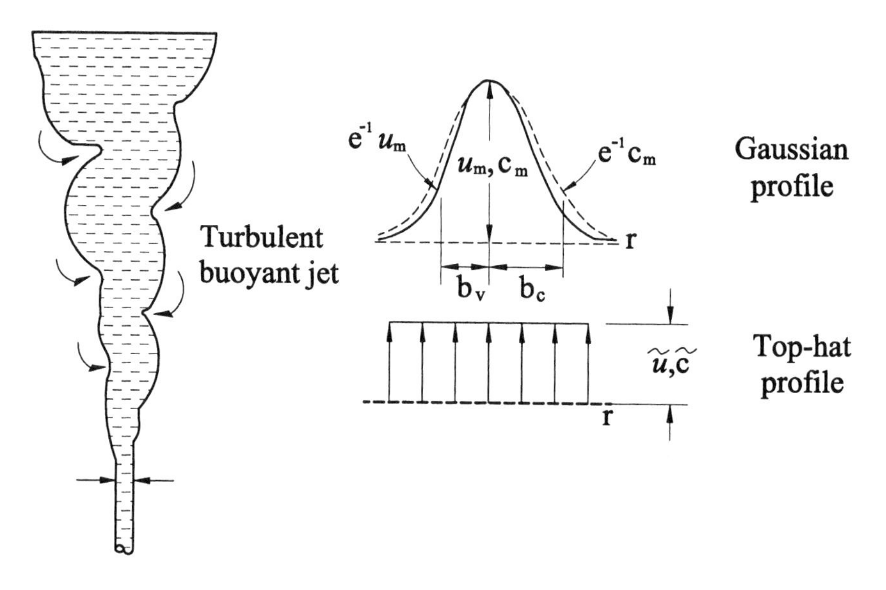

Figure 2.12. Turbulent buoyant jet in stagnant fluid (turbulent mean velocity and concentration distributions described by Gaussian profiles; path averaged jet properties defined by top-hat profiles)

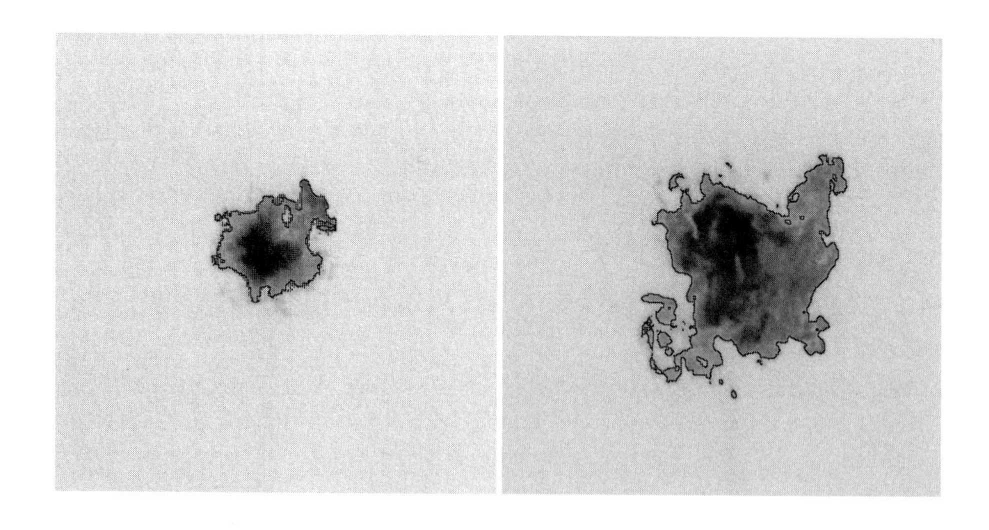

Figure 2.13. LIF image of instantaneous concentration field in round jet cross-section

the source, $x = 0$, $B = R = D/2$, while at the end of the potential core, $x = x_e$, we have $u_m = u_o$, and $U = u_o/2$ respectively (as within the core the x- velocity is u_o). Adopting a linear jet spread hypothesis and invoking momentum conservation between $x = 0$ and $x = x_e$, we then have:

$$\frac{d}{dx}(B - R) = \beta \tag{2.61}$$

$$\pi(u_o/2)^2 B^2 = \pi u_o^2 D^2/4 \tag{2.62}$$

where $\beta = \sqrt{2}\beta_G$ refers to the spread rate defined by the top-hat half-width. The above momentum conservation equation gives $B(x_e) = D$. Integrating from the source, $x = 0$, to the end of the potential core, $x = x_e$, and invoking the conditions $B = R = D/2$ and $B = D; R = 0$ at the respective limits, we have

$$\frac{x_e}{D} = \frac{1}{\beta} \tag{2.63}$$

Substituting, $\beta = \sqrt{2}\beta_G = \sqrt{2} \times 0.114 = 0.16$ into the equation gives $x_e/D = 6.2$ - exactly equal to the observed core length. Note that the spreading rate is independently obtained from the velocity profiles in the zone of established flow. The result is most encouraging; a similar analysis for the two-dimensional jet gives $x_e/d_o = 5.2$ (for $\beta_G = 0.154$). In a later chapter, the same approach will be adopted for the more demanding case of an inclined buoyant jet.

6. SUMMARY

The streamwise momentum flux of a turbulent jet is conserved. In this chapter, we have presented the basic analytical framework for integral modelling of turbulent free jets in a stagnant ambient fluid. The governing equations based on the boundary layer approximations are developed. It is shown that turbulence closure can be achieved either by a jet spreading or an entrainment hypothesis. The characteristic properties of the self-similar jet are derived and the predictions (Table 2.3) are shown to be in excellent agreement with experimental data.

PROBLEMS

2.1 a) By assuming self-similar Gaussian velocity and concentration profiles for a round momentum jet, derive the relations between the jet volume flux Q, kinematic momentum flux M, tracer mass flux Γ, and the centerline velocity u_m, half-width b, and concentration c_m. Determine the ratio of average to centerline dilution if the ratio of concentration to velocity width, λ, is 1.16.

b) Use the entrainment hypothesis to deduce the variation of the jet characteristic properties with downstream distance x in terms of the source fluxes and the entrainment coefficient α. Neglect the initial source flow and the zone of flow establishment.

2.2 a) Explain what is meant by i) the 'Entrainment Hypothesis', and ii) the 'Spreading Hypothesis'.

b) Assume the structure of the flow and tracer concentration in a two-dimensional (slot) turbulent jet with source kinematic momentum flux M_o (in the x-direction) and tracer mass flux Γ_o is schematized by 'top-hat' profiles:

$$u(x,y) = U(x)$$

$$c(x,y) = C(x) \quad y \leq B(x)$$

$$u(x,y) = 0$$

$$c(x,y) = 0 \quad y > B(x)$$

where $U(x)$, $B(x)$, $C(x)$ are the jet average velocity, half-width, and concentration as described by the top-hat profile. Use the entrainment assumption to deduce the characteristic jet properties $U(x)$, $B(x)$, $C(x)$ in terms of M_o and Γ_o, and the entrainment coefficient α. Neglect the initial source flow and the zone of flow establishment.

c) Solve the same problem by using the spreading hypothesis for turbulent closure. Establish the relation between the entrainment coefficient α and the spreading coefficient β for the 2D jet for this profile. Will the same relation hold if Gaussian profiles are assumed instead of top-hat profiles?

2.3 a) By applying a momentum balance to a suitable control volume and invoking the boundary layer approximations, show that the x-momentum equation for a two-dimensional momentum jet (oriented in the x-direction) is:

$$u\frac{\partial u}{\partial x} + v\frac{\partial u}{\partial y} = \frac{1}{\rho}\frac{\partial \tau}{\partial y}$$

where $\tau = \tau_{yx}$ is the shear stress, and the other terms have their usual meaning. Assume the static pressure to be nearly constant throughout the zone of jet diffusion.

Give a physical interpretation to τ for the case of a laminar and a turbulent jet. Show that the jet momentum flux $M(x) = \int_{-\infty}^{\infty} \rho\, u^2\, dy$ is a constant for either case.

b) Using dimensional analysis, deduce the variation of the centerline velocity u_m and average dilution Q/Q_o in a 2D turbulent jet with the downstream distance x, to within a proportionality constant that can be determined by basic experiments.

c) At a distance of 100 m, determine the centerline dilution in a non-buoyant turbulent round jet of initial diameter 1 m.

2.4 Adopting the velocity profiles in the zone of flow establishment (ZFE) of a 2D jet, Eq. 2.1 and Fig. 2.3, show by momentum conservation that $x_o/d_o = (\sqrt{\pi/2\beta})^{-1}$, where $x_o = L_o$ is the length of ZFE, and $\beta = db/dx$ is the jet spread rate. Show that within the ZFE, the inner boundary of the diffusion layer (edge of potential core) is given by $2r/d_o = 1 - x/x_o$. The experimental value of β ranged from 0.154 (Albertson *et al.*1950) to 0.116 (Kotsovinos 1976).

2.5 For a 2D plane jet, experimental measurements (Miller and Comings 1957) show that in the y-momentum equation, the dominant terms in the mixing field of the jet cross-section are the pressure and the transverse turbulent stress; hence the y-momentum equation reduces to $p + \rho\overline{v'^2} \approx p_\infty$, where p_∞= pressure outside the jet flow. Show that the x-momentum equation becomes:

$$u\frac{\partial u}{\partial x} + v\frac{\partial u}{\partial y} = -\frac{\partial \overline{u'^2}}{\partial x} - \frac{\partial \overline{u'v'}}{\partial y} - \frac{1}{\rho}\frac{\partial (p_\infty - \rho\overline{v'^2})}{\partial x}$$

Hence derive the following global x-momentum conservation:

$$\rho\frac{d}{dx}\int_{-\infty}^{\infty}(u^2 + (\overline{u'^2} - \overline{v'^2}))dy = -\int_{-\infty}^{\infty}\frac{\partial p_\infty}{\partial x}\,dy$$

In the case of a uniform external pressure p_∞, and when $\overline{u'^2} \sim \overline{v'^2}$, the momentum balance reduces to the simple form obtained by neglecting all pressure gradients.

$$\frac{d}{dx}\int_{-\infty}^{\infty}u^2\,dy = 0$$

2.6 For the plane jet above, show that the y-momentum equation is given (in conservative form) by:

$$\frac{\partial uv}{\partial x} + \frac{\partial v^2}{\partial y} = -\frac{\partial \overline{u'v'}}{\partial x} - \frac{\partial \overline{v'^2}}{\partial y} + \frac{1}{\rho}\frac{\partial (p_\infty - p)}{\partial y}$$

where $p - p_\infty$ = dynamic pressure. Each of the terms on the left hand side (LHS) of the equation represents convective y-acceleration of the mean flow; its order of magnitude can be estimated by $\sim V^2/b$, where V, b are y-velocity and width scales. On the other hand, the dominant turbulent stress term (second term on the RHS), can be estimated by $\frac{\partial \overline{v'^2}}{\partial y} \sim v_*^2/b$ where v_* is a characteristic scale of the turbulent velocity fluctuation. If experiments show that $v_*^2/u_m^2 \approx 0.06$ near the centre of the jet (u_m = centerline velocity), show that $\frac{\partial v^2}{\partial y} \ll \frac{\partial \overline{v'^2}}{\partial y}$, and hence the approximate y-momentum equation reduces to $(p - p_\infty + \rho\overline{v'^2}) = 0$. (Note that the transverse mean velocity can be estimated by $V \sim u_m\frac{db}{dx} \approx 0.1u_m$).

2.7 Experimental measurements for a plane turbulent jet show that the transverse distribution of turbulent stresses can be well-approximated by the following (Miller and Comings 1957):

$$\frac{\overline{u'^2}}{u_m^2} = 0.17e^{-1.29\eta^2} - 0.10e^{-2.90\eta^2}$$

$$\frac{\overline{v'^2}}{u_m^2} = 0.057e^{-0.76\eta^2}$$

where $\eta = y/b$ and b is the Gaussian half-width based on the $1/e$ velocity point.

Determine the percentage contribution of the turbulence quantities in the momentum flux integral of Prob. 2.5.

2.8 Estimate the position of maximum turbulent shear in the cross-section of i) a plane turbulent jet, and ii) a round turbulent jet. For a plane jet the shear stress can be written as $\tau = -\rho\overline{u'v'} = \rho\nu_t\frac{\partial u}{\partial y}$, where ν_t is the eddy or turbulent kinematic viscosity. The eddy viscosity can be assumed to be constant in the jet cross-section and related to the characteristic velocity $u_m(x)$ and length scale $b(x)$. (This is known as Prandtl's free shear layer hypothesis for turbulence closure). Compare your result with the experimental measurements of Bradbury (1965) and Wygnanski and Fiedler (1969) which show the location of maximum shear at $y = 0.72\ b_g$ and $r = 0.71\ b_g$ for a 2D and 3D jet respectively.

2.9 In the rocket jet shown in Fig. 1.1, estimate the jet Reynolds number assuming a thrust of three million pounds ($13.4\times10^6\ N$), density and viscosity of air $= 1.2\ kg/m^3$ and $2\times10^{-5}\ m^2/s$ respectively.

2.10 The kinetic energy of a turbulent round jet is eventually dissipated into the energy of turbulence. The kinetic energy flux is defined as $E = \int \frac{1}{2}\rho u^3 dA$. As the jet spreads with distance from the source, the kinetic energy flux decreases as x^{-1} due to the conversion of the mean flow energy to turbulent motion. Using the round jet solution, evaluate dE/dx and show that the rate of energy conversion per unit mass of the fluid in a volume equal to $\pi b^2 dx$ is:

$$P = \frac{\beta}{6}\frac{u_m^3}{b}$$

where u_m and b are the centerline velocity and Gaussian radius of the round jet respectively.

The production of the turbulent energy is initially associated with the large scale turbulent motion. Through the nonlinear process of the energy cascade, this energy is passed onto the turbulent motion of smaller and smaller eddies, and finally dissipated at the molecular level as heat.

2.11 Use the jet solution developed in Chapter 2 (Table 2.3) to derive an expression for the stream function $\psi(x,y)$ for the 2D jet, where $\frac{\partial\psi}{\partial y} = u(x,y)$, and $\psi(x,0) = 0$. Sketch a few streamlines to indicate the general flow pattern.

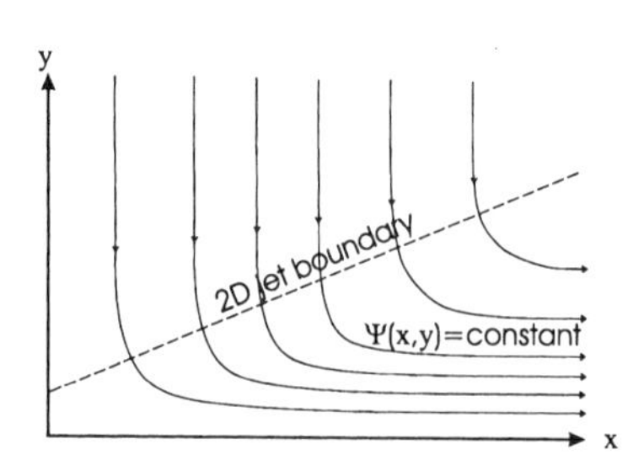

2.12 *Jets for Saving Lakes*:
The hypolimnion of a lake is often low in dissolved oxygen in the summer due to eutrophication and pollution. Sometimes it is desired to raise the oxygen level in the anoxic lower layer while maintaining the stratification (e.g. to keep the low temperature water for fisheries). One way is to inject a highly-oxygenated stream of water into the hypolimnion. The attached figure shows a flow of 1 m^3/s of sidestream withdrawn from a 8 m deep lower layer of a small lake 5 km in length. The sidestream is oxygenated with commercial oxygen by a special device to a DO concentration of 100 mg/L; it is then reintroduced into the hypolimnion and distributed through a number of jet nozzles mounted at 4 m from the bottom. As a reference this system can add around 8 mg/L of DO to a natural water flow of 12 m^3/s.

a) If the jet-induced velocity at the distance where the jet spreads to the bottom, $x = x_1$, is not to exceed 0.1 m/s (to prevent re-suspension of polluted bottom sediment), determine the minimum nozzle diameter and the number of jet nozzles required. Energy considerations dictate that the initial jet velocity should not exceed 6 m/s. Assume the jets act independently of each other and neglect the ambient velocity.

b) For your design estimate the ratio of entrained flow to the jet flow at $x = x_1$. Determine also the distance required for the DO concentration in the jet to drop down to 5 mg/L. Assume zero ambient DO concentration.

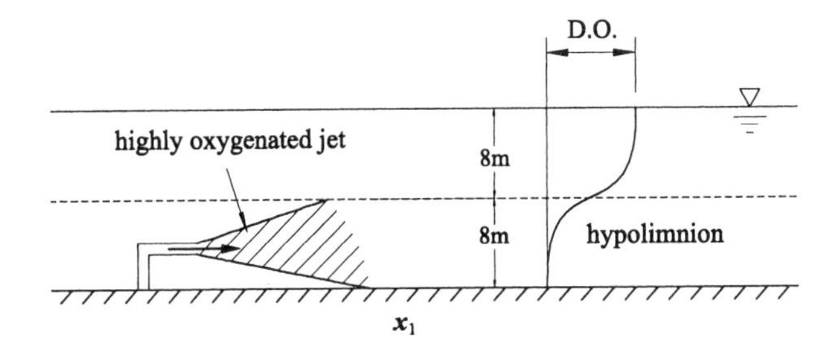

2.13 *Jet Interference*:
a) Consider two identical turbulent free jets located on the y-axis and discharging in proximity to each other; the initial jet velocity and x-momentum of each jet are u_o and M_o respectively. According to Reichardt's inductive theory of turbulence, the jet interaction can be accounted for by superimposing u^2 (where u = turbulent mean velocity), a measure of the x-momentum. In the plane containing the two jets (the $x-y$ plane), the x-momentum is given by:

$$u^2(x,y) = u_m^2(x) \sum_i e^{-2(y-y_i)^2/b^2}$$

where y_i is the location of the jet on the y-axis, and $u_m(x)$, $b(x)$ are the centerline velocity and width of the single jet as given by Table 2.3.

At the location when the two jets merge, when $2b(x) = s$, find the x-velocity at the point M indicated on the figure (mid-way between the two jets). How will your result change if jet velocities (instead of momentum) are additive.

This result can be generalized to predict the velocity for an array of interfering jets (located at mid-depth of hypolimnion) for Prob. 2.12.

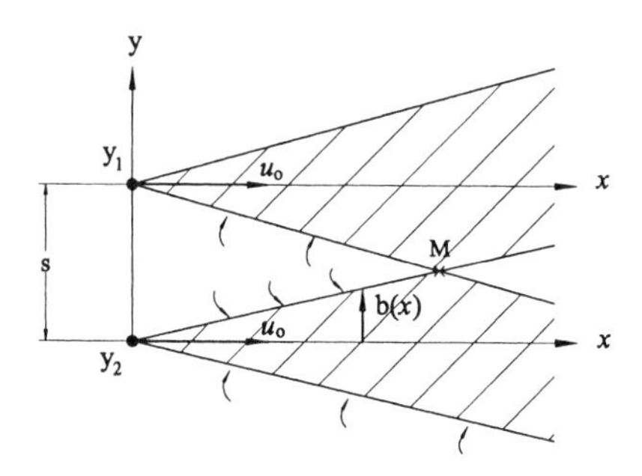

b) Consider the governing x-momentum equation for an axisymmetric jet:

$$\frac{\partial rU^2}{\partial x} + \frac{\partial rUV}{\partial r} = 0$$

where U, V are instantaneous velocities in the x- and radial direction in a cylindrical co-ordinate system. Based on experiments, Reichardt assumes that the rate of transfer of x-momentum in the transverse (radial) is proportional to the transverse gradient in x-momentum:

$$\overline{UV} = -\Lambda(x)\frac{\partial U^2}{\partial r}$$

where $\Lambda(x)$ is a momentum transfer length that is only a function of x. By neglecting turbulent normal stress terms, we then have

$$\frac{\partial u^2}{\partial x} = \frac{\Lambda}{r}\frac{\partial}{\partial r}(r\frac{\partial u^2}{\partial r})$$

Based on self-similarity, show that $M \sim u_m^2 b^2$, and u^2 must have a form of $u^2 = \frac{K}{b^2} f(r/b)$, where K= constant, and $b \sim x$. Hence show that if $\Lambda = \frac{b}{2}\frac{db}{dx}$, the solution to the above equation is

$$u^2 = \frac{K}{b^2} e^{-(r/b)^2}$$

If $\Lambda = kx$ where k=constant, the momentum equation can be transformed to a linear "diffusion" equation in terms of u^2 and a stretched x-coordinate $X = kx^2/2$:

$$\frac{\partial u^2}{\partial X} = \frac{1}{r}\frac{\partial}{\partial r}(r\frac{\partial u^2}{\partial r})$$

Thus at the same X (or x), i.e. if all jets have the same x-dependence for $\Lambda(x)$, the interaction in the transverse $(y - z)$ plane can be obtained by super-imposing u^2.

Chapter 3

TURBULENT BUOYANT PLUMES

Plumes are fluid motions that are produced by continuous sources of buoyancy. Convective currents set up by heated bodies such as a cigarette and a space heater are examples. The fluid in contact with the body attains a higher temperature than its surrounding fluid, rises as a result of its lower density, and in so doing draws ambient fluid radially inwards to mix with the warm fluid in the plume. Whereas the plume generated in this fashion may be laminar near the body, at some distance above it will break up into eddies by virtue of the momentum it derives from the force of buoyancy. Other examples of plumes include sewage effluent from sea outfalls, the localised high temperature fluid discharge from the earth's crust at the bottom of the deep ocean (hydrothermal vents), the injection of concentrated brine (from desalination plants) into sea water, fire plumes, hot gases from smokestacks and volcanic eruptions.

Figures 1.2 and 3.1 show two examples of the turbulent flow in the plume. The turbulent flow in Figure 1.2 is produced by the heat released from a fire. The flow in Figure 3.1 is produced in the laboratory by discharging warm water in the form of a buoyant jet from a tube into a tank of cold water. The fire and the buoyant jet are characterized in the near field by the source densimetric Froude number:

$$Fr = \frac{w_o}{\sqrt{(\Delta\rho_o/\rho_a)gD}} \tag{3.1}$$

where w_o = velocity at the source, $\Delta\rho_o = |\rho_o - \rho_a|$ = density difference between the source fluid and the ambient fluid, ρ_a = ambient density, and D = size of the source. The buoyant jet in the laboratory has a source densimetric Froude number $Fr = 5.6$. In the fire, the initial velocity w_o is zero, hence the densimetric Fr is also zero. Despite the very difference source characteristics, the far field of both plumes are determined by their source buoyancy flux. Any flow produced by a continuous source of buoyancy is expected to behave like a plume at large distances from the source regardless of the generating mechanism, although the plume in the transition between the near field and the far field may be affected by the source densimetric Froude number.

The pioneering work on plumes was due to Rouse, Yih and Humphreys (1952), who studied the convection above line and point sources of heat in air. By assuming similarity of velocity, density deficiency, and shear stress distribution at all heights, they derived the variation of the characteristic plume parameters and confirmed these relationships experimentally. Subsequently, turbulent plumes in stagnant fluid have been extensively studied in the context of buoyant jets for wastewater disposal by e.g. Abraham (1963 a & b), Cederwall (1968), Koh and Brooks (1975), Jirka *et al.*(1975) and Jirka and Harleman (1979). These studies related to wastewater disposal were primarily concerned with the mean concentration field, and relatively few measurements were made for the velocity field. More refined measurements including the turbulence properties were obtained using advanced experimental techniques by Kotsovinos and List (1977) for the plane buoyant jet in water and by Chu *et al.*(1981) and Papanicolaou and List (1988) for the round buoyant jet in water.

The results obtained from the measurements in the buoyant jets and plumes are reviewed in this chapter using the analytical framework based on both Eulerian and Lagrangian methods. The Eulerian method, due to Morton, Taylor & Turner (1956) and Turner (1986), is based on the entrainment hypothesis. The Lagrangian method is based on a spreading hypothesis and the concept of dominant eddy. Both the Eulerian and Lagrangian methods lead to the same asymptotic results that are consistent with experimental observation. However, as it will be demonstration in subsequent chapters, the Lagrangian method generally leads to a more consistent formulation in jet and plume problems in coflows and crossflows.

1. BUOYANCY AND REDUCED GRAVITY

The plume is produced by a continuous source of buoyancy. The flux of the buoyancy in the plume is calculated by considering an element of the buoyant fluid of volume δV and density ρ in an environment of density ρ_a. The weight of the fluid in the element is $W = \rho g \delta V$, which is a downward force. The pressure force on the surface of the fluid element is equal to the weight of *ambient* fluid displaced by the element

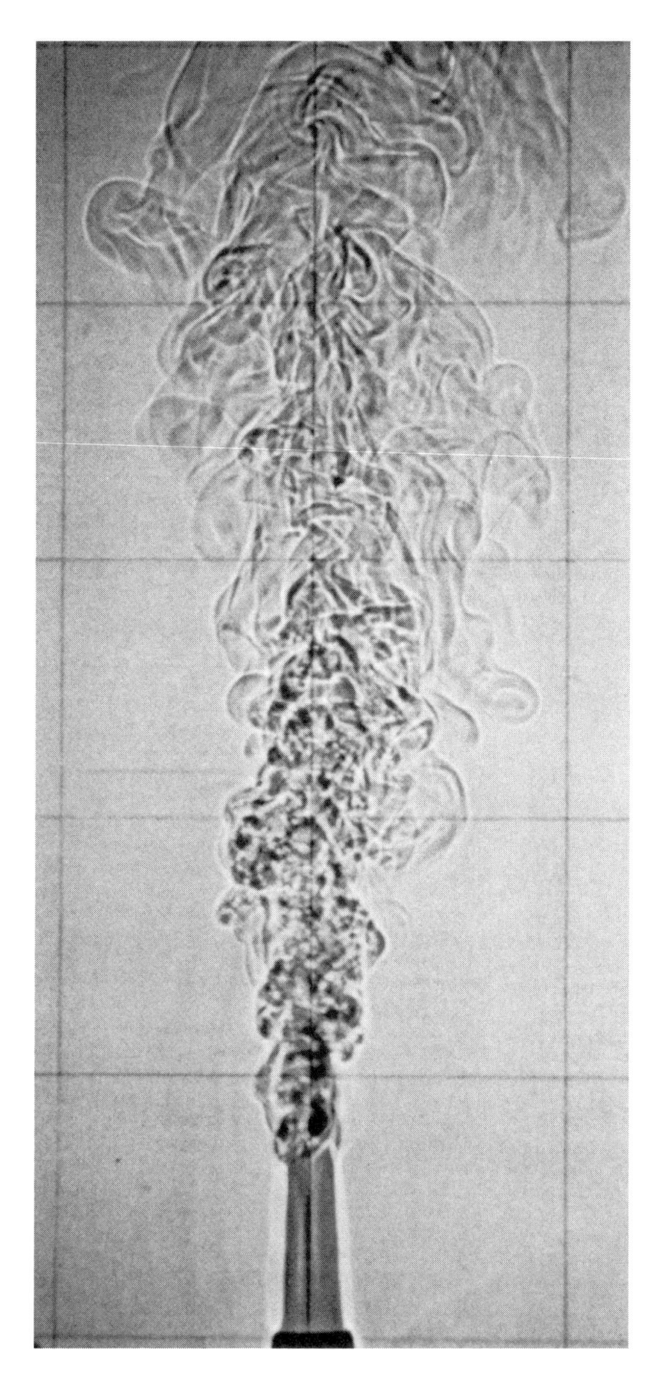

Figure 3.1. Shadow graph of a vertical turbulent buoyant jet in nearly stagnant ambient fluid ($Fr = 5.6$)

is $P = \rho_a g \delta V$, which is an upward force. The net force is the difference between the upward pressure force P and the downward force of the weight W, and that is the buoyancy force $(\rho_a - \rho) g\, \delta V$. The buoyancy force per unit volume of fluid is

$$\text{buoyancy} = (\rho_a - \rho) g$$

The buoyancy force per unit mass of the fluid is

$$\text{reduced gravity} = \frac{(\rho_a - \rho)}{\rho} g$$

The *reduced gravity* is a term introduced to describe the distribution of the force per unit mass of the fluid while the *velocity* is to specify the distribution of momentum per unit mass of the fluid. The momentum in a fluid of non-uniform density is produced continuously by the force of buoyancy. Many physical processes can lead to changes in fluid density and hence give rise to sources of buoyancy; e.g. heating by radiation and conduction, diffusion of solute, or simply the discharging of fluids and mixtures into a receiving environment of different density. In the presence of buoyancy, momentum would be generated at a rate equal to the buoyancy force (Newton's second law of motion).

If the buoyancy in the plume is produced by discharging fluid of density ρ_o at a rate Q_o (volume flux) into an ambient fluid of density ρ_a, the buoyancy flux associated with the discharge would be

$$F_o = (\rho_a - \rho_o) g Q_o. \tag{3.2}$$

If the buoyancy in the plume is produced by a heat source, such as in the case of a fire, the buoyancy flux F_o is related to the heat flux H by the following relation:

$$F_o = g \frac{\gamma_T H}{c_p} \tag{3.3}$$

where

$$\gamma_T = -\frac{1}{\rho}\left(\frac{\partial \rho}{\partial T}\right)$$

is the volume coefficient of thermal expansion (percentage of volumetric change per degree of temperature change) and c_p is the specific heat at constant pressure (change in energy per unit mass of fluid and per degree of temperature change). The thermal expansion coefficient is determined by the equation of state, which for water is given in Appendix A. For ideal gases, the equation of state is $p = \rho RT$, where p = pressure, R = gas constant, and T = absolute temperature. For a given pressure, the fractional changes in density of an ideal gas is negatively proportional to the fractional changes in temperature, i.e., $\delta\rho/\rho = -\delta T/T$. The thermal expansion coefficient for ideal gas is $\gamma_T = 1/T$.

The flux of the buoyancy force in a plume may change due to the variation of the volume expansion coefficient, or due to the variation of the density in a stratified ambient fluid. These effects on the variations are examined in chapter 5 and Appendix A, respectively. However, the buoyancy flux is assumed to be a constant in the consideration of the plume in this chapter.

EXAMPLE 3.1 *The effluent from a desalination plant has a saline concentration of 10% and a flow rate of 10 m³/s. Find the buoyancy flux of this effluent if the salinity in the surrounding sea is 3%.*

SOLUTION: In this case the density of the effluent is heavier than its surrounding sea water. The buoyancy force is downward. Assuming density differences are proportional to salt concentration (see Appendix A), the reduced gravity associated with the source fluid is

$$g'_o = \frac{\text{buoyancy force}}{\text{unit mass}} = \frac{\rho_a - \rho_o}{\rho_o} g = (.03 - .10) \times 9.81 = -0.687 \text{m/s}^2.$$

The buoyancy flux is

$$F_o = \frac{\text{buoyancy force}}{\text{unit time}} = \frac{\text{buoyancy force}}{\text{unit mass}} \times \frac{\text{mass}}{\text{unit time}} = g'_o \rho_o Q_o.$$

The specific buoyancy flux (buoyancy flux per unit density) is

$$\frac{F_o}{\rho_o} \simeq \frac{F_o}{\rho_a} = g'_o Q_o = \underline{-6.87} \text{ m}^4/\text{s}^3$$

In this case, the buoyancy flux is negative because the discharge fluid is heavier than the ambient fluid.

EXAMPLE 3.2 *A fire plume is set up by burning a 0.6 m diameter pool of fuel (heptane) in a large room. Find the buoyancy flux associated with the fire assuming that the heat flux is $H = 500$ kW.*

SOLUTION: The convective heat flux integrated across the plume of cross-sectional area A is

$$H = \int_A \rho c_p (T - T_a) w dA$$

Temperature change is related to density change through the thermal expansion coefficient γ_T by

$$\delta\rho = \frac{\partial \rho}{\partial T} \delta T = -\rho \gamma_T \delta T.$$

Local linearization of the relation between temperature T and T_a gives

$$\frac{\rho - \rho_a}{\rho} \simeq -\gamma_T (T - T_a).$$

For ideal gas, $\gamma_T = 1/T$. Hence, the heat flux is related to the buoyancy flux as follows:

$$H = \int_A \rho c_p T_a \frac{(T - T_a)}{T_a} w \, dA \simeq \frac{\rho c_p T_a}{g} \int_A \frac{\rho_a - \rho}{\rho_a} g \, w \, dA \simeq \frac{c_p T_a}{g} F_o$$

where $F_o = \int_A (\rho_a - \rho) g \, w \, dA$ is the buoyancy flux. (See Equation 3.3, Example 3.3 and also Prob. 3.6.) For air, the specific heat at constant pressure is $c_p = 1.006 \; kJ/kg^oK$. Assuming standard atmosphere with $T = 293^oK$ and $\rho_a = 1.2$ kg/m^3, the *specific* buoyancy flux is

$$\frac{F_o}{\rho_a} = \frac{gH}{\rho_a c_p T} = \frac{9.81 \times 500}{1.2 \times 1.006 \times 293} = \underline{13.9} \, \mathrm{m}^4/\mathrm{s}^3$$

The buoyancy flux is this specific flux times the density of the air.

2. TURBULENT ROUND PLUME

The round (axisymmetric) plume is produced by a steady and continuous source of buoyancy. The source of the buoyancy could be the result of the buoyant fluid discharging from an orifice as the case of the buoyant jet shown in Fig. 3.1, or could be the result of the heat produced by the fire as shown in Fig. 1.2. The problem of the plume is sketched in Fig. 3.2. As in the case of a jet, the objective of the problem is to find the velocity $w(z, r)$ and the tracer concentration $c(z, r)$ field of the plume as a function of the vertical co-ordinate (positive upwards) z and the radial co-ordinate r. Experiments have shown that the plume is a boundary-layer type of flow. The velocity and the concentration profiles in the fully established flow are similar in shape at all heights, and well-described by Gaussian profiles. The problem then reduces to the prediction of the maximum velocity, w_m, the width, b, and the maximum concentration, c_m, as a function of the lengitudinal distance z from the source.

2.1 DIMENSIONAL CONSIDERATIONS

A number of plume properties is derived from dimensional considerations. Neglecting the effects of the initial momentum flux and the size of the source, the width of the plume b is a function of its buoyancy flux at the source, F_o, the fluid density ρ_a, and the elevation from the source, z, as follows:

$$b = \mathrm{fn}(F_o, \rho_a, z). \tag{3.4}$$

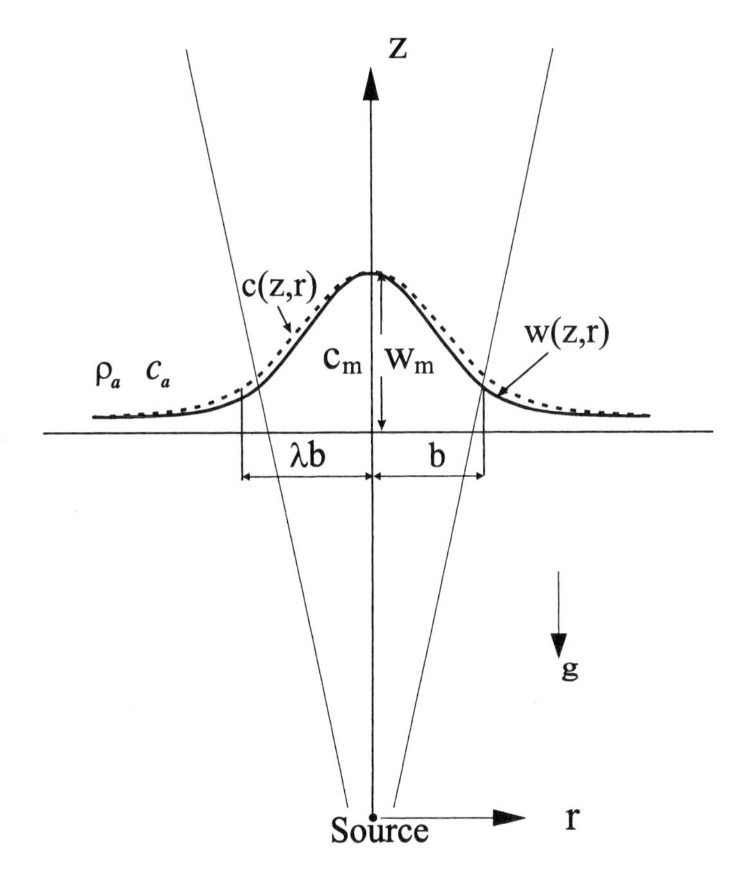

Figure 3.2. Sketch of the mean velocity and concentration profiles in a round plume. The width of the concentration profile, λb, is slightly wider by a factor λ compared with the width of the velocity profile, b.

There are 4 independent variables with 3 basic dimensions and the only dimensionless variable in this functional relation is

$$\Pi = \frac{b}{z}. \tag{3.5}$$

which must be constant. Hence,

$$b = \beta z. \tag{3.6}$$

The width of the plume increase linearly with the distance from the source according this dimensional consideration. The constant β in this linear relation is the spreading rate of the plume.

The method of dimensional analysis can also be applied to find the velocity of the plume. The vertical velocity induced by the plume's buoyancy, w, is a function of its buoyancy flux at the source, F_o, the fluid density ρ_a, the elevation above the source, z, and the radial distance from the plume centerline, r, as follows:

$$w = \text{fn}(F_o, \rho_a, r, z). \tag{3.7}$$

There are 5 independent variables involving 3 basic dimensions in this functional relation. The dimensionless parameters are

$$\Pi_1 = \frac{w}{[F_o/(\rho_a z)]^{\frac{1}{3}}} \quad \text{and} \quad \Pi_2 = \frac{r}{z}. \tag{3.8}$$

These then lead to the following functional form for the velocity profile

$$\frac{w}{w_m} = \text{fn}(\frac{r}{z}), \tag{3.9}$$

where the maximum velocity at the centerline is proportional to the plume characteristic velocity $[F_o/(\rho_a z)]^{\frac{1}{3}}$, i.e.,

$$w_m \sim [\frac{F_o}{\rho_a z}]^{\frac{1}{3}} \tag{3.10}$$

The dimensionless form of the concentration profile

$$\frac{c}{c_m} = \text{fn}(\frac{r}{z}) \tag{3.11}$$

is obtained in a similar manner. The maximum concentration at the centerline (c_m) times the characteristic velocity and flow area, $c_m[F_o/(\rho z)]^{\frac{1}{3}}b^2$ is proportional to the mass flux at the source, Γ_o. Hence,

$$c_m \sim \frac{\Gamma_o}{[F_o/(\rho z)]^{\frac{1}{3}}}. \tag{3.12}$$

Figure 3.3 shows the time-mean velocity and concentration data obtained from the experiment by Papanicolaou and List (1988). A close approximation of these data are the Gaussian profiles

$$\frac{w}{w_m} = \exp[-\frac{r^2}{b^2}] \tag{3.13}$$

$$\frac{c}{c_m} = \exp[-\frac{r^2}{(\lambda b)^2}] \tag{3.14}$$

in which b and λb are the width of the time-mean velocity and concentration profiles, respectively. The reduced gravity, that is the buoyancy force per unit mass of the plume fluid $g' = g(\rho_a - \rho)/\rho_a$, have the same profile as the concentration of any tracer. For a heated jet, the temperature excess, $\Delta T = T - T_a$, is also expected to be Gaussian across the plume. Hence,

$$\frac{g'}{g'_m} = \exp[-\frac{r^2}{(\lambda b)^2}] \tag{3.15}$$

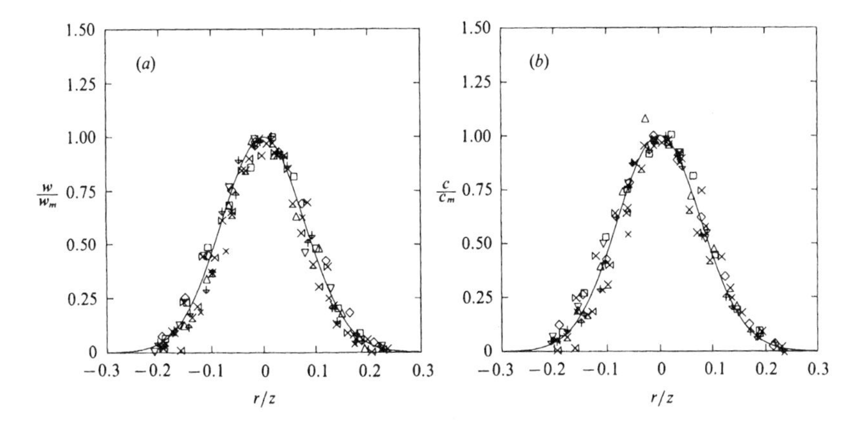

Figure 3.3. Measured mean velocity and concentration profile across a round plume (from Papanicolaou and List 1988, $z/l_s = 8.6 - 62$); w_m, c_m = centerline velocity and concentraton

$$\frac{\Delta T}{\Delta T_m} = \exp[-\frac{r^2}{(\lambda b)^2}] \tag{3.16}$$

The data of Papanicolaou and List (1988) in Fig. 3.3 show the width of the plume increases with distance from the source following the linear relations $b_v = b = 0.105\,z$ and $b_c = \lambda b = 0.112\,z$ with a value of $\lambda = 1.06$ (see also Fig. 3.5). The spreading rate for the velocity profile is different from the rate for the concentration profile; $\lambda = b_c/b_v$ is the ratio of the concentration to velocity width. Some variation in the spreading rate has been observed in different experimental investigations. The average of the experiments by Rouse *et al.* (1952), Morton *et al.*(1956) and Abraham (1963a) gives a spread rate $\beta = db_v/dz = 0.102$ and $\lambda = 1.16$. The plume spread rate of around 0.11 has also been observed in related buoyant plume studies (Lee and Cheung 1991). It appears that the spread rate of the plume is approximately the same as the jet. Over the jet-plume range, the spreading rate varies from $\beta = 0.102 \sim 0.114$. For most practical purposes, a constant spreading rate of $\beta = db_v/dz \simeq 0.108$ may be adopted for both jets and plumes.

Dimensional consistency requires that the width of the round plume increases linearly with distance from the source while the velocity decreases with distance as $(F_o/\rho_a)^{\frac{1}{3}} z^{-\frac{1}{3}}$. The volume flux of a round plume is $Q = \int_0^\infty w 2\pi r\, dr$. Since $w \sim (F_o/\rho_a)^{\frac{1}{3}} z^{-\frac{1}{3}}$ and $b_v \sim z$, the volume flux increase with elevation follows a five-third power law: $Q \sim (F_o/\rho_a)^{\frac{1}{3}} z^{\frac{5}{3}}$. The tracer concentration follows the minus-five-third law: $c \sim \Gamma_o/Q \sim$

$\Gamma_o (F_o/\rho_a)^{-\frac{1}{3}} z^{-\frac{5}{3}}$. If the plume is created by a heat source, the temperature variation will be given by $\Delta T \sim H/(\rho_a c_p)\,(F_o/\rho_a)^{-\frac{1}{3}} z^{-\frac{5}{3}}$.

The above dimensional analysis also shows that the key parameter in the solution is F_o/ρ_a, i.e. the specific buoyancy flux. In most applications we will be concerned with small density differences $\Delta\rho/\rho_a \ll 1$ and working with the specific buoyancy flux $F_o/\rho \simeq F_o/\rho_a$; henceforth unless otherwise stated $F_o = Q_o \frac{\Delta\rho_o}{\rho_a} g \approx Q_o \frac{\Delta\rho_o}{\rho_o} g$ denotes the specific buoyancy flux.

2.2 EULERIAN INTEGRAL MODEL

2.2.1 GOVERNING EQUATIONS

Consider the vertical axisymmetric plume in Fig. 3.2, with the following velocity and scalar concentration excess profiles in the established flow.

$$w(z,r) = w_m(z) e^{-\left(\frac{r}{b}\right)^2}$$

$$c(z,r) = c_m(z) e^{-\left(\frac{r}{\lambda b}\right)^2} \qquad (3.17)$$

where w_m and c_m are the centerline maximum velocity and concentration (excess), and b is the plume radius defined by location with velocity $e^{-1} w_m$. In the axisymmetric co-ordinate system (z, r), the governing equations for turbulent incompressible flow are then:

$$\text{Continuity:} \quad \frac{\partial w}{\partial z} + \frac{1}{r}\frac{\partial}{\partial r}(rv) = 0 \qquad (3.18)$$

$$\text{Momentum:} \quad \rho\left(w\frac{\partial w}{\partial z} + v\frac{\partial w}{\partial r}\right) = -\rho g - \frac{\partial p}{\partial z} - \rho\frac{1}{r}\frac{\partial}{\partial r}(r\overline{w'v'}) \qquad (3.19)$$

$$\text{Mass or heat conservation} \quad w\frac{\partial c}{\partial z} + v\frac{\partial c}{\partial r} = -\frac{1}{r}\frac{\partial}{\partial r}(r\overline{v'c'}) \qquad (3.20)$$

where w, v are the turbulent-mean velocities in the vertical and radial (z, r) directions respectively, and w', v', c' are the velocity and concentration fluctuations. Assuming a hydrostatic pressure distribution, $\frac{\partial p}{\partial z} = -\rho_a g$, we invoke the Boussinesq approximation - that for small density differences, $\frac{\Delta\rho}{\rho} \ll 1$, density differences can be neglected in the governing equations $(\rho \approx \rho_a)$, except in terms multiplied by g. This amounts to assuming *constant mass* but *variable weight.* Eq. 3.19 then becomes

$$w\frac{\partial w}{\partial z} + v\frac{\partial w}{\partial r} = -\frac{(\rho - \rho_a)}{\rho_a}g - \frac{1}{r}\frac{\partial}{\partial r}(r\overline{w'v'}) \tag{3.21}$$

2.2.2 INTEGRAL MODEL EQUATIONS

By applying the integral method (as outlined in Chapter 2) to the continuity equation, Eq. 3.18, and integrating across the plume from $r = 0$ to $r = \infty$, we obtain

$$\frac{d}{dz}\int_0^\infty w\ 2\pi r\ dr = -2\pi r v|_0^\infty = Q_e \tag{3.22}$$

As $r \to \infty$, $v \to 0$, but rv remains finite. The entrainment flow Q_e can be written as $Q_e = 2\pi b v_e$, where v_e is the entrainment velocity at $r = b$. The change in volume flux is due to lateral entrainment. Similarly, the integrated form of the momentum equation, Eq. 3.21, can be shown to be:

$$\frac{d}{dz}\int_0^\infty w^2\ 2\pi r\ dr = \int_0^\infty \frac{\rho_a - \rho}{\rho_a}\ g\ 2\pi r\ dr \tag{3.23}$$

where the right hand side represents the buoyancy force per unit z-length. The buoyant force on a fluid parcel is the gravitational force resulting from non-uniform density. For the infinitesimal plume element, the change in momentum flux is due to the buoyant force associated with a density deficit distribution across the plume.

Assuming $\Delta\rho \propto \Delta c$, the conservation of tracer mass, Eq. 3.20, leads to the conservation of specific buoyancy flux:

$$F = \int_0^\infty w\ \frac{\Delta\rho}{\rho_a}g\ 2\pi r\ dr = \text{constant} = Q_o\frac{\Delta\rho_o}{\rho_a}g \tag{3.24}$$

Substituting the assumed profiles into Eq. 3.22, 3.23, 3.24, and using 3.28, we obtain the governing equations for the integral model in terms of the three unknowns w_m, b, and $\Delta\rho_m$ (or alternatively Q, M, F):

$$\frac{dQ}{dz} = \frac{d}{dz}(\pi w_m b^2) = 2\pi b v_e \tag{3.25}$$

$$\frac{dM}{dz} = \frac{d}{dz}(\frac{\pi}{2}w_m^2 b^2) = \pi\frac{\Delta\rho_m}{\rho_a}g\lambda^2 b^2 \tag{3.26}$$

$$F = (\pi\frac{\lambda^2}{1+\lambda^2}w_m\frac{\Delta\rho_m}{\rho_a}gb^2) = F_o \tag{3.27}$$

2.2.3 ENTRAINMENT HYPOTHESIS

A supplementary condition for the entrainment velocity v_e is required for a unique solution of the above equations (i.e. for closure of the turbulent flow equations). The classical method is to introduce the entrainment hypothesis (Morton *et al.* 1956). It is assumed that the eddies (resulting from a velocity gradient) are characterized by velocities proportional to a characteristic relative velocity. Thus the entrainment flow per unit z-length, Q_e, is given by:

$$Q_e = 2\pi(\alpha w_m)b \tag{3.28}$$

where the entrainment velocity at the nominal boundary of the plume, $r = b$, is assumed proportional to the local centerline velocity, with $v_e = \alpha w_m$, where α = entrainment coefficient, a constant for a round plume.

Defining $Q = \pi w_m b^2$, $M = \frac{\pi}{2} w_m^2 b^2$, F_o = specific buoyancy flux = $\int \frac{\Delta\rho}{\rho} g\ w dA = \frac{\pi\lambda^2}{1+\lambda^2} w_m \frac{\Delta\rho_m}{\rho_a} g b^2$, we have $w_m = \frac{2M}{Q}$, $b = \frac{Q}{\sqrt{M}} \frac{1}{\sqrt{2\pi}}$.
In terms of the set of variables (Q, M, F), the governing equations then reduce to

$$\frac{dQ}{dz} = 2\pi\alpha\sqrt{\frac{2M}{\pi}} \tag{3.29}$$

$$\frac{dM}{dz} = \frac{F_o(1+\lambda^2)}{2} \frac{Q}{M} \tag{3.30}$$

Eq. 3.29 & 3.30 give

$$\frac{d^2}{dz^2}(M^2) = F_o(1+\lambda^2) 2\sqrt{2\pi}\alpha M^{\frac{1}{2}} \tag{3.31}$$

2.2.4 ASYMPTOTIC SOLUTION

The asymptotic solution for a pure plume with zero momentum flux at the source ($M_o = M(0) \approx 0$) is

$$M(z) = az^n \tag{3.32}$$

Substituting this power law form of solution into Eq. 3.31, and equating coefficients of equal power of z, gives:

$$M(z) = (2\pi)^{\frac{1}{3}} [\frac{9\alpha(1+\lambda^2)}{20}]^{\frac{2}{3}} F_o^{\frac{2}{3}} z^{\frac{4}{3}} \tag{3.33}$$

With this solution for M, the solution for Q can be obtained from the continuity equation:

$$Q = \frac{6\alpha}{5}(2\pi)^{\frac{2}{3}}[\frac{9\alpha(1+\lambda^2)}{20}]^{\frac{1}{3}} F_o^{\frac{1}{3}} z^{\frac{5}{3}} = \frac{6\alpha}{5}(2\pi M)^{\frac{1}{2}} z \tag{3.34}$$

The centerline maximum velocity, reduced gravity and the width, w_m, g'_m and b, are uniquely determined by the volume flux, momentum flux and buoyany flux of the plume, Q, M and F_o, respectively. Hence,

$$b = \frac{6\alpha}{5} z \tag{3.35}$$

$$w_m = \frac{5}{3\alpha}[\frac{1}{2\pi}\frac{9\alpha(1+\lambda^2)}{20}]^{\frac{1}{3}} F_o^{\frac{1}{3}} z^{-\frac{1}{3}} \tag{3.36}$$

$$g'_m = \frac{5}{6\alpha\lambda^2}[\frac{20}{9\alpha}]^{\frac{1}{3}}[\frac{1+\lambda^2}{2\pi}]^{\frac{2}{3}} F_o^{\frac{2}{3}} z^{-\frac{5}{3}} \tag{3.37}$$

The solution is solely determined by the driving force F_o - the specific buoyancy flux, with $w_m \sim z^{-\frac{1}{3}}$, $b \sim z$, $\Delta\rho_m \sim z^{-\frac{5}{3}}$, and $M \sim z^{\frac{4}{3}}$, $Q \sim z^{\frac{5}{3}}$. The above solution for the velocity and concentration in a round plume may be compared with those for the round jet. In a round jet, $w_m \sim M_o^{\frac{1}{2}} z^{-1}$, $Q \sim M_o^{\frac{1}{2}} z$ and $c \sim \Gamma M_o^{-\frac{1}{2}} z^{-1}$. Jets are efficient devices for mixing on a small scale but the energy cost is expensive when a large volume of fluid is involved. In the atmosphere and the oceans, mixing is dependent on buoyancy. For a buoyant plume, the kinetic energy flux continually increases, with $E \sim w_m^3 b^2 \sim z$, as the potential (buoyant) energy is continually converted into kinetic energy with the ascent of the plume. If a local Reynolds number $Re(z)$ is defined as $Re = \frac{w_m b}{\nu}$, we see that $Re \sim z^{\frac{2}{3}}$. Thus the plume is guaranteed to be turbulent at some distance above the source, even if it is initially laminar.

Unlike the case of a jet, the momentum flux in a plume continually increases, with a much slower velocity decay. The fact that $w_m \to \infty$, $\Delta\rho_m \to \infty$ as $z \to 0$ reflects the mathematical singularity of the point plume solution. For finite sources with $Q_o, M_o \neq 0$, the flow will be described by these equations corresponding to a virtual source below the actual source. For $Q(z) \gg Q_o$, $M(z) \gg M_o$, z will be considerably larger than the difference in location between the actual and virtual source.

Using the entrainment hypothesis , a linear spread of the plume is predicted. This implies that we could have solved the plume problem by a constant spreading rate hypothesis, $db/dz = \beta$, as in Chapter 2. The plume spread rate must be related to the entrainment coefficient as:

$$\beta_p = \frac{6}{5}\alpha_p \tag{3.38}$$

where the subscript 'p' denotes plume values.

For a plume width defined by the Gaussian profile, the experimentally observed β value is in the range of 0.102 to 0.112. Morton *et al.*(1956) recommended a value $\alpha = 0.093$ based on their experimental observation of the plume in a density stratified ambient. From the experiments by Rouse *et al.* (1952) and Abraham (1963a), the experimental constants α, λ for a round plume can be determined to be $\alpha = 0.085$ (corresponding to a plume spread rate of $\beta = 0.102$), and $\lambda = 1.16$. On the other hand, the experiments of Papanicolaou and List (1988) which covered a larger distance downstream of the source, give $\beta = 0.105$ (implying $\alpha = 0.088$), and $\lambda = 1.06$; a value of $\lambda = 1.19$ is however suggested for use over entire jet-plume range. The plume spread rate of around 0.11 has also been observed in related buoyant plume studies (Lee and Cheung 1991). For the pure plume, if we adopt $\alpha = 0.088$ and $\lambda = 1.19$, the plume solution as summarized in Table 3.1 can be obtained. The round plume solution is very useful in many environmental problems.

Given the values $\alpha \approx 0.088$ and $\lambda = 1.19$, the above asymptotic formulae for the plume become

$$w_m = 4.71\, F_o^{\frac{1}{3}} z^{-\frac{1}{3}},\; g'_m = 10.46\, F_o^{\frac{2}{3}} z^{-\frac{5}{3}},\; Q = 0.163\, F_o^{\frac{1}{3}} z^{\frac{5}{3}} \tag{3.39}$$

As the reduced gravity, the maximum concentration of the tracer at the centerline of the plume follows the same minus five-third power law

$$c_m = 10.46\, \Gamma_o\, F_o^{-\frac{1}{3}} z^{-\frac{5}{3}}. \tag{3.40}$$

where Γ_o = flux of tracer mass. If the plume is created by a heat source, the maximum temperature excess (above its ambient) will be given by

$$(\Delta T)_m = (T_m - T_a) = 10.46\, \frac{H}{\rho_a c_p}\, F_o^{-\frac{1}{3}} z^{-\frac{5}{3}}. \tag{3.41}$$

where H = flux of heat. The symbol F_o in these equations denotes the *specific* buoyancy flux.

EXAMPLE 3.3 <u>*Volume of smoke produced by a fire*</u>

A fire plume is set up by burning a 0.6 m diameter pool of fuel (heptane) in a 10m×10m×10m large room. If the source heat release rate is H = 500 kW, estimate the maximum plume velocity, temperature excess, and mass flux entrained by the plume at a height of z = 10 m above the heat source. Estimate the time for the smoke of the fire to fill the room. Density of air at standard atmospheric condition is $\rho_a = 1.2\ kg/m^3$.

SOLUTION: Experiments have shown that the smoky hot gas flow above the heat source of a fire can be approximated by a round plume

(Heskestad 1984). For air, the specific heat at constant pressure is $c_p = 1.006\ kJ/kg^oK$. Assuming standard atmosphere with $T = 293^oK$ and $\rho_a = 1.2$ kg/m^3, the specific buoyancy flux can be evaluated using Eq. 3.3 as follows:

$$F_o = \frac{gH}{\rho_a c_p T} = \frac{9.81 \times 500}{1.2 \times 1.006 \times 293} = 13.87\,\mathrm{m}^4/\mathrm{s}^3$$

Using the plume solution in Eqs. 3.39, 3.40, and 3.41, and the solution summarized in Table 3.1, the maximum velocity at $z = 10$ m is $w_m = 4.71 F_o^{1/3} z^{-1/3} = 4.71 \times 13.9^{1/3}/10^{1/3} = \underline{5.3}$ m/s. The maximum excess temperature $(T_m - T_a)$ at the same level is given by

$$(\Delta T)_m = 10.46 \frac{H}{\rho_a c_p} F_o^{-1/3} z^{-5/3} = \underline{38.8}^o\mathrm{C}$$

The volume flux induced by the fire plume at this height is

$$Q = 0.163\, F_o^{1/3} z^{5/3} = 0.163 \times 13.87^{1/3} \times 10^{5/3} = \underline{18.2}\ \mathrm{m}^3/\mathrm{s}.$$

The mass flow rate is $\rho_a Q = 1.2 \times 18.2 = \underline{21.8}$ kg/s. In well-developed fires, the convective heat flux usually accounts for a major fraction of the total heat release rate. Plume theory is useful in the design of extraction fans or sprinkler systems for shopping malls and atria. The hot and smoky gases generated by the fire rise to the top of the room and form a buoyant smoke layer below the ceiling. In the above example, the smoke of the fire is produced at a rate equal to the plume's volume flux $Q = 18.2\ \ \mathrm{m}^3/\mathrm{s}$. If this volume flux is maintained, the time for the smoke to fill the room of 10^3 m^3 would be $t = 10^3/18.2 = \underline{55}$ s.

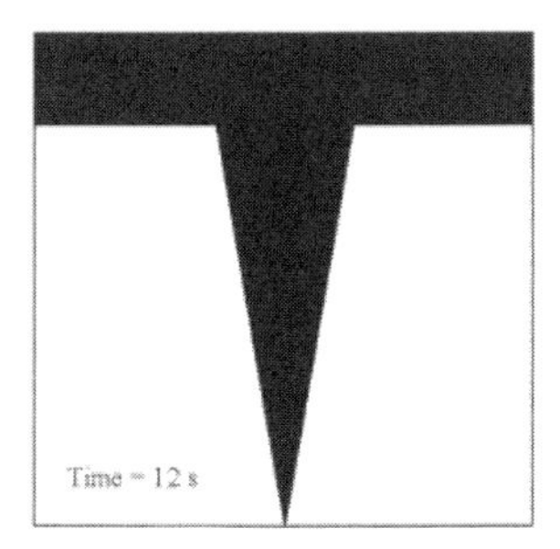

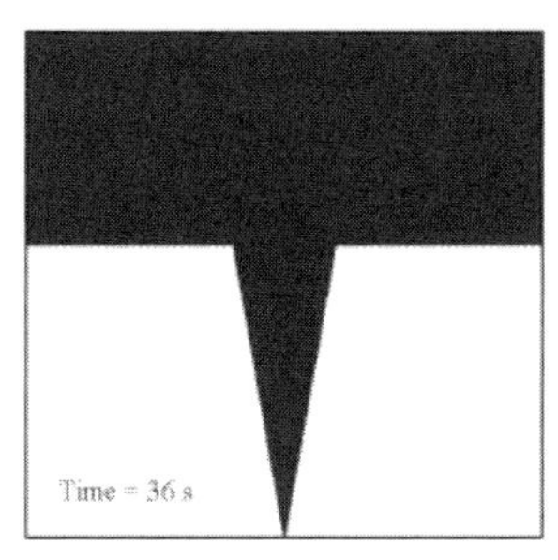

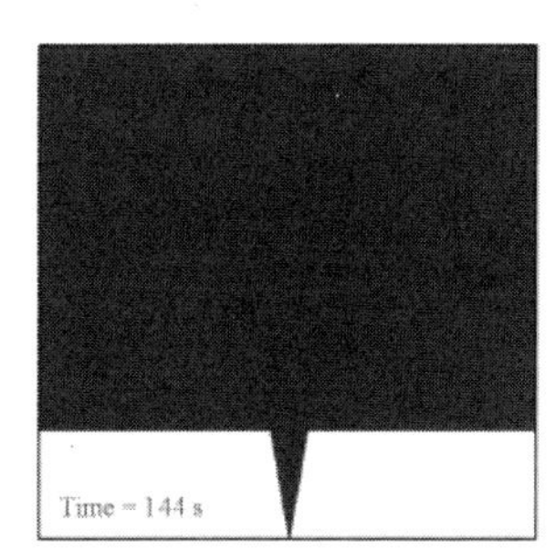

Figure 3.4. Smoke produced by 500 kW fire in a 10m×10m×10m room at time $t =$ 12 s, 36 s and 144 s. The plume boundary in this figure is defined by $\tilde{b} = 0.17z$ and the interface by Eq. 3.42

As the interface of the smoke layer moves down from the ceiling, the volume flux $Q(z)$ decreases with elevation of the interface z. (See Fig.

3.4 and also Fig. 5.15 in chapter 5 for the downward movement of the interface.) The downward velocity of the interface, $-dz/dt$, is equal to $Q(z)/A$, i.e.,

$$-\frac{dz}{dt} = \frac{Q(z)}{A}$$

where A is area of the interface or the floor area of the room. Integrating the equation over the period of time from time $t = 0$ to t as the elevation of the interface moves down from the height of the ceiling where $z = z_c$, gives

$$\int_o^t dt = -\int_{z_c}^{z} \frac{A\,dz}{Q(z)} = -\int_{z_c}^{z} \frac{A\,dz}{0.163\,F_o^{1/3}\,z^{5/3}}$$

that is

$$t = [\frac{3A}{2 \times 0.163 F_o^{\frac{1}{3}}\, z_c^{\frac{2}{3}}}]\,[(\frac{z_c}{z})^{\frac{2}{3}} - 1] \tag{3.42}$$

The formula gives $t = 300$ s (5 minutes) for a floor area of $A = 100$ m^2, interface height $z = 1$ m, ceiling height $z_c = 10$ m, and specific buoyancy flux $F_o = 13.9\,\mathrm{m}^4/\mathrm{s}^3$. According to this calculation, the smoke of the fire will fill 90% of the 10m $\times$ 10m $\times$ 10 m large room in a very short time, of the order of minutes! Note that the effect of ceiling boundary and extraction fans on the plume dynamics have been tacitly ignored in this worst case scenario.

2.2.5 DENSIMETRIC FROUDE NUMBER

The local densimetric Froude number of the plume based on the centerline values of the velocity and reduced gravity is

$$\mathrm{Fr_L}(z) = \frac{w_m}{\sqrt{\frac{\Delta\rho_m}{\rho_a} g b}} \tag{3.43}$$

Based on the asymptotic solution, it can be shown that for a round plume,

$$\mathrm{Fr_L}(z) = \mathrm{Fr}_p = (\frac{5}{4})^{\frac{1}{2}} \frac{\lambda}{\sqrt{\alpha}} = \text{ constant,} \tag{3.44}$$

which has a numerical value of $\mathrm{Fr}_p = 4.5$ if $\alpha = 0.09$ and $\lambda = 1.19$. The constancy of $\mathrm{Fr_L}$ indicates a constant ratio of inertia to buoyancy forces throughout the plume. The rate of production of turbulent energy per unit mass of fluid from buoyant forces is $P \sim g' w_m$, while the rate of turbulent energy dissipation is $\epsilon \sim w_m^3/b$. Thus the densimetric Froude number can also be interpreted as the ratio of turbulent energy dissipation to production, with $\mathrm{Fr_L}^2 \sim \frac{w_m^2}{g'b} \sim \epsilon/P$; the constancy of the Froude number represents a fixed ratio of dissipation to production.

For any buoyant jet with both momentum and buoyancy, it can be shown that the buoyant jet will approach the asymptotic Froude number Fr_p as it rises and mixes with the surrounding fluid, regardless of whether Fr is greater or less than Fr_p (Prob. 3.8). For example, for a vertical buoyant jet, if the discharge densimetric Froude number is significantly greater than the plume value, $Fr \gg \mathrm{Fr}_p$, the effluent may initially behave like a turbulent jet, and the local densimetric Froude number would approach the asymptotic value from above (see Fig. 3.1). Alternatively, in the case of the plume produced by a fire, since the velocity w_m is initially zero, the densimetric Froude number $\mathrm{Fr_L}$ increases from zero at the source and asymptotically approaches the plume value of $\mathrm{Fr}_p = 4.5$ far from the source (see Fig. 1.2).

While the spread rate for both a jet and a plume is about the same, the entrainment coefficient can vary by as much as 60 percent over the jet-plume range, from $\alpha_j = 0.057$ for the round jet to $\alpha_p \approx 0.09$ for the round plume. A dimensional analysis of the entrainment velocity as a function of the local variables, $v_e = \mathrm{fn}(w_m, g'_m, b)$, would show the dependence of the entrainment coefficient α on the local densimetric Froude number $\mathrm{Fr_L}$. This dependence can also be deduced from the governing equations (Prob. 3.14). In buoyant jet models the variation of entrainment coefficient is typically incorporated in a form which gives the correct values in the limiting cases of a jet or plume:

$$\alpha = \alpha_j + (\alpha_p - \alpha_j)(\frac{\mathrm{Fr}_p}{\mathrm{Fr_L}})^2 \tag{3.45}$$

In many practical applications, the plume solution serves a good approximation to strongly buoyant discharges (small Fr) with finite Q_o and M_o. For these cases the dilution can be inferred from Eq. 3.37. By definition we have ($F = F_o$)

$$F = Q\frac{\Delta\rho_m}{\rho_a}g\frac{\lambda^2}{1+\lambda^2} = Q_o\frac{\Delta\rho_o}{\rho_a}g \tag{3.46}$$

The relative centerline maximum concentration (inverse of centerline dilution, $S_m = c_o/c_m$) is then given by

$$\frac{c_m}{c_o} = \frac{\Delta\rho_m}{\Delta\rho_o} = \frac{5}{6\alpha\lambda^2}(\frac{20}{9\alpha})^{1/3}(\frac{1+\lambda^2}{2\pi})^{\frac{2}{3}} Q_o F_o^{-1/3} z^{-\frac{5}{3}} \tag{3.47}$$

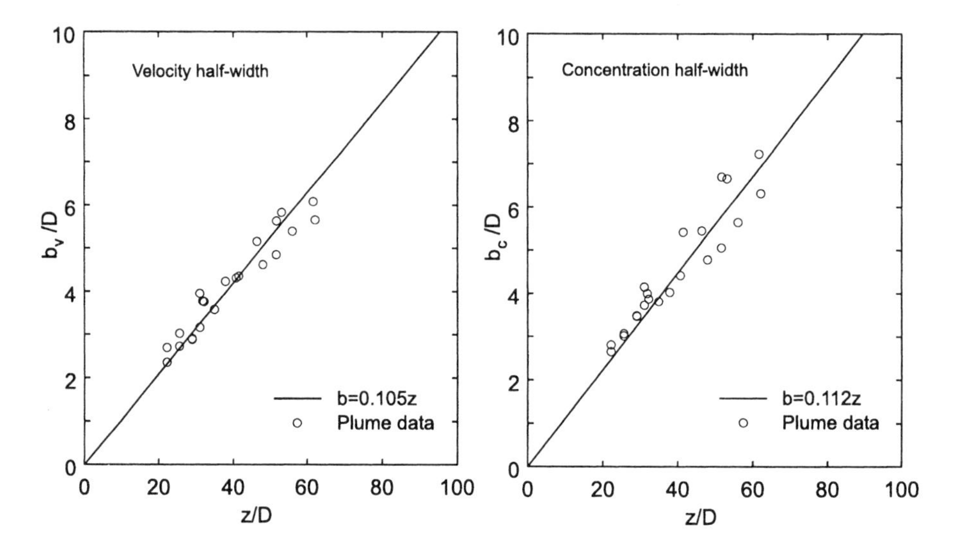

Figure 3.5. Variation of plume half-width defined by velocity (b_v) and concentration (b_c) with vertical distance from source (data from Papanicolaou and List 1988)

2.2.6 EXPERIMENTS

Fig. 3.3 shows measured radial profiles of velocity and scalar concentration at different elevations above a round plume; it is seen that both the velocity and concentration are self-similar and can be well-approximated by Gaussian distributions. The velocity half-width (and the concentration half-width) increases linearly with distance (Fig. 3.5). The data include the plume experiments of Papanicolaou and List (1988).

The plume solution (Eq. 3.39 or Table 3.1) obtained by the Eulerian integral method may be compared with the following formulae obtained from the laboratory experiments:

Papanicolau and List (1988) - tracer concentration and velocity measurements in water

$$w_m = 3.85\ F_o^{\frac{1}{3}} z^{-\frac{1}{3}}, c_m = 11.0\ \Gamma_o\ F_o^{-\frac{1}{3}} z^{-\frac{5}{3}}, Q = 0.14\, F_o^{\frac{1}{3}} z^{\frac{5}{3}} \quad (3.48)$$

where Γ_o = tracer mass flux at the source, and c_m = tracer mass concentration at the centerline.

Rouse, Yih and Humphreys (1952) - temperature and velocity measurements in air

$$w_m = 4.70\ F_o^{\frac{1}{3}} z^{-\frac{1}{3}}, \Delta T_m = 11.0\ \frac{H_o}{\rho_a c_p} F_o^{-\frac{1}{3}} z^{-\frac{5}{3}} \quad (3.49)$$

where H_o = heat flux at the source, c_p = specific heat, and ΔT_m = maximum temperature excess at the plume centerline.

George, Alpert and Tamanini (1977) - temperature measurements in air

$$\Delta T_m = 9.1\ \frac{H_o}{\rho_a c_p} F_o^{-\frac{1}{3}} z^{-\frac{5}{3}} \tag{3.50}$$

Lee and Cheung (1991) - temperature measurements in water

$$\Delta T_m = 10.0\ \frac{H_o}{\rho_a c_p} F_o^{-\frac{1}{3}} z^{-\frac{5}{3}} \tag{3.51}$$

Chu, Senior and List (1981) - temperature measurements in water

$$\Delta T_m = 10.9\ \frac{H_o}{\rho_a c_p} F_o^{-\frac{1}{3}} z^{-\frac{5}{3}} \tag{3.52}$$

Baines (1983) - direct measurement of the volume flux of a salty plume through a stratified interface

$$Q = 0.123\, F_o^{\frac{1}{3}} z^{\frac{5}{3}} \tag{3.53}$$

(see also Baines and Chu (1996) for a review of the available experimental data).

The plume theory predicts a velocity decay coefficient of 4.71; this is in excellent agreement with the measured value of 4.7 by Rouse, Yih and Humphreys (1952), but a bit higher than the observed 3.85 by Papanicolaou and List (1988). On the other hand, the predicted centerline concentration of $c_m = 10.46\ Q_o c_o F_o^{-1.3} z^{-1/3}$ compares favorably with measured coefficients of 11.0 (Rouse *et al.* 1952), 10.9 (Chu *et al.* 1981), 9.1 (George *et al.* 1977), and 10.0 (Lee and Cheung 1991). The coefficient of 0.163 for the predicted volume flux also compare favorably with the experimental value of 0.14 by Papanicolaou and List (1988).

Chu *et al.* (1981) have carried out turbulence measurements in a plume; their data show that turbulent fluctuations in a plume are in general much more intense than in a jet. For example, the instantaneous maximum and minimum temperatures are approximately 2.7 and 0.53 times that of the turbulent mean concentration at the centerline. In Fig. 3.6 the measured concentration turbulent intensity across a plume is compared with those of a jet. Whereas it is customary to see a twin peak of turbulence intensity, around 0.2 to 0.25, at the edge of the jet (corresponding to the position of maximum shear), in a plume the rms fluctuation is distinctly higher, with a maximum intensity of around 0.4.

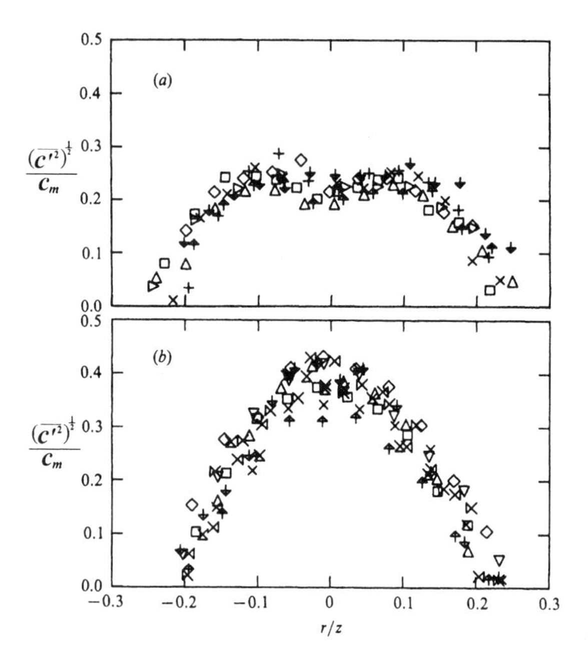

Figure 3.6. Radial profile of turbulence intensity of concentration a) across a jet; and b) across a plume. (from Papanicolaou and List 1988)

The velocity turbulent intensities in a plume are however similar to a jet, with maximum axial velocity turbulent intensities of around 0.25, and similar shapes. In view of this larger plume concentration intermittency, it is perhaps not surprising that, although the spreading rate of plumes and jets are approximately the same, plumes often appear to be more 'blobby' than jets (cf Fig. 3.1 and 2.1). The larger turbulence intensity of a plume is responsible for its greater entrainment coefficient.

EXAMPLE 3.4 <u>*Vertical round plume*</u>
Consider the round jet in Example 2.3 again, with however the effect of buoyancy included. This is a discharge with jet diameter $D = 0.15$ *m, depth* $H = 14$ *m, discharge velocity* $w_o = 1$ *m/s, and initial relative density difference* $\frac{\Delta\rho_o}{\rho_a} = 0.025$*. The jet discharge densimetric Froude number is* $Fr = \frac{1}{\sqrt{9.81 \times 0.025 \times 0.15}} = 5.2$.

SOLUTION: The initial volume, specific momentum and buoyancy fluxes Q_o, M_o and F_o are:

$$Q_o = \frac{\pi(0.15^2)}{4}(1) = 1.767 \times 10^{-2}\ m^3/s$$

$$M_o = \frac{\pi(0.15^2)}{4}(1^2) = 1.767 \times 10^{-2}\ m^4/s^2$$

$$F_o = \frac{\pi(0.15^2)}{4}(1)(0.025)9.81 = 4.334 \times 10^{-3}\ m^4/s^3$$

Assuming Q_o and M_o are negligible, the centerline concentration can be calculated from Table 3.1 (Eq. 3.47) as:

$$\frac{c_m}{c_o} = \frac{10.46 \times \pi(0.15^2)/4}{(4.334 \times 10^{-3} \times 14^5)^{1/3}} = \underline{0.0138}$$

– the inverse of which is the centerline dilution, $S_m = \underline{73}$. Therefore, $\frac{Q}{Q_o} \gg 1$. Also Table 3.1 gives $\frac{M}{M_o} = 19.3 \gg 1$. Thus our starting assumption of negligible Q_o, M_o is justified.

Similarly, the maximum velocity at surface is obtained (Table 3.1) as:

$$w_m(z = H) = 4.71\ (\tfrac{4.334 \times 10^{-3}}{14})^{1/3} = \underline{0.32}\ m/s$$

Comparing the numerical results for the round plume and the round jet (with and without buoyancy) for this same design, we see that in the presence of buoyancy, the dilution is enhanced considerably. This can also be seen analytically by contrasting the solution for $Q(z)$. From Table 3.1 and Table 2.3, we have:

round plume	round jet
$Q = 0.26M^{1/2}z$	$Q = 0.28M^{1/2}z$

Whereas the momentum flux $M(z)$ = constant in a jet, it increases continuously with z in a plume, resulting in a greater total induced flow.

2.3 EFFECT OF INITIAL MOMENTUM: VERTICAL BUOYANT JET

Practical discharges possess both initial momentum and buoyancy. We have seen that jet behaviour is governed by the momentum flux, while plume behaviour is governed by the buoyancy flux. For a buoyant jet with initial momentum flux M_o and buoyancy flux F_o, we would expect that initially the jet momentum may be important. Since the kinetic energy of the jet-induced motion decreases with distance, we would expect buoyancy will eventually dominate the mixing. Considering the effect of jet momentum and buoyancy in turn, the jet momentum-induced velocity at a distance z can be estimated by $w_j \approx 7M_o^{1/2}z^{-1}$ (Table 2.3), while the buoyancy-induced velocity is around $w_p \approx 4.7F_o^{1/3}z^{-1/3}$. Since w_p decays much slower than w_j, the two effects will be about the same when $w_j \approx w_p$:

$$z \sim \frac{M_o^{3/4}}{F_o^{1/2}}$$

This is exactly the momentum length scale $l_s = \frac{M_o^{3/4}}{F_o^{1/2}}$ we previously introduced (Chapter 1) via pure dimensional reasoning. For $z/l_s \gg 1$, we expect the buoyant jet behaves like a plume, while for $z/l_s \ll 1$, we expect the discharge to behave like a momentum jet. For a vertical buoyant jet, the precise transition from a jet to a plume is offered by the experiments of Papanicolaou and List (1988). Fig. 3.7 shows a plot of normalized measured volume flux as a function of dimensionless distance z/l_s from the source. Using the jet and plume solution for the volume flux Q, it can be shown that for the jet region, $Q^* = \frac{Q(x)F_o^{1/2}}{M_o^{5/4}} \sim z/l_s$, while for plumes, $Q^* \sim (z/l_s)^{5/3}$ (Prob. 4.6). The data shows that the buoyant jet behavior can be very well-correlated with this length scale; the dilution (and other variables) can be well-predicted with the plume solution for $z/l_s \geq 5$, and by the jet solution for $z/l_s \leq 2$. Except for cases when the source dimension is important ($z/D \sim 1$), this length scale provides a good means of defining jet/plume behaviour.

Based on Fig. 3.7 and other experimental data, a single formula to describe the average dilution $\overline{S}$ throughout the jet-plume range of a vertical round buoyant jet has been suggested (Wright and Wallace 1991):

$$\frac{\overline{S}l_Q}{z} = 0.29\ (1\ +\ 0.16(z/l_s)^2)^{1/3} \tag{3.54}$$

where z=vertical distance above the source and $l_Q = Q_o/M_o^{1/2}$. It can be shown that this dilution formula is consistent with the limiting cases of a pure jet and plume (Prob. 4.7). For a round buoyant jet, it can be shown that the average dilution $\overline{S} = Q/Q_o$ and centerline minimum dilution $S_m = c_o/c_m$ are related by $\overline{S}/S_m = \frac{1+\lambda^2}{\lambda^2} \approx 1.7$ (Prob. 2.1).

EXAMPLE 3.5 <u>*Mixing of vertical round buoyant jet*</u>
Consider a round jet discharge with diameter $D = 0.1$ *m, discharge velocity* $w_o = 1$ *m/s, initial relative density difference* $\frac{\Delta\rho_o}{\rho_a} = 0.025$*, and*

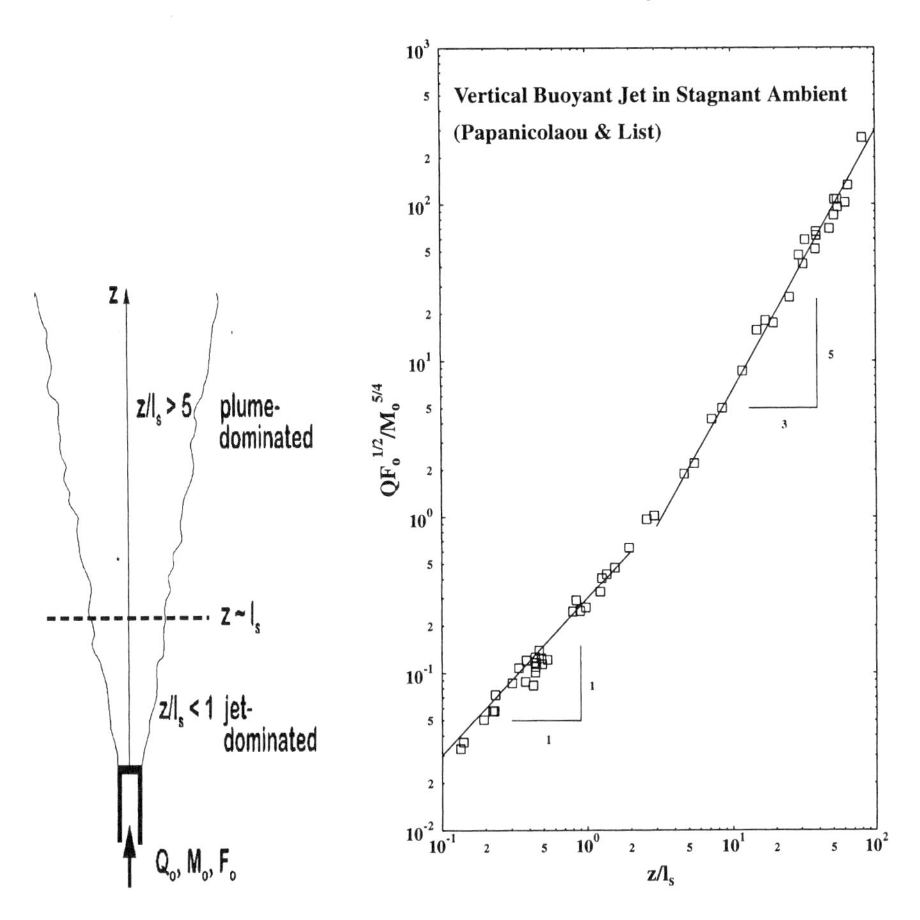

Figure 3.7. Measured volume flux in a vertical turbulent buoyant jet plotted against dimensionless distance from the origin (data from Papanicolaou and List 1988)

discharging in depth $H = 12$ m. The jet discharge densimetric Froude number is $Fr = \frac{1}{\sqrt{9.81 \times 0.025 \times 0.1}} = 6.4$. Estimate the maximum velocity and minimum dilution at $z = 5$ m and $z = 1$ m from the jet origin.

SOLUTION: The initial volume, specific momentum and buoyancy fluxes Q_o, M_o and F_o are:

$$
\begin{aligned}
Q_o &= \frac{\pi(0.1^2)}{4}(1) = 7.854 \times 10^{-3}\ m^3/s \\
M_o &= \frac{\pi(0.1^2)}{4}(1^2) = 7.854 \times 10^{-3}\ m^4/s^2 \\
F_o &= \frac{\pi(0.1^2)}{4}(1)(0.025)9.81 = 1.9262 \times 10^{-3}\ m^4/s^3
\end{aligned}
$$

The momentum length scale can be computed to be $l_s = \underline{0.6 \text{ m}}$. At $z = 5$ m, $z/l_s = 8.3$; the mixing is plume like at this location. We can then use the plume solution (Table 3.1) to compute the centerline velocity as $w_m = \underline{0.35}$ m/s and the centerline dilution as $S_m = \underline{21}$. Similarly, at $z = 1$ m, $z/l_s = 1.7$, the mixing is jet like. The jet solution can then be used to give $w_m = \underline{0.62}$ m/s, $S_m = \underline{2}$. The use of Eq. 3.54 would have resulted in $S_m = \underline{22}$ and $S_m = \underline{2.2}$ for the two cases respectively.

EXAMPLE 3.6 *Chemical waste gas discharge*
A chemical laboratory discharges a waste gas (sulphur dioxide) from a 0.2 m diameter smokestack located on the roof top of a building. The jet velocity is $w_o = 5$ m/s, and the temperature of the waste gas is $T_o = 33^oC$. If the temperature of the outside air is $T_a = 23^oC$, determine if the plume dilution would be adequate at a vertical height of 10 m above the roof top. The gas concentration at the source is $c_o = 1000$ ppm; the maximum allowable concentration for 8-hour exposure is 10 ppm (i.e. requiring a dilution of 100).

SOLUTION: For small density and temperature differences, we can assume $\frac{|\rho_a - \rho_o|}{\rho_o} \approx \frac{|T_a - T_o|}{T_o}$, where T_o, T_a are the absolute temperature of the jet and ambient air respectively (Prob. 3.5). The initial relative density difference of the buoyant waste discharge is hence $\Delta\rho_o/\rho_a \approx (33 - 23)/(273 + 23) = \underline{0.0338}$. The jet densimetric Froude number is hence $Fr = 5/\sqrt{0.0338 \times 9.81 \times 0.2} = \underline{19.4}$.

The initial volume, specific momentum and buoyancy fluxes Q_o, M_o and F_o are:

$$\begin{aligned} Q_o &= \frac{\pi(0.2^2)}{4}(5) = 0.157\ m^3/s \\ M_o &= \frac{\pi(0.2^2)}{4}(5^2) = 0.785\ m^3/s \\ F_o &= \frac{\pi(0.2^2)}{4}(5)(0.0338)9.81 = 0.052\ m^4/s^3 \end{aligned}$$

Compute the length scales as $l_Q = \sqrt{\pi/4} \times 0.2 = 0.177$ m, $l_s = 0.785^{3/4}/\ 0.052^{1/2} = 3.66$ m. At a height of 10 m, $z/l_s = 2.73$, the buoyant jet is in the transition region from a jet to a plume. Eq. 3.54 can be used to compute the average dilution:

$$\frac{\overline{S} \times 0.177}{10} = 0.29\ (1 + 0.16 \times 2.73^2)^{1/3} = 0.377$$

with $\overline{S} = 0.377 \times 10/0.177 = \underline{21.3}$. The centerline dilution can be obtained by $S_m = \overline{S}/1.7 = \underline{12.5}$. Thus the plume dilution due to buoyancy is about 12, much less than the required dilution of 100. In the real situation, the effect of wind (crossflow) would act to bring down the concentration further (see Chapter 7 to 10).

2.4 BUOYANCY REDUCTION DUE TO DENSITY-TEMPERATURE NONLINEARITY

In the foregoing development, we have assumed a linear density-concentration dependence. In general, density differences are linearly proportional to salinity differenes (see Appendix A); e.g. for freshwater discharges into the ocean. It is then legitimate to write $\frac{(\rho-\rho_a)}{\rho} \approx \gamma_S(S - S_a)$ where $\gamma_S = \frac{1}{\rho}\frac{\partial \rho}{\partial S}|_{S_a,T_a}$. The conservation of buoyancy flux is a direct consequence of this assumed linearity. However the variation of water density with temperature is quite nonlinear (Fig. 3.8); a linear approximation, $\Delta\rho/\rho \approx \gamma_T(T - T_a)$, is good only for small ΔT of a few degrees. In laboratory experiments, it is often convenient to use heat as a source of buoyancy. Due to the nonlinear equation of state, conservation of the heat flux does not guarantee buoyancy flux conservation. For large initial temperature excess, there can be a substantial loss of buoyancy within a short distance from the source – a point that can be noted in many previous experimental studies.

EXAMPLE 3.7 <u>*Heated water jet: buoyancy conservation*</u>
A thermal plume in a laboratory experiment is produced by using a heated discharge with an initial temperature difference of 40° Centigrade. The discharge is designed to simulate a plume with initial relative density difference of around 0.015. The plume has the following characteristics:

initial velocity	$w_o = 0.1075\ m/s$
source temperature	$T_o = 61.0^oC$
ambient temperature	$T_a = 21.0^oC$
jet diameter	$D = 0.75\ cm$
initial density difference	$\frac{\Delta\rho_o}{\rho_a} = 0.01527$

where the initial density difference has been determined with the use of equation of state (Appendix A). This is a plume like discharge with $Fr = 3.2$.

At a distance of 0.1 m above the source, compute the centerline temperature and density deficit, and comment on the conservation of buoyancy flux.

SOLUTION: The initial volume and specific buoyancy fluxes Q_o, and F_o can be computed to be:

$$Q_o = \frac{\pi(0.0075^2)}{4}(0.1075) = 4.7492 \times 10^{-6}\ m^3/s$$
$$F_o = \frac{\pi(0.0075^2)}{4}(0.1075)(0.01537)9.81 = 0.7161 \times 10^{-6}\ m^4/s^3$$

The centerline dilution, $S_m = \frac{\Delta T_o}{\Delta T_m}$, can then be computed to be (from Table 3.1):

$$S_m = 0.0956\ \frac{(0.7161 \times 10^{-6})^{1/3}(0.1)^{5/3}}{4.7492 \times 10^{-6}} = \underline{3.9} \tag{3.55}$$

Hence the centerline (maximum) temperature at $z = 0.1\ m$ is $T_m = T_a + \Delta T_o/S_m = 21.0 + 40.0/3.9 = 31.3^oC$. If the $\rho(T)$ relation were linear, the buoyancy corresponding to this dilution would be simply $\Delta\rho_m/\rho(linear) = \Delta\rho_o/\rho/S_m = 0.0153/3.9 = 0.00396$. However, from the equation of state, the centerline temperature of 31.3^oC results in an actual centerline density deficit of $\Delta\rho_m/\rho(actual) = \underline{0.00275}$. Thus at only 0.1 m above the source, where the discharge can be shown to be behaving like a pure plume, only about 0.69 of the original buoyancy flux remains (see also Appendix A)! This point can be of some importance in designing experiments and interpreting experimental data. Fortunately, in the prediction of dilution using the plume formula, $S \sim F_o^{1/3}$, such that the dilution is at most over-predicted by $0.69^{-1/3} = 1.13$, or around 13 percent. For non-vertical heated jets, this loss of buoyancy results in a *lower* observed jet trajectory compared to the case if the buoyancy is created using salinity difference. In some cases, the jet trajectory can also be sensitive to ambient temperature (Prob. 4.8).

3. LAGRANGIAN APPROACH FOR PLUME MODELLING

The Eulerian integral model in the previous section is obtained from the basic Eulerian equations of fluid motion. At this junction we take a detour and introduce an alternative Lagrangian approach to model buoyant jets. Consider a round plume discharge as shown in Fig. 3.9, with initial velocity $w_o = U_o$. The marked material volume issuing from the source over a time interval Δt has a streamwise length of $U_o\Delta t$,

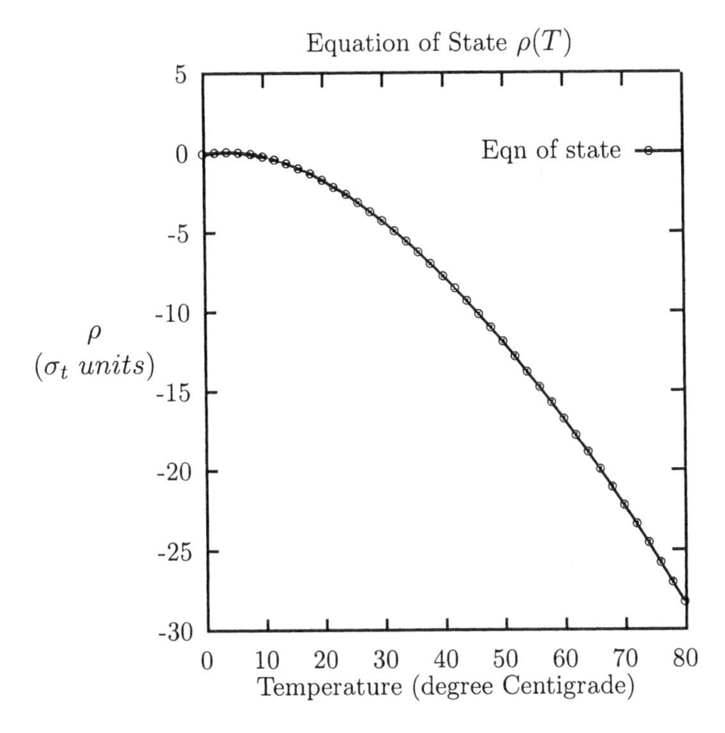

Figure 3.8. Density of water as a function of temperature

cross-section area $A_o = \pi D^2/4$, and velocity U_o. This material volume then has initial mass, momentum, and buoyancy as follows:

$$\begin{array}{llll} \text{mass } Q & = & \rho A_o (U_o \Delta t) & [\text{kg}] \\ \text{momentum } M & = & \rho A_o (U_o \Delta t) U_o & [\text{kg-m/s}] \\ \text{buoyancy force } F & = & \Delta \rho_o g A_o (U_o \Delta t) & [\text{N}] \end{array}$$

Instead of using the Eulerian equations, the motion of this material volume is followed with time. This plume element mixes with its surrounding by turbulent entrainment, and in so doing, its width, mass, momentum, and species concentration (buoyancy) change correspondingly. In general, at any time t (or elevation z) in the established flow, the plume element is characterized by

$$\begin{array}{lll} \text{mass } Q & = & \rho A (U \Delta t) \\ \text{momentum } M & = & \rho A (U \Delta t) U \\ \text{buoyancy force } F_B & = & \Delta \rho g A (U \Delta t) \end{array}$$

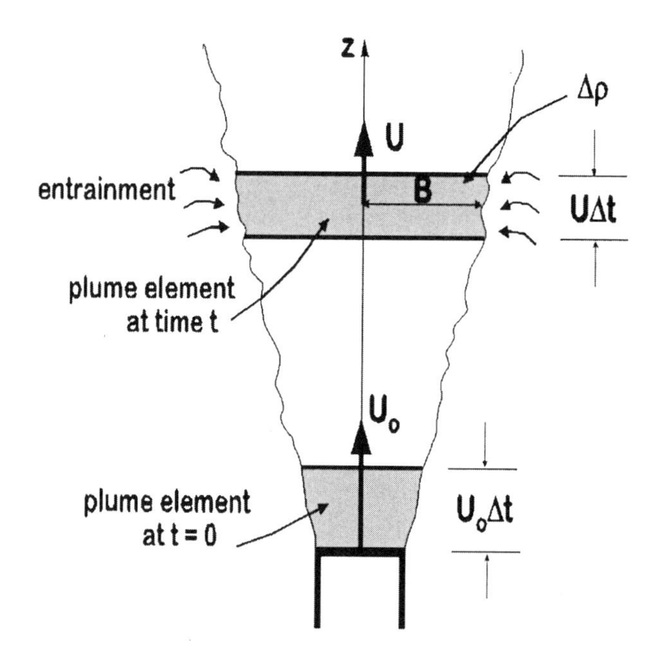

Figure 3.9. Lagrangian formulation for the motion of a plume element

where B, $U\Delta t$ = half-width and thickness of plume element, $A = \pi B^2$, $U, \Delta\rho$ = average velocity and density deficit respectively (this is the 'top-hat' profile introduced in Chapter 2). Viewed in this manner, the mass, momentum, and buoyancy fluxes at time t are proportional to the mass, momentum, and buoyancy of the corresponding plume element. Applying Newton's Law to this material volume, the rate of change of the vertical momentum of the plume element is equal to the buoyancy force acting on it:

$$\frac{dM}{dt} = F_B$$

$$\frac{d}{dt}(\rho A(U\Delta t)U) = \Delta\rho g A(U\Delta t) = \rho\frac{F_o}{U}(U\Delta t) \tag{3.56}$$

or in a more general form,

$$\frac{d}{dt}(AU^2) = \int u\frac{\Delta\rho}{\rho_a}g dA = F_o \tag{3.57}$$

Integration yields an exceedingly simple solution for the kinematic momentum flux $M = AU^2$:

$$M(t) = F_o t \; + \; M_o \tag{3.58}$$

where M_o is the initial momentum flux and F_o the specific buoyancy flux. In the presence of buoyancy, the momentum of the plume element increases linearly with *time*; significant fluid motions can hence be generated by apparently small density differences. Coupled with a plume spreading hypothesis, we show how the plume solution can be easily obtained.

First, the spreading hypothesis, $db/dz = \beta$, can be written in a more general Lagrangian form (multiplying throughout by U):

$$U\frac{dB}{dz} = U\beta \tag{3.59}$$

where $\beta = \frac{dB}{dz} = \sqrt{2}\frac{db}{dz}$. Since $U\frac{dB}{dz} = \frac{dB}{dt}$ is the rate of change of plume radius with time (following the material volume), the changes in properties of the plume element are then governed by the following two equations (for $M_o = 0$):

$$\frac{dB}{dt} = \beta U \tag{3.60}$$

$$\pi U^2 B^2 = F_o t \tag{3.61}$$

Noting that $U = (1/\beta)dB/dt$, and $z = (1/\beta)B$, the solution can then be obtained by simple integration:

$$B = \left(\frac{4\beta}{3}\right)^{1/2}\left(\frac{F_o}{\pi}\right)^{1/4} t^{3/4} \tag{3.62}$$

$$U = \left(\frac{3}{4\beta}\right)^{1/2}\left(\frac{F_o}{\pi}\right)^{1/4} t^{-1/4} \tag{3.63}$$

$$z = \left(\frac{4}{3\beta}\right)^{1/2}\left(\frac{F_o}{\pi}\right)^{1/4} t^{3/4}$$

$$t = \left(\frac{3\beta}{4}\right)^{2/3}\left(\frac{\pi}{F_o}\right)^{1/3} z^{4/3} \tag{3.64}$$

This leads to the following expressions for the volume and momentum of the plume element (corresponding to the Eulerian volume and momentum fluxes):

$$Q = \pi U B^2 = \sqrt{\pi}\beta\left(\frac{3\sqrt{\pi}\beta}{4}\right)^{1/3} F_o^{1/3} z^{5/3} \tag{3.65}$$

$$M = F_o t = \left(\frac{3\sqrt{\pi}\beta}{4}\right)^{2/3} F_o^{2/3} z^{4/3} \tag{3.66}$$

Noting the correspondence between the top-hat and Gaussian profiles, $\beta = \sqrt{2}\beta_G = \sqrt{2}\ \frac{6\alpha}{5}$, Eq. 3.65 and 3.66 are exactly the same as the plume solution developed in the previous section for $\lambda = 1$.

The underlying assumption of the Lagrangian approach is that there is negligible streamwise mixing between the plume elements at consecutive times. In the above development, we have tacitly assumed $\lambda = 1$, i.e. neglecting differences in velocity and concentration width. Nevertheless, the model gives very satisfactory results. Adopting $\beta = \sqrt{2} \times 0.105 = 0.149$, the predicted coefficients for the volume and momentum fluxes (Eq. 3.65 and 3.66) are 0.153 and 0.34 respectively, and that of centerline velocity (Eq. 3.64, with $w_m = 2U$) is 4.4. Knowing the volume flux, the centerline concentration can be obtained by mass conservation, with a predicted coefficient of 11.1. Compared to the measured values of Q and c_m (0.14 and 11.0), the Lagrangian model gives the same degree of agreement with data as the Eulerian formulation. The treatment of $\lambda > 1$ and related details can be found in Chu (1994).

The Lagrangian approach is physically appealing, and the solution can be much simpler to obtain and interpret. Unlike the Eulerian framework, the Boussinesq approximation is not a requirement in the Lagrangian formulation (see also Prob. 5.5 and Prob. 10.7). More important, the extension to the complex and practically important case of a buoyant jet in crossflow is relatively straightforward. The use of the Lagrangian approach for complex buoyant jet problems was first proposed by Chu (1977, 1985) in a line impulse model. A fundamental formulation and its interpretation in terms of 'dominant eddies' (Chu 1994) is developed in the later chapters. A Lagrangian model (JETLAG) based on the heuristic approach outlined above (Lee and Cheung 1990) will be described in Chapter 10.

EXAMPLE 3.8 *Plume versus jet mixing*

Buoyant fluid is discharged at a rate of 0.3 m^3/s through a 0.6 m diameter opening on the sea floor of the deep ocean. Find the specific momentum flux, M, the volume flux, Q, the average dilution, $\overline{S} = Q/Q_o$, and the reduced gravity, g', at elevation 30 m, 60m, and 120 m above the source. The density of the sea water is 2.5% heavier than the effluent at the source.

SOLUTION: Calculations are carried out first for the momentum flux and buoyancy flux at the source. The cross section area of the source is

$A_o = \pi D^2/4 = 0.283$ m^2. The velocity at the source is $V_o = Q_o/A_o =$ 1.06 m/s. The specific momentum flux is

$$M_o = Q_o V_o = 0.318 \text{ m}^4/\text{s}^2$$

The reduced gravity at the source is $g'_o = 9.81 \times 0.025 = 0.245$ m/s^2. The specific buoyancy flux is

$$F_o = g'_o Q_o = 0.0736 \text{ m}^4/\text{s}^3$$

The length scale defined by the buoyancy and momentum fluxes is

$$l_s = \frac{M_o{}^{\frac{3}{4}}}{F_o{}^{\frac{1}{2}}} = 1.56 \text{ m}$$

Within a short distance away from the source (when $z >> l_s$), momentum would become negligible and buoyancy would be the dominant effect. Neglecting the initial momentum at the source, the time for the material volume to move a distance $z = 30$ m, according to Eq. 3.64, is

$$t = (\frac{3\beta}{4})^{2/3}(\frac{\pi}{F_o})^{1/3} z^{4/3} = \underline{75.6}\text{s}.$$

where $\beta = 0.149$. Over this period of time, the momentum is increased by the buoyancy force to

$$M = F_o t = 0.0736 \times 82.5 = \underline{5.6} \text{ m}^4/\text{s},$$

that is 18 times greater than the initial momentum flux at the source, $M_o = 0.318$ m^4/s^2. The radius of the plume at this position ($z = 30$ m) is

$$B = \beta z = \underline{4.8}\,\text{m}.$$

The corresponding upward velocity of the plume element is

$$W = \sqrt{\frac{M}{\pi B^2}} = 0.298\,\text{m/s}.$$

The volume flux at this time is

$$Q = \pi W B^2 = \underline{18.8} \text{ m}^3/s$$

At $z = 30$ m above the source, the dilution of the plume is $\overline{S} = Q/Q_o =$ $\underline{62}$. The table below summarizes the calculations.

Plume Model					Jet Model		
z (m)	t (s)	M ($\mathrm{m^4/s^2}$)	Q/Q_o	g' ($10^{-2}\mathrm{m/s^2}$)	t (s)	M ($\mathrm{m^4/s^2}$)	Q/Q_o
30	76	5.6	62	0.394	240	0.318	17
60	191	14.0	198	0.124	960	0.318	34
90	327	24.1	389	0.063	2160	0.318	51
120	480	35.3	628	0.039	3940	0.318	68

It is clear from these calculations that buoyancy is the dominant effect. The dilution would be only $Q/Q_o = 68$ at $z = 120$ m if mixing were due to the jet momentum alone. With buoyancy, the momentum flux at 120 m is $M = 35.3\ \mathrm{m^4/s^2}$ that is 111 times the initial momentum flux at the source. Enormous momentum is generated by the minute difference in density between the buoyant fluid and its surrounding sea.

In the absence of buoyancy, the jet momentum would have to be dramatically increased to obtain the same dilution. Since $Q \sim \sqrt{M_o}$ in a jet, to acheive a dilution of 628 by the jet momentum alone, the momentum should be increased by $(628/68)^2 = 9.2^2$ times. For the same Q_o, this would produce a source velocity $V_o = 1.06 \times 9.2^2 = 89.7$ m/s; the diameter of the source opening would have to be $d_o = 6.5$ cm (a 9.2 times reduction in the size of the opening). The power consumption to produce such a high-velocity jet would be:

$$\frac{1}{2}\rho V_o^2 Q_o = \frac{1}{2}(1000) \times 89.7^2 \times 0.3 \text{ Nm/s} = 1207 \text{ kW}$$

The pumping cost to maintain such a jet would be prohibitively costly (in addition to possible bed erosion and other problems).

4. NEGATIVELY BUOYANT JETS

In many situations, negatively buoyant fluids are discharged upwards into surrounding fluid of lighter density. An example is the discharge of concentrated brine from desalination plants. Waste gas exhaust from air-conditioned buildings in the hot summer is another example. Fig. 3.10 shows the rise of the negatively buoyant jet to a maximum height and then the falling back of the curtain of the negatively buoyant fluid. The initial momentum flux is not negligible in the negatively buoyant jet. The negative buoyancy acts to oppose the initial jet momentum until the height of the maximum rise, z_m, after which the dense fluid will fall back due to its greater density. This case provides an example to demonstrate the versatility of the Lagrangian method.

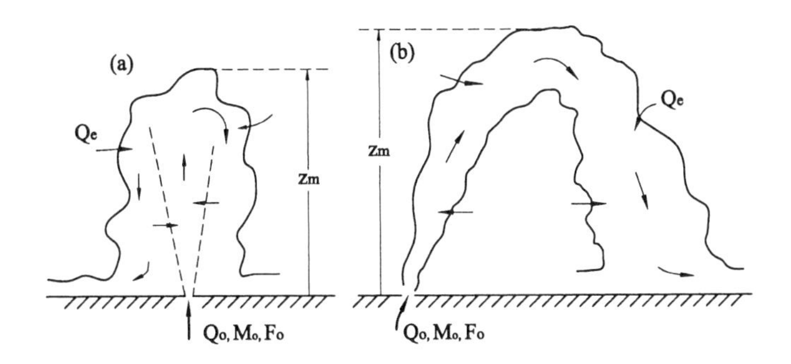

Figure 3.10. Negatively buoyant plume produced by upward momentum flux

Applying the Lagrangian method, the momentum equation gives:

$$\frac{dM}{dt} = -F_o \tag{3.67}$$

Since the density deficiency flux is conserved, we have $M(t) = M_o - F_o t$, and when the vertical momentum vanishes at $z = z_m$, $t = M_o/F_o$. Applying the spreading hypothesis $dB/dt = \beta U$ again, we have ($M = \pi U^2 B^2$):

$$B\frac{dB}{dt} = \beta U B = \frac{\beta}{\sqrt{\pi}}(M_o - F_o t)^{1/2} \tag{3.68}$$

from which the solution can be easily found to be:

$$B(t^*) = (\frac{4\beta}{3\sqrt{\pi}})^{1/2}\ l_s\ (1 - (1 - t*)^{3/2})^{1/2} \tag{3.69}$$

where $t^* = F_o t/M_o$ varies between 0 and 1 from the source to the maximum rise height, and $l_s = \frac{M_o^{3/4}}{F_o^{1/2}} = (\frac{\pi}{4})^{1/4} FrD$ is the jet momentum length scale. Thus the maximum height of rise is given by:

$$z_m = \frac{1}{\beta}B(t* = 1) = \sqrt{\frac{2}{3\beta}}FrD = 1.98\ FrD \tag{3.70}$$

where $Fr = u_o/\sqrt{|g_o'|D}$, and a value of $\beta = 0.17$ is used as suggested by Baines and Chu (1996). The predicted rise coefficient of 1.98 is somewhat greater than the experimental coefficient of 1.74 according to Turner (1966). In the experiment, the falling negatively buoyant fluid is re-entrained into the rising jet. The re-entrainment, however, can be avoided by tilting the dense plume at a small angle (say 7°) from the vertical as shown in Fig. 3.10. Without the interaction of the upward flow with the falling curtain, as would be expected, the rise height is larger. Measurements of the slightly inclined fountains give $z_m = 2.04FrD$ (Baines *et al.* 1990), in remarkable agreement with the prediction of the simple Lagrangian model (Eq. 3.70).

It is instructive to compute the jet volume flux for this case, $Q = \pi UB^2$. Substituting $U = \frac{1}{\beta}\frac{dB}{dt}$, and noting that $B = \beta z$, it can be shown that the volume flux is:

$$Q(t^*) = \sqrt{\pi}\beta\ M_o^{1/2} z(1-t^*)^{1/2} = 0.30\ M_o^{1/2} z(1-t^*)^{1/2} \tag{3.71}$$

Close to the source, $z \ll z_m$ or $t^* \ll 1$, the jet volume flux varies linearly with elevation as for the pure jet (cf $Q = 0.28M_o^{1/2}z$ for round jet). However, the volume flux actually decreases towards zero at the maximum rise height, $z = z_m$ or $t^* = 1$. Denoting $\tau = 1 - t^*$, we have $Q \sim z\tau^{1/2} \sim [(1-\tau^{3/2})\tau]^{1/2}$. It follows then that the positive entrainment is a maximum at τ_c, where $\tau_c^{3/2} = 2/5$, i.e. $\tau_c = 0.543$ or $t_c^* = 0.457$. At this critical time, the maximum amount of 'clean' fluid flow entrained by the dense plume is given by:

$$Q_c = \sqrt{\pi}(\frac{4\beta}{3\sqrt{\pi}})^{1/2}[(1-\tau_c^{3/2})\tau_c]^{1/2}\ M_o{}^{1/2} l_s \tag{3.72}$$

The maximum average dilution $\overline{S} = Q_c/Q_o$ can be shown to depend only on the jet densimetric Froude number:

$$\overline{S} = (8\beta/3)^{1/2}(3/5 \times 0.543)^{1/2} Fr = 0.384\ Fr \tag{3.73}$$

The centerline minimum dilution is then given by ($\overline{S}/S_m \approx 1.7$)

$$S_m = 0.23\ Fr \tag{3.74}$$

Since entrainment is the incorporation of the ambient fluid into the turbulent source fluid, by definition Q should always increase. The fact that Q decreases suggests that there is "detrainment" - i.e. the turbulent fluid is peeling off from the jet to form the falling part of the dense plume.

EXAMPLE 3.9 <u>Dense gas plume</u>

Consider the waste gas discharge from the chemical laboratory in Example 3.6 again, with $D = 0.2m$, and the jet velocity is $w_o = 5$ m/s. In the hot summer, however, the exhaust from an air-conditioned building has a temperature lower than the ambient. Assume now the temperature of the waste gas is $T_o = 23^oC$. If the temperature of the outside air is

$T_a = 33^oC$, estimate the maximum height of rise of the waste gas in a calm atmosphere and the minimum dilution achieved.

SOLUTION: This is a negatively buoyant jet with the effluent denser than the surrounding environment; the other parameters are similar to Example 3.6 with the buoyancy reversed. The effluent discharged upwards has a relative density difference of $\Delta\rho_o/\rho_a \approx (23 - 33)/(273 + 23) = -\underline{0.0338}$. The jet densimetric Froude number is hence $Fr = 5/\sqrt{|0.0338| \times 9.81 \times 0.2} = \underline{19.4}$. The maximum height of rise is $z_m = 1.74 Fr D = 1.74 \times 19.4 \times 0.2 = \underline{6.8}$ m. The minimum dilution can be estimated as that achieved at the level of maximum positive entrainment, with $S_m \approx 0.23 Fr = 0.23 \times 19.4 = \underline{4.5}$.

The dense plume will eventually sink back to the discharge level and spread over on the roof or ground. From the air pollution point of view, this is important as the pollutant may contaminate the area near the discharge point. Since the most concentrated part of the plume mixes further beyond t_c and on its descent, the dilution is expected to be somewhat greater than estimated. More details on dense plumes can be found in Baines and Chu (1996).

5. TURBULENT LINE PLUME

The flow generated by a line source of buoyancy is called a line plume or plane plume. The flow in this line plume is governed by the buoyancy flux per unit length of the source. There have been only a few studies of the line plumes. Our present knowledge is mainly based on the studies of Rouse *et al.*(1952) who studied convection over line fires, and more recently by Kotsovinos and List (1977) who carried out detailed measurements with a plane heated water jet. Chu and Baines (1989) measured the entrainment flow rate into a plane plume using a direct method. The axial velocity and tracer concentration profiles in a plane plume are found to be self-similar and can be represented by Gaussian profiles as in the plane jet. The formulation for the two dimensional case is similar in almost every respect to the axisymmetric plume.

Consider the flow generated by a line source of buoyancy (located at $z = 0$) in an otherwise stagnant ambient fluid of constant density ρ_a. The motion is sustained by a steady constant buoyancy flux $F_o = \int_{-\infty}^{\infty} \frac{\Delta\rho}{\rho} gw \, dy$ at the origin, $z = 0$. The continuity, momentum, and tracer mass conservation equations for the plume are (in the $y - z$ plane, where z=upward vertical co-ordinate from source) respectively:

$$\frac{\partial w}{\partial z} + \frac{\partial v}{\partial y} = 0 \tag{3.75}$$

$$w\frac{\partial w}{\partial z} + v\frac{\partial w}{\partial y} = -\frac{\partial \overline{w'v'}}{\partial y} - \frac{(\rho - \rho_a)}{\rho_a}g \tag{3.76}$$

$$w\frac{\partial c}{\partial z} + v\frac{\partial c}{\partial y} = -\frac{\partial \overline{v'c'}}{\partial y} \tag{3.77}$$

An integral model can be obtained from the above equations by suitable turbulent closure. By assuming a linear equation of state, an analytical solution can be obtained. In particular, using the Lagrangian model with a spreading hypothesis, $dB/dt = \beta W$, the following solution can be easily obtained in terms of the 'top-hat' velocity and half-width W, B, and the spreading rate, $\beta = \frac{dB}{dz}$ (Prob. 3.13).

$$B = (\frac{\beta}{\sqrt{2}})^{2/3} F_o^{1/3} t$$

$$W = (2\beta)^{-1/3} F_o^{1/3} \tag{3.78}$$

For the line plume, it is interesting to note the plume velocity is a constant, implying that $z = Wt$. The volume and momentum fluxes (per unit length of source), and the centerline concentration can then be obtained as (see Prob. 3.12 for relation between average and centerline variables for plane buoyant jets):

$$Q(z) = (2\beta)^{2/3} F_o^{1/3} z$$

$$M(z) = (2\beta)^{1/3} F_o^{2/3} z$$

$$c_m(z) = (2\beta)^{-2/3} \sqrt{\frac{1+\lambda^2}{\lambda^2}} Q_o F_o^{-1/3} z^{-1} \tag{3.79}$$

An analysis of the earlier experiments of Rouse *et al.*(1952) gives $\beta_G = 0.147$ and $\lambda = 1.24$ (Jirka and Harleman 1979). More detailed velocity and temperature measurements by Kotsovinos and List (1977) give $\beta_G = 0.116$ and $\lambda = 1.35$. This suggests a spread rate of $\beta = \sqrt{\frac{\pi}{2}}\beta_G$ $= \sqrt{\frac{\pi}{2}} \times 0.116 = 0.145$ and $\lambda = 1.35$. A recent review of data by Chu (1994) has suggested a spreading rate $\beta = dB/dz = 0.17$. Adopting $\beta = 0.145$, the solution in terms of the fluxes and centerline variables can be obtained; it is summarized in Table 3.1.

For the line plume, the volume flux or dilution varies linearly with distance. The predicted momentum coefficient (0.74) and centerline concentration (2.68) are in excellent agreement with the measured values of Rouse *et al.*(0.72 and 2.6 respectively). It is also consistent with Kotsovinos' measured concentration coefficient of 2.38. The predicted coefficient for the volume flux is 0.47. Based on direct measurements of the entrainment flow rate into a plane buoyant jet, Chu and Baines (1989) have obtained the following asymptotic formulae for the plane plume:

$$Q = 0.48\,F_o^{1/3} z \tag{3.80}$$

The formula proposed by Kotsovinos and List (1977) to fit their data is

$$Q = 0.34\,F_o^{1/3} z \tag{3.81}$$

The classical experiment of the plane plume by Rouse *et al.*(1952) gives the formula

$$Q = 0.57\,F_o^{1/3} z \tag{3.82}$$

There is considerable variation in the results obtained by the different investigations. The measured centerline dilution of $S_m = 0.42 F_o^{1/3} z$ given by Kotsovinos and List (1977) is also not consistent with their measured average dilution coefficient of 0.34. The nature of the plane plume is sensitive to the ambient condition (see Baines and Chu, 1996). The measured spreading characteristics may depend on this condition. Clearly, further experimental data are needed.

Plane Buoyant Jet - For a two-dimensional slot buoyant jet with initial unit discharge Q_o m^2/s, specific momentum flux $M_o = Q_o w_o$m^3/s^2, and specific buoyancy flux $F_o = Q_o g'_o$ m^3/s^3, a momentum length scale similar to that of the round buoyant jet can be defined as:

$$l_s(2D) = \frac{M_o}{F_o^{2/3}} \tag{3.83}$$

The momentum length scale is a measure of the distance over which the initial jet momentum is important. The dilution data of Kotsovinos and List (1977) shows that the buoyant jet flow is plume-like for $z/l_s \geq 4$.

For practical application an important equation for dilution prediction is that of a line plume in a stagnant unstratified ocean. The experimental result of Rouse (1952) is often used for the prediction of centerline dilution:

$$S_m = \frac{c_o}{c_m} = 0.38 \frac{F_o^{1/3} z}{Q_o} \tag{3.84}$$

The above results are for purely two-dimensional plumes. Roberts (1979, 1980) studied the mixing of a line pume of finite length in a current. For near stagnant conditions, the minimum surface dilution measured in his experiments give:

$$S_m = 0.27 \frac{F_o^{1/3} H}{Q_o} \tag{3.85}$$

for a depth of H. Compared with Eq. 3.84, the smaller dilution coefficient reflects the effect of blocking of the surface layer. Dilution charts to account for the effect of an unstratified current can be found in Roberts (1980).

EXAMPLE 3.10 *Thermal discharge from slot diffuser*
An oil-fueled steam electric generating station is located on one bank of a tidal river. Condenser cooling water is withdrawn from the river (at a flowrate of $Q_o = 20$ m^3/s*) into an upstream intake channel, and discharged back into the river at an excess temperature of* 20° *C. It is proposed to discharge the thermal effluent vertically through a submerged slotted pipe (a 2D slot diffuser) running along the canal bank, at* $H = 10$ *m below the free surface. The regulatory environmental standard stiputates that the temperature in the upper 5 m layer cannot exceed 28° C. In addition, to minimize any offshore momentum that may be hazardous to navigation, the maximum allowable transverse velocity (perpendicular to river axis) is 0.5 m/s.*

If the maximum ambient temperature that can occur in the year is 23° C, determine the required dilution of the diffuser discharge so that the temperature standard will be met all the year around. Determine the diffuser length and discharge velocity that will satisfy all the environmental constraints. Assume a worst case scenario of zero river velocity. The ambient water has a typical salinity of 32 ppt. Typical channel depths at the site are about 12-15 m. To minimize head losses, the initial jet velocity should not exceed 2 m/s.

SOLUTION: The thermal discharge from the slotted diffuser can be modelled as a two-dimensional vertical buoyant plume. For the highest

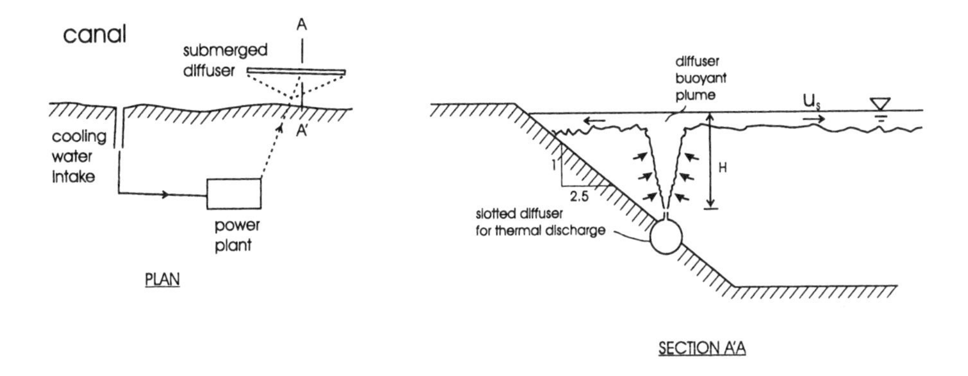

Figure 3.11. Slot plume thermal discharge

ambient temperature $T_a = 23^o$C and and a maximum allowable surface temperature of 28 ^{o}C, the required minimum dilution is $S_m = \Delta T_o/\Delta T_m = 20/(28-23) = \underline{4}$. For a plane buoyant jet, the ratio of minimum to average dilution is given by $\sqrt{\pi\lambda^2/(1+\lambda^2)} = 0.8$. The required average dilution is hence $\overline{S} = 4/0.8 = \underline{5.0}$.

Under stagnant ambient conditions, the vertical plume rises to the surface and spreads into two symmetrical layers. For a 2D plume it can be shown the upper layer thickness is approximately 5/6 H (Jirka and Harleman 1979). For a salinity of S = 32 ppt and the temperatures given, the initial density difference can be determined to be (Appendix A) $\Delta\rho_o/\rho = 0.0071$. The average dilution achieved by the time the thermal effluent enters the surface layer is then (Table 3.1):

$$\overline{S} = 5 = 0.465 F_o^{\frac{1}{3}}(\frac{5}{6}H)/Q_o = 0.465(0.0071 \times 9.81)^{\frac{1}{3}}(\frac{5}{6} \times 10)/Q_o^{\frac{2}{3}}$$

The required unit discharge is then $Q_o = 20.0/L_d = 0.18$ m^2/s, corresponding to a diffuser length of $L_d = 20/0.18 = \underline{111}$ m. Adopting a vertical discharge velocity of $U_o = 2.0$ m/s, the slot width is then $d_o = 20/111/2. = \underline{0.09}$ m. When the mixed thermal effluent reaches the surface, it flows away symmetrically in two directions. The transverse velocity of the mixed flow can be estimated to be $U_s = 0.5\overline{S}Q_o/(H/6) = (5/2) \times 0.18/1.67 = \underline{0.27}$ m/s. Thus the navigation contraint is also met. For this case, the unit momentum flux is $M_o = 0.18 \times 2 = 0.36$ m^3/s^2, and the buoyancy flux is $F_o = 0.18 \times 0.0071 \times 9.81 = 0.125$ m^3/s^3, giving a momentum length scale of $l_s = M_o/F_o^{2/3} = 1.44$ m. Thus $z/l_s = 5.8$, and the use of the 2D plume solution is well-justified.

An alternative to the use of a slot diffuser would be to discharge in the form of a large number of round jets from a multiport diffuser

(Prob. 3.3). If located in sufficient depth, wastewater discharges from closely spaced jets on a submerged multiport diffuser tend to form line plumes before they surface or form a trapped spreading layer, and the dilution in stagnant water can often be estimated by the line plume formula (e.g. Fischer *et al.*1979, Roberts *et al.*1989). It can be shown that for a given diffuser length, optimal dilution can be achieved using a line plume (Prob. 3.10).

6. SUMMARY

The development of plane and round plumes, and the positive and negative buoyant jets, are analysed in this chapter using both the Eulerian method and the Lagrangian method. The entrainment coefficient of a plume differs greatly from that of a jet, but the spreading rate for the plume is practically the same as the rate for the jet. The Lagrangian method differs from the Eulerian method in the manner in which the velocity and concentration profiles are specified. In the Eulerian method, the plume is typically described in terms of the centerline variables, and the solution obtained by specifying the Gaussian velocity and concentration profiles. Such an assumption is not required in the Lagrangian formulation which adopts a 'top-hat' profile for the velocity of the dominant eddies. The most important distinction between the Lagrangian and Eulerian methods lies in the formulation of the turbulent entrainment hypothesis. The entrainment coefficient defined from a Eulerian point of view such as the one given in Eq. 3.45 is flow dependent. In the positively buoyant jet this entrainment coefficient α_G varies from a jet value of 0.057 to a plume value of 0.09. The entrainment coefficient may become negative in the negatively buoyant jet and is equal to zero in the far field region of the coflowing jet (see Chapter 6). However, the spreading coefficient β (defined in a Lagrangian reference frame following the dominant eddies) is relatively constant and not dependent on the flow. The Lagrangian method is simple and convenient in its implementation because the important results can be obtained without actually specifying the velocity and concentration profiles, which are not generally known for more complex turbulent flows. In the subsequent chapters, the Lagrangian method will be adopted to study the inclined buoyant jet, jets and plumes in stratified fluid, coflowing jets, puffs and thermals, and for arbitrary-inclined buoyant jet in non-uniform stratified crossflow.

Table 3.1. Summary of plume properties (Based on Gaussian Profiles)

Parameter	Round plume	Plane plume
Plume Volume Flux $Q = \int w dA$	Dimensions L^3T^{-1} $Q = \pi w_m b^2$	L^2T^{-1} $Q = \sqrt{\pi} w_m b$
Specific momentum Flux $M = \int w^2 dA$	L^4T^{-2} $M = 0.38\ F_o^{\frac{2}{3}} z^{\frac{4}{3}}$	L^3T^{-2} $M = 0.743\ F_o^{\frac{2}{3}}$
Specific buoyancy flux $F = \int w \frac{\Delta\rho}{\rho} g dA$	L^4T^{-3} $F = \frac{\pi\lambda^2}{1+\lambda^2} w_m g_m' b^2$	L^3T^{-3} $F = \sqrt{\frac{\pi\lambda^2}{1+\lambda^2}} w_m g_m' b$
Maximum time-averaged velocity w_m	$w_m = 4.71\ F_o^{\frac{1}{3}} z^{-\frac{1}{3}}$	$w_m = 2.26\ F_o^{\frac{1}{3}}$
Plume width b	$b = 0.105\ z$	$b = 0.116\ z$
Maximum time-averaged tracer concentration c_m	$c_m = 10.46 Q_o C_o F_o^{-\frac{1}{3}} z^{-\frac{5}{3}}$	$c_m = 2.68 Q_o C_o F_o^{-\frac{1}{3}} z^{-1}$
Average Dilution $\overline{S} = Q/Q_o$	$\overline{S} = 0.163\ F_o^{\frac{1}{3}} z^{\frac{5}{3}} Q_o{}^{-1}$	$\overline{S} = 0.465\ F_o^{\frac{1}{3}} z Q_o{}^{-1}$
Plume spreading angle β	$\beta_G = 0.105$ $(\beta = \frac{6}{5}\alpha)$	$\beta_G = 0.116$ $(\beta = \frac{2}{\sqrt{\pi}}\alpha)$
Entrainment coefficient α	$\alpha_G = 0.088$	$\alpha_G = 0.103$
Ratio of concentration to velocity width, λ	$\lambda = 1.19$	$\lambda = 1.35$
Plume densimetric Froude number	$\mathrm{Fr}_p = (\frac{5}{4})^{1/2} \frac{\lambda}{\sqrt{\alpha}} = 4.50$	$\mathrm{Fr}_p = (\frac{\pi}{2})^{1/4} \sqrt{\frac{\lambda}{\alpha}} = 4.06$

PROBLEMS

3.1 The vertical centerline velocity of a round plume in near stagnant fluid is given by $w_m = CF_o^{1/3}z^{-1/3}$, where F_o and z are the specific buoyancy flux and elevation above source respectively, and $C = 4.71$. Show that the average travel time t_H from the source to the free surface can be estimated by $t_H = 3/(2C)F_o^{-1/3}H^{4/3}$, where H is the depth of the plume. Calculate the travel time for the discharge in Prob. 4.1 for $U_a = 0$. Comment on your results.

How will the result change for the same discharge in an ambient current of $U_a = 1$ m/s (see Chapter 8). Justify any assumptions made in your estimate.

3.2 Consider the discharge of fresh water from a round nozzle into ambient salt water in the laboratory, with $D = 0.015$ m, $w_o = 0.07$ m/s, $\frac{\Delta\rho_o}{\rho_a} = 0.03$. Assuming that the plume will be turbulent if the local Reynolds number defined by $Re = \frac{w_m b}{\nu}$ exceeds 2000, estimate the distance over which the plume may be laminar. This is a consideration in the design of laboratory experiments of buoyant discharges.

3.3 Merging plumes

Wastewater plumes are often discharged into rivers or coastal waters through a multiple port outfall laid along the bottom (a multiport diffuser). Wastewater is discharged as a number of equally-spaced vertical buoyant jets into water of depth H. Field observations showed that as long as the depth of discharge below the surface was less than three times the distance between adjacent outlets, the measured dilution at the surface is the same as that of a single jet (Rawn and Palmer (1930), *Trans. ASCE*, Vol.94, 1036).

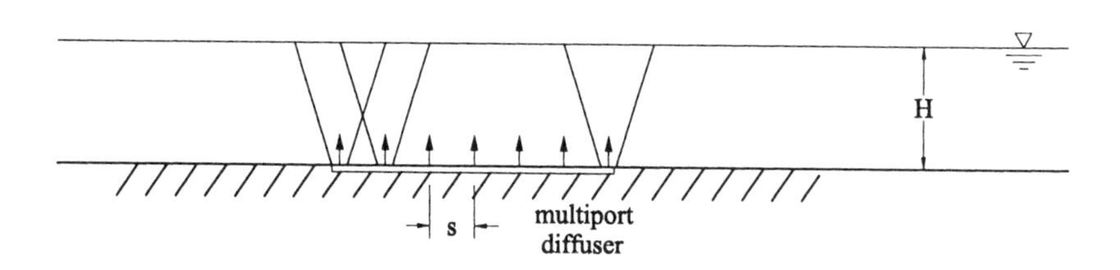

Use jet and plume theory to establish a criterion for the optimal spacing (s) between adjacent buoyant jets. Compare your result with the observations. Assume that the optimal dilution is given by the case when the jets do not interfere.

3.4 For a heated water jet discharge into a receiving water of constant ambient temperature T_a, show that the source heat flux is given by

$$H = c_p \int \rho w \Delta T dA = c_p \rho_a Q_o \Delta T_o$$

where $\Delta T = T - T_a$, and c_p = specific heat of water. Show that the predicted centerline temperature excess ΔT_m in a round thermal plume (Eq. 3.51) can be written as:

$$S_m = \frac{\Delta T_o}{\Delta T_m} = 0.1 \frac{F_o^{1/3} z^{5/3}}{Q_o}$$

For a discharge of $Q_o = 0.1\ m^3/s$ and $\Delta T_o = 10^oC$, calculate the heat flux H (kW) if $c_p = 4.18\ kJ/kg - {}^oC$ (at 20^oC).

3.5 Consider an atmospheric plume that satisfies perfect gas law, $p = \rho RT$, where p= pressure, R=gas constant, and T=absolute temperature. Show that at constant pressure we have:

$$\frac{|\Delta T|}{T} \approx \frac{|\Delta \rho|}{\rho}$$

Show also the volume coefficient of thermal expansion (percentage change of volume per degree Kelvin) is given by $\gamma_T = 1/T$.

3.6 For a buoyant gas plume arising from a heat source, the convective heat flux is given by $H = c_p \int \rho w \Delta T dA$. Show that for small temperature excess $\Delta T/T \ll 1$ and assuming constant pressure, the heat flux is related to the specific buoyancy flux F_o as:

$$H = \frac{c_p \rho_a T}{g} \int w g' dA = \frac{c_p \rho_a T}{g} F_o$$

For the standard atmosphere, with $R = 287 J/kg^oK, p = 101.3$ kPa, $c_p = 1.006\ kJ/kg^oK$, show that

$$H\ (kW) \approx 36.2\ F_o\ (m^4/s^3)$$

3.7 A desalination plant discharges concentrated brine (specific gravity 1.032) into the adjacent sea ($\rho_a = 1025$ kg/m^3) of depth 15 m. The dense effluent contains copper which can be toxic to marine phytoplankton. It is proposed to design a multiport diffuser to mix the effluent (flow $Q_o = 0.6$ m^3/s) sufficiently to reach a safe copper concentration by dilution. Physical model tests show that maximum trajectory and hence dilution are obtained by having the jet nozzles discharging at 60° to the horizontal .

The experiments show that for $\phi_o = 60^o$, the maximum height of rise z_m and the minimum dilution at the top of the jet are correlated with the jet discharge parameters as $z_m = 2.04\ FrD$, and $S_m = 0.5\ Fr$ respectively.

If the required dilution of the copper in the effluent is $S = 15$, and the jet nozzle diameter is $D = 0.1$ m, determine the jet velocity and the number of jets required. Assume non-interfering jets. Determine the maximum height of rise; will the dense effluent impact on the surface?

3.8 *Asymptotic densimetric Froude number of vertical round buoyant jet*
Consider a vertical round buoyant jet with initial velocity w_o, relative density difference $\Delta\rho_o/\rho_a$, and jet diameter D, discharging in the z-direction in ambient fluid of constant density ρ_a. The initial jet densimetric Froude number is $Fr = w_o/\sqrt{g_o' D}$. The velocity and concentration (density-deficiency) distribution can be assumed to be self-similar and Gaussian, with velocity $w(z,r) = w_m e^{(-r^2/b^2)}$, and density deficiency, $\Delta\rho(z,r) = \Delta\rho_m e^{-r^2/(\lambda b)^2}$, where $\Delta\rho = \rho_a - \rho$.

An integral model based on the entrainment hypothesis results in the following governing equations for the centerline maximum velocity and density deficit, u_m and $\Delta\rho_m$, and jet-width b:

$$
\begin{aligned}
\frac{d}{dz}(w_m b^2) &= 2\alpha w_m b \\
\frac{d}{dz}(\frac{w_m^2 b^2}{2}) &= g'\lambda^2 b^2 \\
\frac{d}{dz}(w_m g' b^2) &= 0
\end{aligned}
$$

where $g' = \Delta\rho_m/\rho_a$ and $\alpha=$ entrainment coefficient.

It can be shown from the governing equations of the integral model that the local densimetric Froude number defined by
$\mathrm{Fr_L} = \frac{w_m}{\sqrt{\Delta\rho_m/\rho_a g b}} = \frac{w_m}{\sqrt{g' b}}$ satisfies

$$
\frac{d\mathrm{Fr_L}}{dz} = \frac{2\alpha}{\mathrm{Fr_L} b}(\frac{5\lambda^2}{4\alpha} - \mathrm{Fr_L}^2)
$$

a) Show that the continuity equation and the momentum equation results in the following expressions:

$$
\begin{aligned}
2w_m^2 b\frac{db}{dz} + w_m b^2 \frac{dw_m}{dz} &= 2\alpha w_m^2 b \\
w_m^2 b\frac{db}{dz} + w_m b^2 \frac{dw_m}{dz} &= g'\lambda^2 b^2
\end{aligned}
$$

Hence show that

$$
\frac{db}{dz} = 2\alpha - \frac{\lambda^2}{\mathrm{Fr_L}^2}
$$

b) From the momentum equation, and using the result in (a), deduce that:

$$
\begin{aligned}
w_m \frac{dw_m}{dz} &= \frac{g'\lambda^2 b^2 - w_m^2 b db/dz}{b^2} \\
&= 2g'(\lambda^2 - \alpha \mathrm{Fr_L}^2)
\end{aligned}
$$

c) From the buoyancy conservation equation, show that $d(g'b)/dz = -\frac{g'}{w_m} d(w_m b)/dz$. Since from the momentum equation, we have $w_m b \frac{d}{dz}(w_m b) = g'\lambda^2 b^2$, show that:

$$
\frac{d}{dz}(g'b) = -\frac{g'\lambda^2}{\mathrm{Fr_L}^2}
$$

d) The local densimetric Froude number is a measure of the ratio of the inertia to buoyancy forces, with $\mathrm{Fr_L}^2 = w_m^2/(g'b)$. We then have:

$$
\begin{aligned}
\frac{d\mathrm{Fr_L}^2}{dz} &= \frac{(g'b)\frac{dw_m^2}{dz} - w_m^2 \frac{d}{dz}(g'b)}{(g'b)^2} \\
&= \frac{2\alpha}{\mathrm{Fr_L} b}(\frac{5\lambda^2}{4\alpha} - \mathrm{Fr_L}^2)
\end{aligned}
$$

The above expression shows that the vertical buoyant jet attains an asymptotic densimetric Froude number of $\mathrm{Fr}_p = \sqrt{\frac{5\lambda^2}{4\alpha}}$. If $Fr \leq \mathrm{Fr}_p$, $d\mathrm{Fr_L}/dz \geq 0$, and the local Froude number will increase. Conversely, if $Fr \geq \mathrm{Fr}_p$, $d\mathrm{Fr_L}/dz \leq 0$, and the local Froude number will decrease. The asymptotic value of $\mathrm{Fr_L}$ for a vertical buoyant jet with any Fr is Fr_p.

Determine the value of Fr_p using the values of $\alpha = 0.088$ and $\lambda = 1.2$ for a round plume.

3.9 a) Under what conditions will wastewater plumes from a multiport diffuser resemble a "line plume ".

b) Consider a turbulent line plume (freshwater) in otherwise still sea water of constant density (and at the same temperature). The initial buoyancy flux is $F_o = Q_o g'_o$, where Q_o is the initial volume flux per unit diffuser length. Deduce from first principles the conservation of specific buoyancy flux at any elevation above the source (cf Eq. 3.24).

c) Using dimensional reasoning, arrive at a prediction (to within a constant) of the centerline velocity $w_m(z)$ and the average dilution $Q(z)/Q_o$ of a line plume in terms of the important physical parameters. Determine also the plume width and the local densimetric Froude number. Comment on your results.

For a fixed wastewater flow and depth, how will the dilution depend on the diffuser length?

3.10 Pure plume solutions are very useful for estimating the initial dilution of buoyant discharges. The average dilution for a round and line plume at elevation z above the source are given by:

$$S_{3D} = 0.156 \, \frac{F_o^{1/3} z^{5/3}}{Q_o}$$

$$S_{2D} = 0.535 \, \frac{F_o^{1/3} z}{Q_o}$$

Consider a submerged multiport diffuser of given length L in depth H. It is desired to handle a design flow Q_d and two opposing port arrangements are proposed:

- very closely spaced ports to simulate a line plume
- N ports spaced to avoid jet interference before the plumes arrive at the surface

a) Show that the ratio of the average dilution (total flow entrained) at any z for the two cases is

$$R = \frac{S_{row\ of\ N\ ports}}{S_{line\ plume}} = 0.29 \, (zN/L)^{2/3}$$

b) If $w = 2b(z)$ is taken as the nominal half-width of each round buoyant jet (what percentage of $Q(z)$ is accounted for by this definition?), calculate R at the surface, $z = H$. What about R for $z < H$? For a round plume, $db/dz = \beta_G = 0.105$.

3.11 *Turbulent line plume*

a) The velocity and tracer concentration in a *two-dimensional plume* are confirmed to be self-similar and well-approximated by Gaussian profiles. Employing the entrainment hypothesis and a control volume analysis, deduce the governing equations of a line plume in terms of the centerline velocity $w_m(z)$, plume width $b(z)$, and centerline density deficit $\Delta\rho_m(z)$. Neglect initial volume and momentum fluxes.

b) Adopting the values of the entrainment coefficient $\alpha = 0.103$ and the ratio of concentration/velocity width $\lambda = 1.35$, deduce the relations for the variation of the characteristic properties of a line plume.

3.12 For a plane buoyant jet, the volume, momentum, and tracer mass fluxes can be obtained in terms of either the average properties (U, B, C) or the centerline variables of the Gaussian profile (u_m, b, C_m), where the terms have their usual meaning. Show by mass, momentum, and tracer mass conservation that:

$$\begin{aligned} Q = 2UB &= \sqrt{\pi} u_m b_g \\ M = 2U^2 B &= \sqrt{\frac{\pi}{2}} u_m^2 b_g \\ QC &= \sqrt{\frac{\pi\lambda^2}{1+\lambda^2}} u_m C_m b_g \end{aligned}$$

Hence show that the average and centerline properties are related by: $U = u_m/\sqrt{2}$, $B = \sqrt{\pi/2} b_g$, and $C = \sqrt{\lambda^2/(1+\lambda^2)} C_m$.

3.13 *Lagrangian model for line plume in stagnant fluid*

a) Assume the line plume cross-section can be represented by an average velocity U and a half-width B. Using a Lagrangian plume spreading hypothesis, $dB/dt = \beta U$, develop the governing equations for the plume element as a function of *time*.

b) Derive the line plume equations, Eq. 3.78, by simple integration.

c) Noting that, $w_m = \sqrt{2}U$, $b_g = B/(\sqrt{\pi}/2)$, obtain the prediction of centerline velocity and centerline concentration. For $\beta_G = 0.116$ and $\lambda = 1.35$ evaluate the coefficients in the expressions. Compute the local densimetric Froude number of the plume.

3.14 Dependence of entrainment coefficient on Froude number

When a discharge has both momentum and buoyancy, the entrainment coefficient α is in general not a constant but a function of the local jet densimetric Froude number $\mathrm{Fr_L}$. The dependence of α on $\mathrm{Fr_L}$ can be developed from the mechanical energy equation and is illustrated herein for an arbitrarily inclined plane buoyant jet in stagnant fluid (Jirka and Harleman 1973).

Consider a two-dimensional buoyant jet with velocity $[u, v]$ (s, n) in the streamwise (s) and normal (n) co-ordinates respectively. In this natural co-ordinate system (s, n) following the jet trajectory, the governing continuity and axial momentum equations can be shown to be (invoking the boundary layer approximations):

$$\frac{\partial u}{\partial s} + \frac{\partial v}{\partial n} = 0$$

$$u\frac{\partial u}{\partial s} + v\frac{\partial u}{\partial n} \quad = \quad g' \sin\phi - \frac{\partial \overline{u'v'}}{\partial n}$$

where $g' = \Delta\rho g$, and ϕ= local inclination of jet trajectory to horizontal.

a) Show that by multiplying the continuity equation with u^2 and the momentum equation with u, the mechanical energy equation for the two-dimensional buoyant jet can be obtained as follows:

$$\frac{\partial u^3}{\partial s} + \frac{\partial u^2 v}{\partial n} \quad = \quad 2\,u\frac{\Delta\rho}{\rho_a} g \sin\phi - 2u\frac{\partial \overline{u'v'}}{\partial n}$$

b) Assume Gaussian profiles for u and $\Delta\rho$, with $u(s,n) = u_m e^{-n^2/b^2}$, and density deficiency, $\Delta\rho(s,n) = \Delta\rho_m e^{-n^2/(\lambda b)^2}$. Assume that the shear stress is also self-similar (see e.g. Mih and Hoopes 1972), $\overline{u'v'} = u_m^2 f(n/b)$, where $f(n/b)$ is a universal but unknown function.

The integral model equations for continuity and axial momentum can be shown to be:

$$\frac{d}{ds}(u_m b) \quad = \quad \frac{2}{\sqrt{\pi}}\alpha w_m$$

$$\frac{d}{ds}(u_m^2 b) \quad = \quad \sqrt{2} g' \lambda b \, \sin\phi$$

By integrating the mechanical energy equation across the buoyant jet, show that the entrainment function must be of the form

$$\alpha = \sqrt{3} I_1 + (\sqrt{2} - \sqrt{\frac{3}{1+\lambda^2}}) \sin\phi \, \frac{\sqrt{\pi}\lambda}{\mathrm{Fr_L}^2}$$

where $I_1 = \int_{-\infty}^{\infty} f(n/b) d(n/b)$. Determine the constant $\alpha_1 = \sqrt{3} I_1$ using available data on jets and plumes.
(Hint: Use the identity $\frac{d}{ds}(u_m^3 b) = 2u_m \frac{d}{ds}(u_m^2 b) - u_m^2 \frac{d}{ds}(u_m b)$)

3.15 For an unbounded 2D heated gas plume, show from the plume solution in Table 3.1 that the mass flux of the plume is given by:

$$M_a = C_m \rho_a \, H^{1/3} L^{2/3} z$$

where L is the lateral length of the line plume, H = heat flux, and ρ_a is the ambient density. Evaluate the dimensionless entrainment coefficient C_m.

Experiments were performed for a laboratory scale model of a line fire (length L). For $H = 8.2$ kW, $L = 1.0$m, find the mass flow rate in the fire plume at a height of 2 m above the source.

3.16 Settled sewage from a population of about 3 million is to be discharged from a 1200 m long submarine diffuser located in a harbour. The design sewage flow is 20 $\mathrm{m^3/s}$, and the depth is 12 m. Assuming a line plume and $\Delta\rho_o/\rho = 0.025$, estimate the minimum surface dilution in near stagnant water. Will the initial dilution be adequate?

Chapter 4

INCLINED BUOYANT JET IN STAGNANT ENVIRONMENT

The analysis of straight jets and plumes in the previous chapters has provided us with a basic understanding of turbulence and mixing in jets and plumes, and some tools to obtain reliable estimates in asymptotic cases. However, for many engineering problems it is far more convenient and often necesssary to use a general integral jet model. In this chapter we develop such a model for the case of inclined buoyant jets. The formulation for the inclined buoyant jets is an extension to the problems of the straight jets and plumes, and is a prelude to the general formulation of the jet-and-plume problems to include the effects of density stratification and crossflow in the later chapters.

Figure 4.1 shows the inclined buoyant jet in an otherwise still fluid. Momentum and buoyancy are discharged at a rate equal to M_o and F_o, respectively and at an angle ϕ_o to the horizontal. Figure 4.2 shows a buoyant jet in a model test for a wastewater discharge (Wah Fu Outfall, Hong Kong) at an inclination of 20 degrees to the horizontal. Close to the source, jet momentum dominates. The jet spreads linearly as it entrains the ambient fluid. The buoyant force acts in the vertical direction of the gravity. It changes the vertical component of the momentum at a rate equal to the buoyant force. The horizontal component of the momentum flux is constant. The initial specific momentum and buoyancy fluxes, M_o and F_o, are the two parameters characterising the source. The length scale

$$l_s = \frac{(M_o)^{\frac{3}{4}}}{(F_o)^{\frac{1}{2}}} \tag{4.1}$$

defines the near field and the far field. Since $M_o = \pi D^2 V_o^2/4$ and $F_o = \pi D^2 V_o g'_o/4$,

$$l_s = (\frac{\pi}{4})^{\frac{1}{4}} D Fr_o \tag{4.2}$$

where D is the diameter of the source and $Fr = V_o/\sqrt{g'_o D}$ is the densimetric Froude number. The buoyancy effect becomes comparable to that due to the initial momentum at distance $z = l_s$ from the source. Eventually at large distances from the source, $z \gg l_s$, the buoyant jet behaves like a plume. The strength of the jet is directly related to the value of its densimetric Froude number, Fr. The greater the value of the densimetric number the stronger the flow is determined by the jet's initial momentum. The jet turns rapidly into a plume when Fr is small.

The jet path is in general unknown *a priori* and it is governed by the initial discharge angle and the balance of inertia and buoyancy. The flow and concentration distributions (and hence the buoyancy distribution) are coupled and have to be solved simultaneously by numerical integration of the governing equations.

The inclined buoyant jet in stagnant fluid has been extensively studied by the integral method in connection with environmental applications (e.g. Brooks 1972; Koh and Brooks 1975; Jirka *et al.*1975; Fischer *et al.*1979). In a model of this type, the governing boundary-layer equations for continuity, momentum, and tracer mass conservation can be written in a local, 'natural' co-ordinate system (s, r), where s is a streamwise co-ordinate along the jet centerline (defined by e.g. point of maximum concentration), and r the radial co-ordinate away from the centerline. As the shear flow is induced by jet momentum and buoyancy, locally the mixing is analogous to straight jets and plumes. The same boundary-layer approximation and assumptions of self-similar Gaussian profiles for velocity and concentration in the jet cross-section (Fig. 4.1) can be made. The governing equations can be integrated across the buoyant jet to yield a set of differential equations in terms of the centerline variables, or the mass, momentum, and tracer mass (buoyancy) fluxes. Given the initial conditions, the numerical integration of the resulting system of equations gives a prediction of the jet trajectory, width, velocity, and tracer concentration (dilution).

In most of the models reported in the literature, turbulent closure is formulated using the entrainment hypothesis (e.g. Fan 1967; Fan and Brooks 1969; Ayoub 1971; Hirst 1972; Schatzmann 1979). As noted in the previous chapter, the entrainment coefficient α is not a constant but varies as the turbulent flow changes from a jet to a plume. The dependence of α on the local densimetric Froude number was recognised and a relation formulated for vertical buoyant jets (Fox 1970; List and Imberger

1973). As an alternative, models based on the spreading hypothesis have also been formulated (Abraham 1963a & b; Jirka *et al.* 1975; Lee and Jirka 1981; Wood 1993); the equivalence of the two approaches for the limiting cases has also been discussed. Different variants of these models have been extensively applied to wastewater outfall design and related environmental applications. For example, predictions of initial dilution

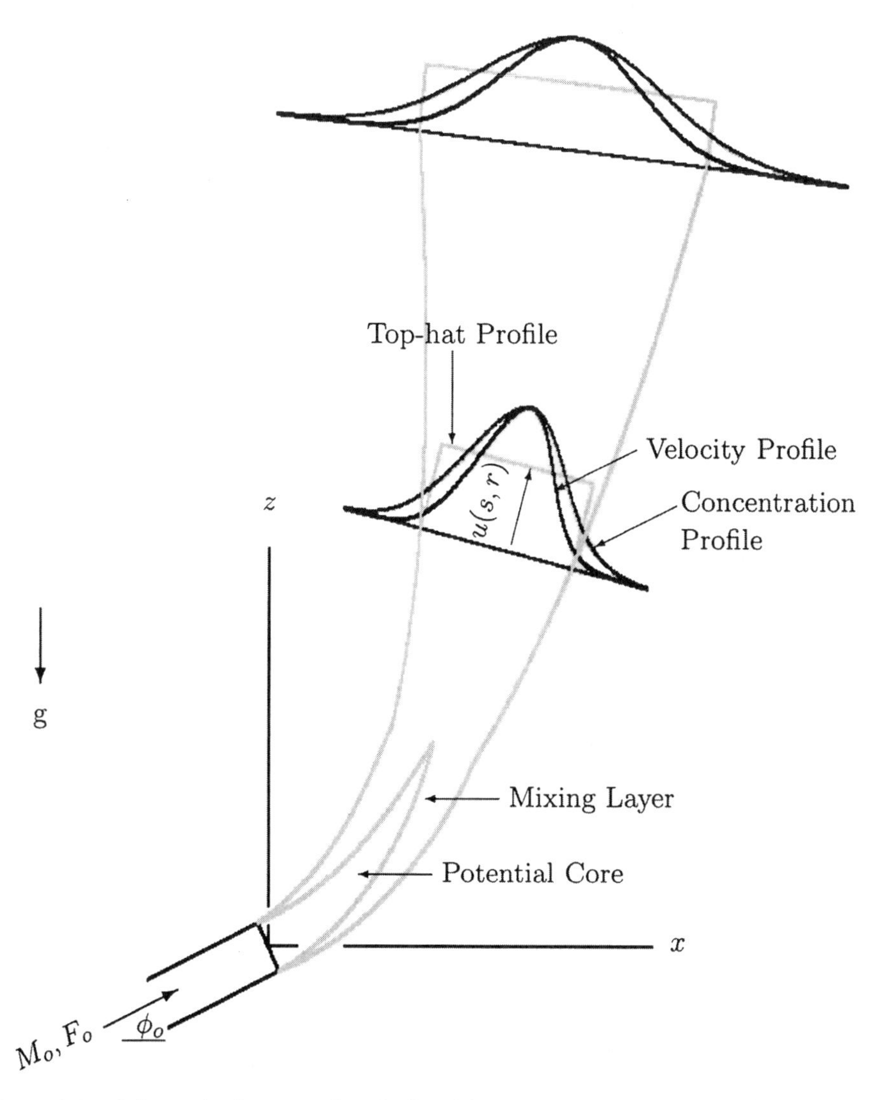

Figure 4.1. Schematic diagram of an inclined buoyant jet in stagnant ambient fluid. Gaussian velocity and concentration profiles, and the top-hat profile, are shown at two cross sections along the jet trajectory.

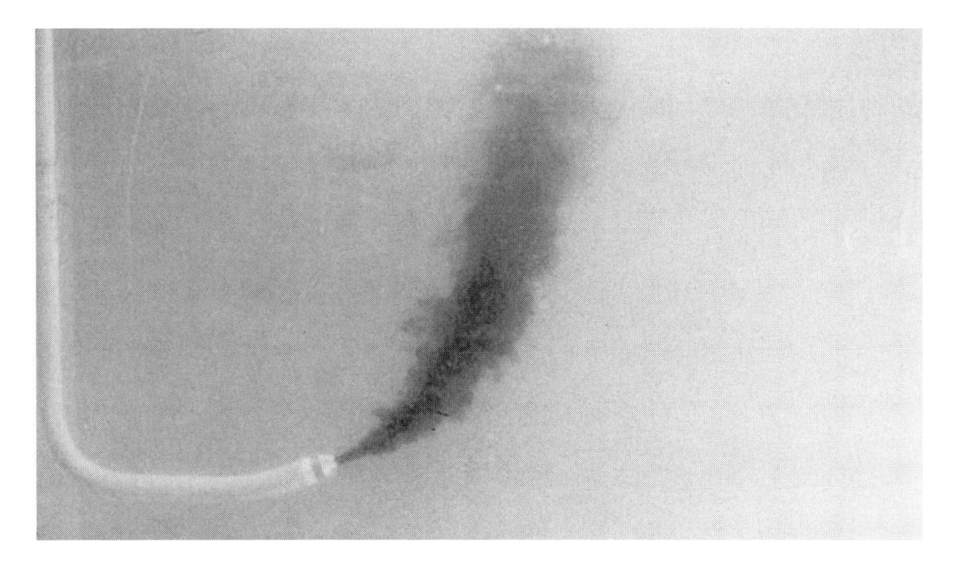

Figure 4.2. Observed buoyant jet trajectory in a model test: $Fr = 6.8$, $\phi_o = 20^o$

contours as a function of Fr and dimensionless depth y/D were generated from the numerical model of Fan and Brooks (1969), and widely used as design charts. Backed by extensive laboratory experiments, the dilution predictions of Abraham (1963) and Cederwall (1968) were also extensively used in practice. In particular, the USEPA model UPLUME (Muellenhoff *et al.* 1985) is an elaborate computer implementation of Abraham's (1963b) theory based on the spreading hypothesis.

For a general numerical calculation, provided the correct dependence of α on the local densimetric Froude number is incorporated, both the spreading or entrainment hypothesis will give similar and equally acceptable results. Nevertheless, the great majority of the models either neglect the potential core, or assume a straight jet of fixed core length for coupling source conditions and those at the start of the zone of established flow (ZEF) where Gaussian profiles can be assumed. In many applications, this may be a reasonable approximation, as the distance from source to the point of interest (e.g. free surface or trapping level) may be much larger than the source dimension (jet diameter), $z \gg D$. Since far away from the source, all discharges behave like plumes, neglecting the potential core will not introduce significant error. Nevertheless, there are many situations where the source dimension or initial core length cannot be conveniently ignored. In the general case, it would be highly desirable to be able to predict the path and length of the potential core.

In this chapter we present a Lagrangian model for an inclined buoyant jet in stagnant fluid, using a spreading hypothesis following the motion of

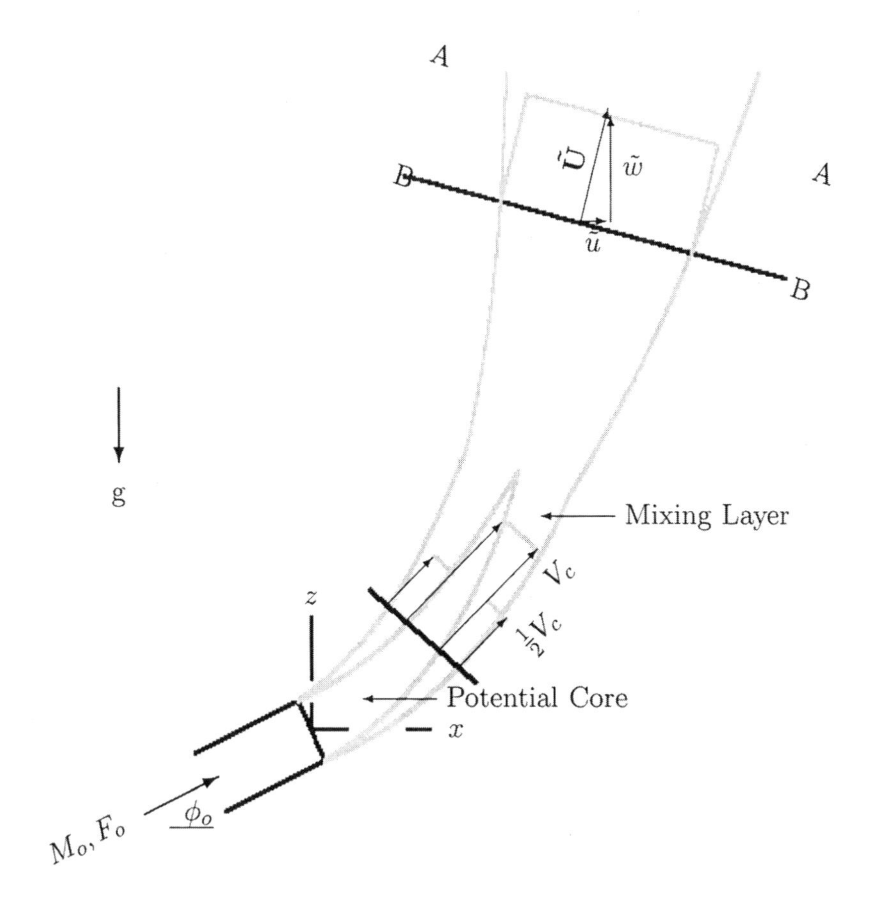

Figure 4.3. Schematic diagram of an inclined buoyant jet in stagnant ambient fluid. The top-hat profiles for the velocity of the dominant eddies are shown at a cross section in the ZFE and a cross section in the ZEF.

the dominant eddy in the buoyant jet. The constancy of spread rate over the jet-plume range is employed to advantage. The formulation is made first for the zone of established flow (ZEF). The treatment of the zone of flow establishment (ZFE) in the potential core is then described along with its coupling with the flow in the ZEF. The connection between the model and Eulerian formulations is discussed. The model predictions are illustrated and compared with experimental data.

1. LAGRANGIAN MODEL FOR BUOYANT JET IN STAGNANT FLUID

The formulation is illustrated for the round buoyant jet in a stagnant fluid of uniform density (Fig.4.3). At sufficiently high Reynolds number, the fluid emerging from the circular nozzle of a round jet would immediately begin to mix with the ambient fluid. A turbulent mixing layer is formed surrounding a core of essentially irrotational fluid from the nozzle. As the mixing layer grows by entrainment of the surrounding fluid, the area of the core decreases correspondingly. The velocity and concentration in the ZFE depend on the potential core and mixing layer surrounding the core. In the potential core, the velocity and concentration, V_c and C_c, are unaffected by the mixing process. In the mixing layer, the velocity and concentration are $V_c/2$ and $C_c/2$. These are the average velocity and concentration of the dominant eddies in the mixing layer as envisaged in the dominant-eddy hypothesis proposed by Chu (1994). Entraining fluid from the potential core, and fluid from the stagnant environment, the dominant eddies in the mixing layer move forward with a velocity equal to $\frac{1}{2}V_c$ that is the average of the velocities of the fluid on either side of the mixing layer. Beyond the ZFE, the velocity and concentration of the dominant eddies are $\tilde{U}$ and $\tilde{C}$, respectively. The governing equations are derived first for the ZEF. This is followed by the treatment of the ZFE in a subsequent section.

1.1 ZONE OF ESTABLISHED FLOW (ZEF)

We adopt a Lagrangian approach following the motion of the dominant eddies (Chu 1994). The formulation is carried out using a material volume of fluid that is defined by the release of momentum and buoyancy from the source over a period of one time unit. The initial momentum and buoyancy force associated with this material volume is equal to the momentum flux M_o and buoyancy flux F_o. At a subsequent time, this material volume as schematised in Figure 4.3 is bounded between section AA and section BB. The fluid associated with section AA left the source one unit of time earlier than those associated with section BB. If the average velocity of the plume element in the direction of the flow (s-direction) is $\tilde{U} = \sqrt{\tilde{u}^2 + \tilde{w}^2}$, the length of this material volume would be equal to the velocity $\tilde{U}$ times one time unit and that is $\tilde{U}$. If the radius of the cylinderical volume is $\tilde{b}$, the cross section area will be $\pi\tilde{b}^2$ and the volume of the element AA-BB would be $\pi\tilde{b}^2|\tilde{U}|$. Momentum and buoyancy associated with the material volume are $\mathbf{M} = (\pi\tilde{b}^2|\tilde{U}|)\tilde{\mathbf{U}}$ and $F = (\pi\tilde{b}^2|\tilde{U}|)\tilde{g}'$, respectively. In an ambient fluid of uniform density, the buoyancy force associated with the material volume is a constant and is equal to the buoyancy flux at the source as follows:

$$F = (\pi\tilde{b}^2|\tilde{U}|)\tilde{g}' = F_o \tag{4.3}$$

(assuming without loss of generality a linear equation of state). The momentum flux (that is the momentum assciated with the material element AA-BB) has two components. The horizontal component is constant:

$$M_x = (\pi \tilde{b}^2 |\tilde{U}|)\tilde{u} = M_o \sin\phi_o \tag{4.4}$$

since the buoyancy force acts in the vertical direction. The vertical component is changed at a rate equal to the buoyancy force; i.e.,

$$\frac{dM_z}{dt} = \frac{d}{dt}[(\pi \tilde{b}^2 |\tilde{U}|)\tilde{u}] = F_o \tag{4.5}$$

where d/dt is the material derivative for the rate with time in a Lagrangian co-ordinate system moving with the dominant eddies.

A Lagrangian spreading hypothesis is introduced for turbulent closure. The dominant eddies in the buoyant jet are assumed to spread at a rate proportional to the relative velocity between the dominant eddies and its surroundings:

$$\frac{d\tilde{b}}{dt} = \beta|\tilde{U}| \tag{4.6}$$

where β is the speading coefficient which is assumed to be a constant equal to the value adopted for straight jets and plumes. The path of the jet defined by the co-ordinates $\tilde{x} = x(t)$ and $\tilde{z} = z(t)$ are related to the velocity components of the dominant eddies as follows:

$$\frac{d\tilde{x}}{dt} = \tilde{u} \tag{4.7}$$

$$\frac{d\tilde{z}}{dt} = \tilde{w} \tag{4.8}$$

The six unknowns, $\tilde{b}, \tilde{u}, \tilde{g}', \tilde{w}, \tilde{x}$ and $\tilde{z}$, are determined by numerical integration of the sets of the four ordinary differential equations, Equations 4.5, 4.6, 4.7, 4.8, and the two algebraic relations, Equations 4.3 and 4.4, for constant F and constant M_x.

In a steady flow, the Lagrangian operator is related to the Eulerian advection operator by

$$\frac{d}{dt} = \tilde{U}\frac{d}{ds} \tag{4.9}$$

where $s=$ distance along the jet trajectory. With this relation, the above system of ordinary differential equations can be cast into Eulerian form as follows:

$$\frac{d\tilde{b}}{ds} = \beta \tag{4.10}$$

$$\frac{dM_z}{ds} = \frac{F_o}{U} \tag{4.11}$$

$$\frac{dx}{ds} = \cos\phi \tag{4.12}$$

$$\frac{dz}{ds} = \sin\phi \tag{4.13}$$

where $\phi = \tan^{-1}(M_z/M_x)$. This system of ordinary differential equations are in standard (canonical) form and can be solved using standard numerical techniques, e.g. a 5th order Runge-Kutta scheme. The above equations can also be obtained by integrating the equations of motion in natural co-ordinates across the buoyant jet, and can be extended to account for ambient density stratification (see Chapter 5 and Prob. 5.1). The Lagrangian formulation however allows a straight-forward extension of the procedure to include effects such as unsteadiness, density stratification and crossflow. The general derivation for some of these effects using the Lagrangian method are considered in the later chapters.

1.2 THE POTENTIAL CORE (ZFE)

Most of the existing integral models ignore the presence of the potential core by assigning a virtual origin for the jet which is assumed to be located at a distance $s_e \simeq 6.2D$ from the source. In many cases a straight jet behaviour is postulated close to the source; a potential core of fixed length 6.2 D is assumed. The initial conditions for the governing equations in the zone of established flow (ZEF) are then obtained at $s = s_e$ by momentum conservation (Prob. 4.2). In many practical applications, particularly for buoyancy-dominated jets, the initial mixing of the flow around the core is significant and the mixing process cannot be correctly represented by a point source model. We present here a model of the potential core, and the mixing layer surrounding the core, as shown in Figure 4.4. The radius of the core is R. The width of the mixing layer surrounding the core is r. The core velocity is V_c and is uniform across the core. The dominant eddies in the mixing layer moves with a velocity equal to $\frac{1}{2}V_c$ which is the average of the velocity in the core, V_c, and the velocity in the stagnant environment. Hence,

$$\frac{ds}{dt} = \frac{1}{2}V_c. \tag{4.14}$$

where s is advection distance of the dominant eddies. In a co-ordinate moving with the dominant eddies, the size of the eddies (that is the

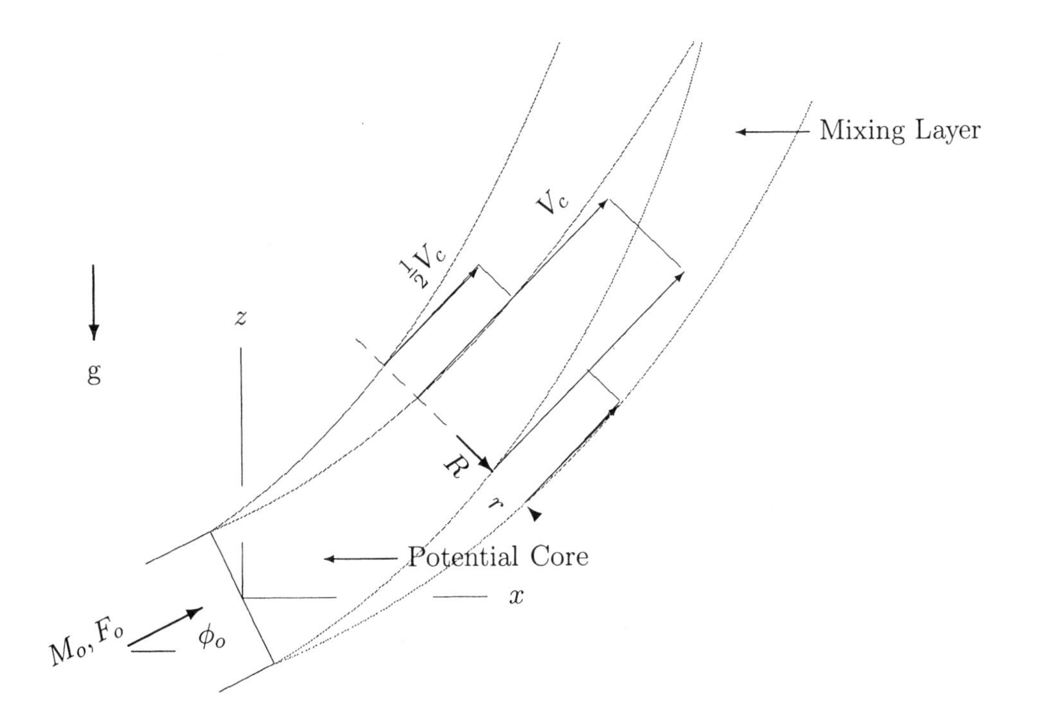

Figure 4.4. Development of the mixing layer around the potential core in the ZEF. The radius of the core is R and the size of the dominant eddies in the mixing layer is r.

thickness of the mixing layer), r, increases at a rate proportional to the velocity difference across the mixing layer; i.e.,

$$\frac{dr}{dt} = \beta \frac{1}{2} V_c. \tag{4.15}$$

The ratio of the above relations gives the spreading rate of the mixing layer

$$\frac{dr}{ds} = \beta. \tag{4.16}$$

This rate for the mixing layer is known to be the same as the jets and the plume; that is $\beta = \sqrt{2}\beta_G \approx 0.17$ (cf Eq. 6.15 and Fig. 2.9). With this spreading rate and an overall conservation of momentum relation, the length of the potential core can be computed. For a pure momentum jet, the core length ($s_e \simeq 6.2D$) is fixed. For a buoyant jet, the core length depends on buoyancy force and also on the direction of the source momentum relative to this buoyancy force. In general, the velocity in the core, V_c, is not constant but varies with the elevation from the source, z, according to the Bernoulli equation

$$\frac{1}{2}V_c^2 - g_o' z = \frac{1}{2}V_o^2. \tag{4.17}$$

where g_o' is the reduced gravity and V_o the jet discharge velocity. The gravity term is negative in this equation since the buoyancy force is in the upward direction. The average concentration and buoyancy across the mixing layer are $\frac{1}{2}c_o$ and $\frac{1}{2}g_o'$ respectively. The specific momentum flux in the core is $\pi R^2 V_c^2$. The specific momentum flux associated with the mixing layer is $\pi[(R+r)^2 - R^2](\frac{1}{2}V_c)^2$. The total momentum flux is

$$M = \pi R^2 V_c^2 + \pi[(R+r)^2 - R^2](\frac{V_c}{2})^2$$

$$= \pi V_c^2 [R^2 + \frac{1}{2}Rr + \frac{1}{4}r^2] \tag{4.18}$$

Similar to the analysis of the ZEF, the horizontal momentum flux is constant, i.e.,

$$M_x = M_o \cos\phi_o, \tag{4.19}$$

while the vertical momentum flux increases with the height at a rate equal to the buoyant force:

$$\frac{dM_z}{ds} = \pi R^2 g_o' + \pi[(R+r)^2 - R^2]\frac{g_o'}{2}$$

$$= \pi g_o' [R^2 + Rr + \frac{1}{2}r^2] \tag{4.20}$$

The path of the potential core is defined by

$$\frac{dx}{ds} = \cos\phi \tag{4.21}$$

$$\frac{dz}{ds} = \sin\phi \tag{4.22}$$

We thus have a system of 4 differential equations (Equations 4.16, 4.20, 4.21 and 4.22) and two algebric relation (Equations 4.17 and 4.19) for the 6 unknowns in terms of $r(s)$, $R(s)$, $M_x(s)$, $M_z(s)$, $x(s)$, $z(s)$. At any longitudinal location s, we have $M = \sqrt{M_x^2 + M_z^2}$, and $\phi = \sin^{-1}(M_z/M)$. Once $M(s)$ and $r(s)$ are calculated, the radius of the core $R(s)$ can be obtained. To facilitate calculations, we non-dimensionalise all lengths with the source diameter D and momentum fluxes by the initial momentum flux $M_o = V_o^2 \pi D^2/4$. The dimensionless governing equations are then:

$$\frac{d\hat{r}}{d\hat{s}} = \beta \tag{4.23}$$

$$\frac{d\hat{M}_x}{d\hat{s}} = 0 \tag{4.24}$$

$$\frac{d\hat{M}_z}{d\hat{s}} = \frac{2}{Fr^2}(2\hat{R}^2 + 2\hat{R}\hat{r} + \hat{r}^2) \tag{4.25}$$

$$\frac{d\hat{x}}{d\hat{s}} = \cos\phi \tag{4.26}$$

$$\frac{d\hat{z}}{d\hat{s}} = \sin\phi \tag{4.27}$$

The dimensionless momentum flux along the jet, after eliminating V_c, is

$$\hat{M} = (1 + 2Fr^{-2}\hat{z})(4\hat{R}^2 + 2\hat{R}\hat{r} + \hat{r}^2) \tag{4.28}$$

The development of the potential core can then be determined by numerical integration of the system of non-linear relations, Equation 4.23 to 4.27. The initial conditions at the source are: $(\hat{M}_x, \hat{M}_z)_o = (\cos\phi_o, \sin\phi_o)$, $\hat{R} = 0.5$, $\hat{r} = 0$ and $(\hat{x}, \hat{z}) = (\hat{x}_o, \hat{z}_o)$. At any $\hat{s}$, given the magnitude of $\hat{M}$, the radius of the core $\hat{R}$ is obtained from the solution of the quadratic equation, Equation 4.28; i.e.,

$$\hat{R} = \frac{-\hat{r}}{4} + \frac{1}{4}\sqrt{4\frac{\hat{M}}{(1 + 2Fr^{-2}\hat{z})} - 3\hat{r}^2} \tag{4.29}$$

The core radius R reduces with distance from the nozzle. The core reaches its end when the radius of the core $\hat{R}$ is reduced to zero. At the end of the ZFE, the computed location and orientation, velocity, width, and concentration then become the initial conditions for the calculations in the ZEF. The half-width or radius of the jet at the end of the potential core $\tilde{b}_e$ is the same as the width of the mixing layer, r_e. The initial streamwise velocity and angle at $s = s_e$ are $V_c/2$ and ϕ_e respectively, from which $(M_x, M_z)_e$ can be determined.

2. NUMERICAL SOLUTION

2.1 JET TRAJECTORY AND POTENTIAL CORE DEVELOPMENT

Fig. 4.5 and 4.6 show typical examples of the computed jet trajectory including the potential core development for a selected range of jet densimetric Froude numbers Fr and discharge angles. For comparision with available experimental data, the results are normalized using the length scale $l_s = \frac{M_o^{\frac{3}{4}}}{F_o^{\frac{1}{2}}} = (\frac{\pi}{4})^{\frac{1}{4}} FrD$. Similarly we can define time and velocity scales based on M_o and F_o. The time scale is

$$t_s = \frac{M_o}{F_o} = \frac{\frac{\pi}{4} D^2 V_o^2}{\frac{\pi}{4} D^2 V_o g_o'} = \frac{V_o^2}{g_o' D} \frac{D}{V_o} = Fr^2 \frac{D}{V_o} \tag{4.30}$$

and the velocity scale is

$$u_s = \frac{l_s}{t_s} = \frac{F_o^{\frac{1}{2}}}{M_o^{\frac{1}{4}}} = (\frac{\pi}{4})^{\frac{1}{4}} Fr^{-1} V_o \tag{4.31}$$

Fig. 4.5 shows an inclined buoyant jet with $Fr = 4$ and $Fr = 1$ discharged at $\phi_o = -20$ to 45°. It can be seen that the potential core can occupy a significant portion of the jet path before it turns vertical by virtue of buoyancy; it is interesting that the trajectory of a horizontal plume with $Fr = 1$ is quite different from a vertical plume. The resemblance of the computed jet path for $Fr = 4, \phi_o = 20$ to that of the experiment for a similar Fr should also be noted (Fig. 4.2). The generality of the present model is demonstrated in Fig. 4.6 which shows the predicted initial mixing of a vertical buoyant jet for $Fr = 20 - 0.1$. It can be seen that for a momentum-dominated discharge, $Fr = 20$, the potential core length is about $z_e/D \approx 6$, but the core length decreases to less than 5 for $Fr = 2$, and smaller than 3 for $Fr = 0.1$. The examples show that depending on the value of the densimetric Froude number and the initial orientation of the jet, the development of the core can be significantly altered by the buoyancy forces.

For a discharge from a nozzle, the limiting condition for which a full flow from the nozzle can be defined is $Fr \approx 1$ (Wilkinson 1988); smaller Fr (e.g. lower jet velocity) will cause intrusion of the ambient fluid into the nozzle. Thus for practical aquatic discharges only $Fr \geq 1$ need to be considered. On the other hand, the case of a pure thermal source of buoyancy (i.e entirely temperature-induced) has a $Fr \approx 0$ ($u_o \approx 0$). For these extreme cases, the acceleration of the source fluid (or fluid near the source) can lead to contraction of the core such as the example shown in Figure 4.6 ($Fr = 0.1$). The densimetric Froude number is zero in the case of plume produced by a fire; the core of the fire is contracting and highly unstable. A swirl is often observed in such strongly buoyant flows as the surrounding fluid is drawn into the low pressure region of the core.

The results of the numerical integration for the jet trajectory obtained for the horizontal buoyant jet ($\phi_o = 0$) are given in Figure 4.7. The dimensionless vertical coordinate of the jet path, $z/(FrD) = (\pi/4)^{\frac{1}{4}}(z/l_s)$,

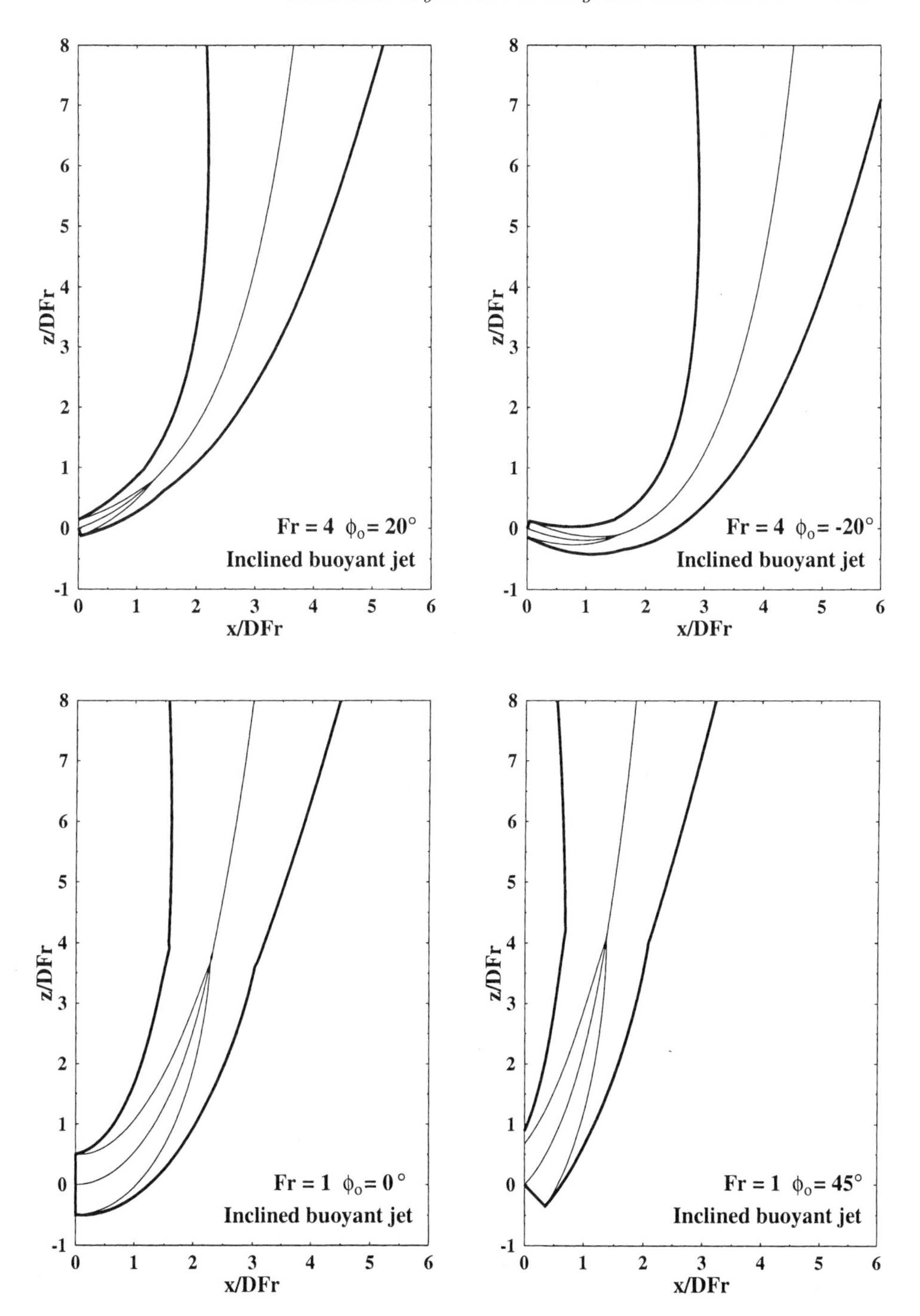

Figure 4.5. Computed inclined buoyant jet trajectory showing development of potential core, $Fr = 4, 1$ and different discharge angles $\phi_o = -20^o - 45^o$. The jet path is shown in dimensionless co-ordinates $x/(FrD)$ and $z/(FrD)$.

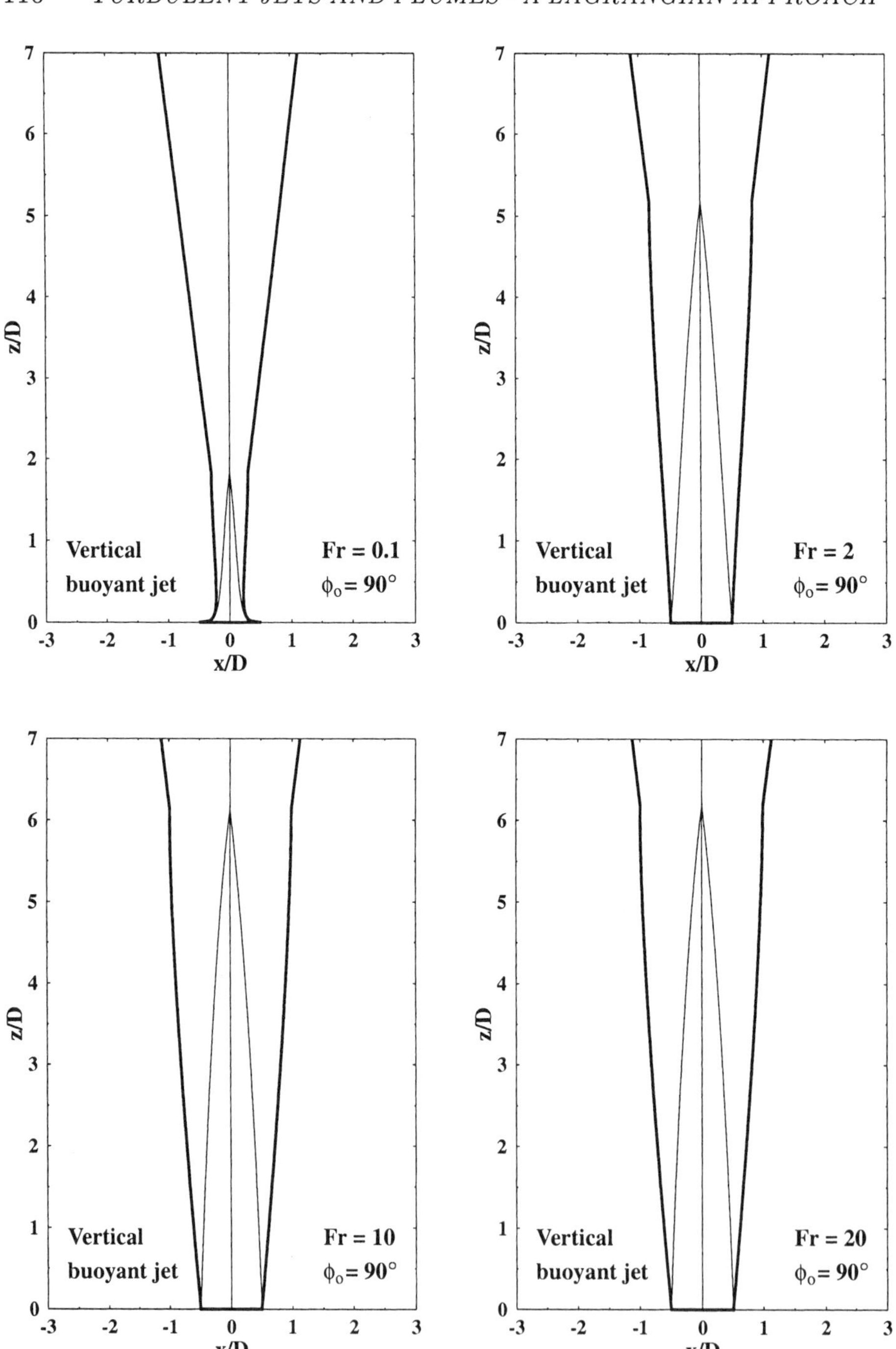

Figure 4.6. Potential core development of a vertical buoyant jet for different densimetric Froude number Fr; $Fr = 0.1 - 20$

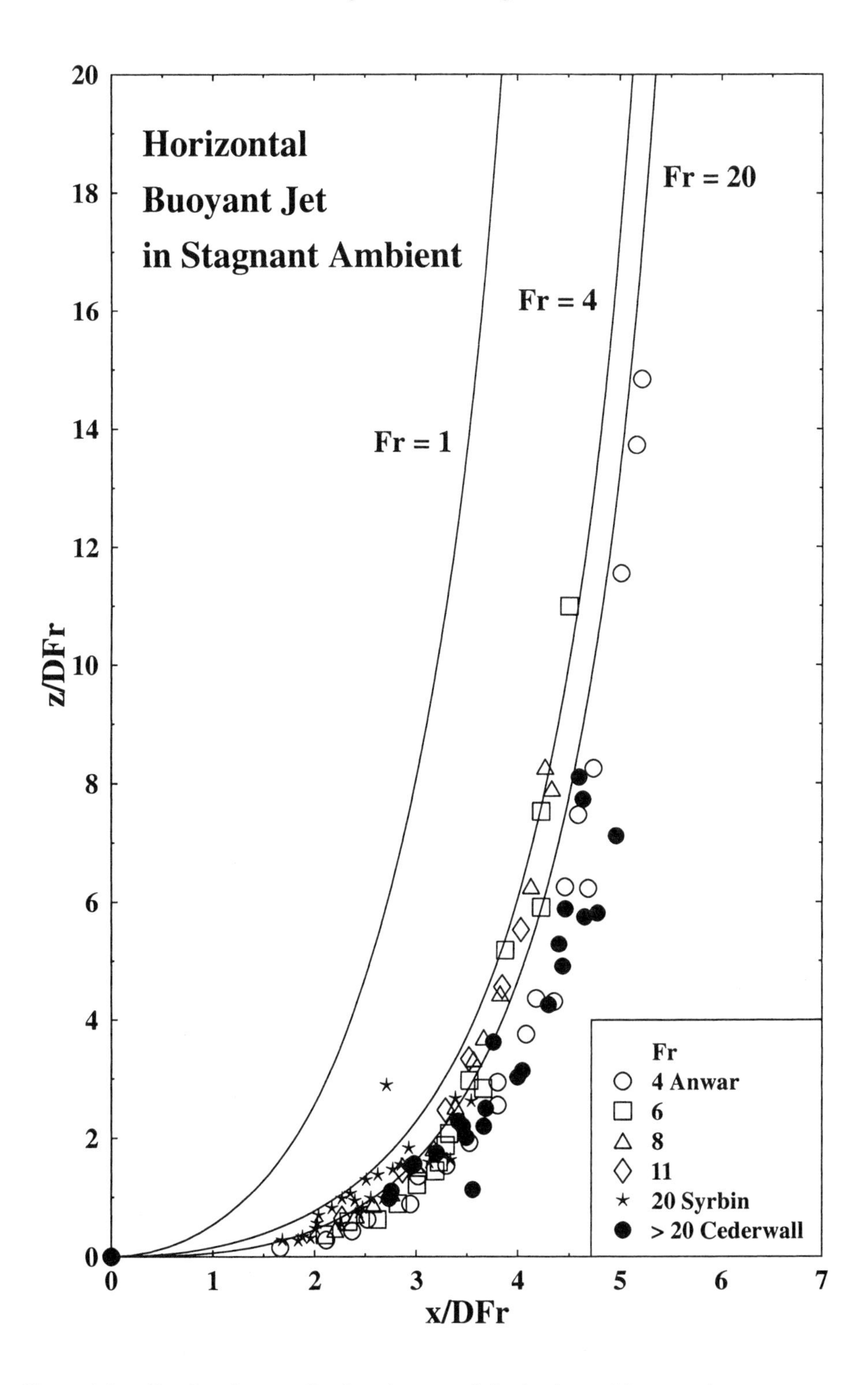

Figure 4.7. Predicted normalized trajectory of the horizontal buoyant jet

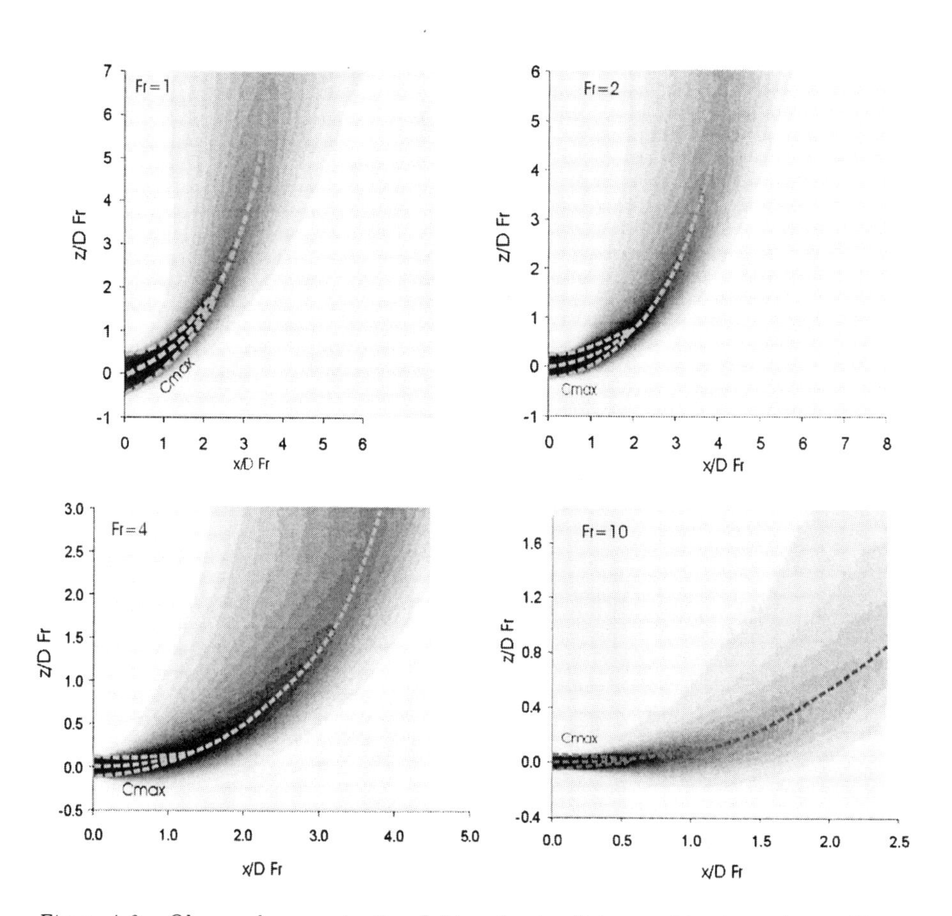

Figure 4.8. Observed concentration field and potential core of horizontal round buoyant jet in stagnant fluid

is plotted against the horizontal coordinate, $x/(FrD) = (\pi/4)^{\frac{1}{4}}(x/l_s)$. Model predictions obtained for three densimetric Froude numbers, $Fr =$ 1,4 and $Fr = 20$, are also shown in the figure as lines in comparison with corresponding experimental data obtained by Anwar (1969), Syrbin and Lyakhovsky (1936), and Cederwall (1968). Plotted on this scale, it is seen that, supported by the data, the dimensionless trajectories for $Fr \geq 4$ differ very little; the trajectories for $Fr \geq 20$ collapse to the same curve.

Figure 4.8 shows the observed trajectory and potential core of a turbulent horizontal round buoyant jet in stagnant ambient. These experiments were performed with a negatively buoyant jet, with the initial relative density difference created by salt, $\frac{\Delta\rho_o}{\rho} = 0.025$; at a jet densimetric Froude number of $Fr = 1, 2, 4, 10$. The jet diameter is $D = 33$ mm for $Fr = 1$, $D = 15$ mm for $Fr = 2, 4$, and $D = 5$ mm for $Fr = 10$. A fluorescent dye tracer, Rhodamine 6B, is added to the discharge and the tracer concentration field in the centerline plane made visible by the laser-induced fluorescence (LIF) technique (see Chapter 6). It can be seen the predicted potential core length and shape are well-supported by the experiments, with the core length $s_e/D \approx 3$ for $Fr = 1$ and $s_e/D \approx 4 - 5$ for $Fr = 2 - 4$ respectively.

2.2 DILUTION

The mass flux of a conservative tracer is constant. If the average concentration (mass per unit volume) in the material volume AB is $\tilde{c}$, the mass flux would be

$$\Gamma = \tilde{U}\pi\tilde{b}^2\tilde{c} = \Gamma_o. \tag{4.32}$$

where $\Gamma_o = V_o \frac{\pi}{4} D^2 c_o$ is the mass flux at the source. The ratio of the concentration at a particular section of the jet to the source concentration is

$$\frac{\tilde{c}}{c_o} = \frac{\frac{\pi}{4} D^2 V_o}{\pi \tilde{b}^2 \tilde{U}} = \frac{Q_o}{Q} = \frac{1}{\tilde{S}} \tag{4.33}$$

where $\tilde{S} = c_o/\tilde{c} = Q/Q_o$ is the average dilution, defined as the ratio of the jet volume flux Q and the source volume flux Q_o. In dimensionless form, we have

$$\tilde{S} = \frac{c_o}{\tilde{c}} = (\frac{4}{\pi})^{\frac{1}{4}} Fr \pi \hat{b}^2 \hat{U} \tag{4.34}$$

where $\hat{b} = \tilde{b}/l_s$, and $\hat{U} = \tilde{U}/u_s$. For a Gaussian velocity and concentration profile, defined by velocity half-width (based on the $1/e$ centerline velocity point) b_v and concentration half-width $b_c = \lambda b_v$, where $\lambda \approx 1.2$, application of a mass balance (Prob. 2.1) gives the relation between the centerline dilution (minimum dilution based on the maximum concentration in the jet cross-section) and the average dilution:

$$S_m = \frac{c_o}{c_m} = \frac{\lambda^2}{1+\lambda^2}\frac{Q}{Q_o} = \frac{\lambda^2}{1+\lambda^2}\tilde{S} \tag{4.35}$$

Hence,

$$S_m = (\frac{4}{\pi})^{\frac{1}{4}}\ Fr\ \pi \hat{b}^2 \hat{U} \frac{\lambda^2}{1+\lambda^2} \tag{4.36}$$

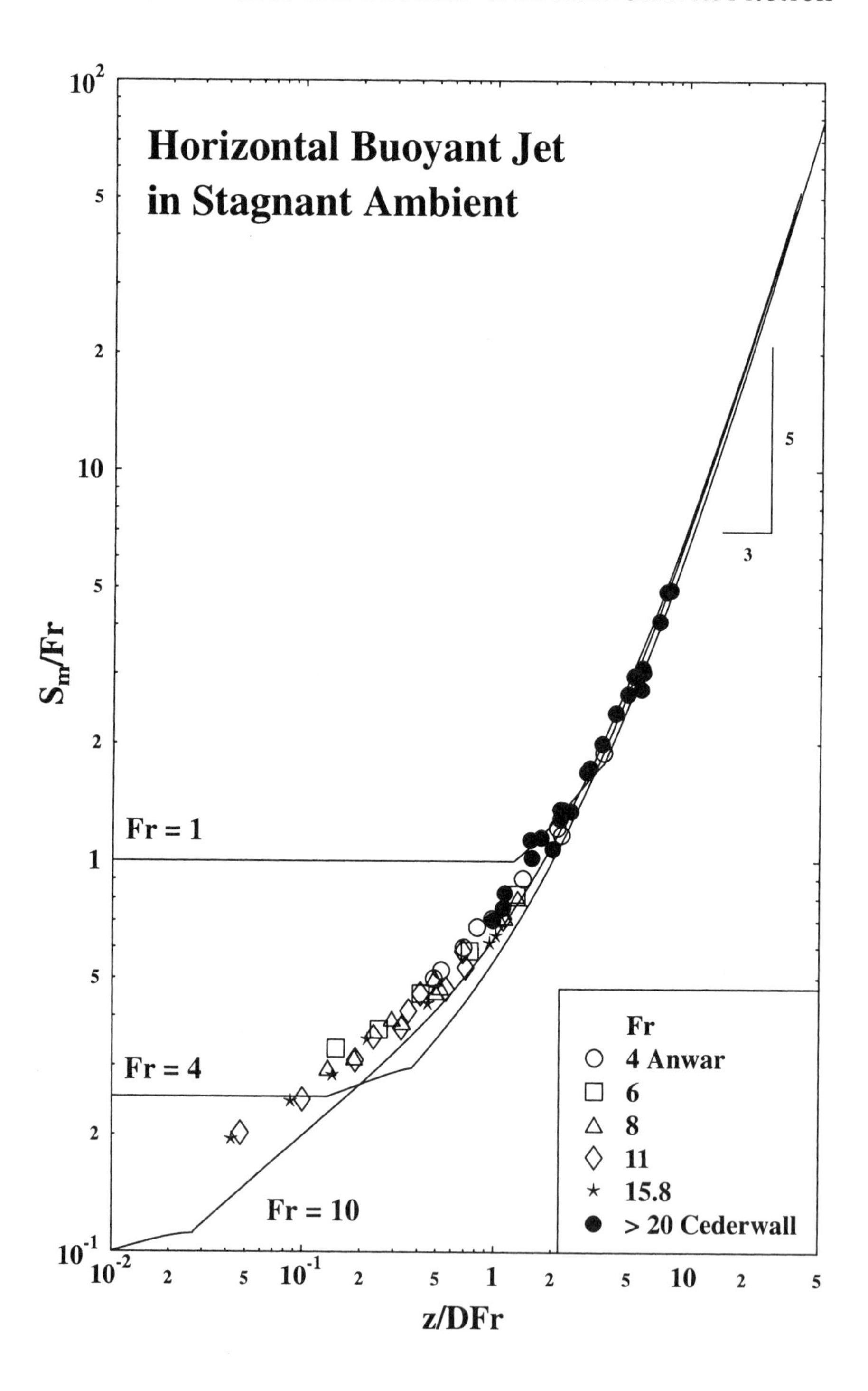

Figure 4.9. Predicted centerline (minimum) dilution of the horizontal buoyant jet, S_m/Fr, as a function of $z/(FrD)$.

The model predictions of this minimum dilution ratio, S_m, divided by the densimetric Froude number Fr is plotted as a function of the dimensionless elevation $z/(FrD)$, and compared with the experimental data of the same range of densimetric Froude number in Figure 4.9. In the potential core, the concentration c_m is a constant equal to c_o and the minimum dilution $S_m = c_o/c_m = 1$. The horizontal lines in the figure, $S_m/Fr = 1/4 = 0.25$ and $S_m/Fr = 1/10 = 0.1$, are the values associated with the potential core for $Fr = 4$ and 10, respectively. On the whole the model predictions of the dilution follow quite closely the trend of the experimental data but appears to be slightly below the values observed in the experiment.

It should be noted that a secondary circulation is produced on the plane perpendicular to the path of the jet due to buoyancy. This circulation, in the form of a vortex pair, promotes turbulent entrainment and may cause a further dilution of the jet that has not been accounted for by the present model (see Chapter 8 on advected line thermal).

For the same source densimetric Froude number Fr, a horizontal buoyant jet will result in a longer jet trajectory to rise to a given depth; hence the dilution is greater for a horizontal jet than a corresponding vertical jet. Figure 4.10 compares the predicted dilution obtained for a horizontal jet and a vertical jet. The dilution of the horizontal jet is clearly the higher up to a dimensionless depth of $z/DFr \approx 10$.

Based on experiments and numerical model calculations, Cederwall (1968) has given a widely used equation for predicting centerline dilution for a horizontal round buoyant jet in still fluid.

$$S_m = 0.54\ Fr(0.38\frac{z}{FrD} + 0.66)^{5/3} \quad \text{for} \quad \frac{z}{FrD} > 0.5 \tag{4.37}$$

$$S_m = 0.54\ Fr(\frac{z}{FrD})^{7/16} \quad \text{for} \quad \frac{z}{FrD} \leq 0.5 \tag{4.38}$$

It can be shown the above equation reduces to the point plume equation as $z/(FrD) \to \infty$. For $z/D \geq 10$, the present model predictions agree closely with the experimentally derived Cederwall equation (not shown).

2.3 BOUNDARY EFFECTS

The theoretical model generally works well for free buoyant jets in stagnant ambient. However, if the jet nozzle is located very close to the bed, the pressure reduction due to the confinement of the entrainment flow demand may cause the plume to cling to the bottom boundary (Coanda effect). For nearly horizontal buoyant jets inclined at angle ϕ_o to vertical, a simple criterion for bottom attachment is given by Jirka and Doneker (1991) as $\tan\phi_o \leq 0.2 - z_o/l_s$, where z_o = height of nozzle centerline

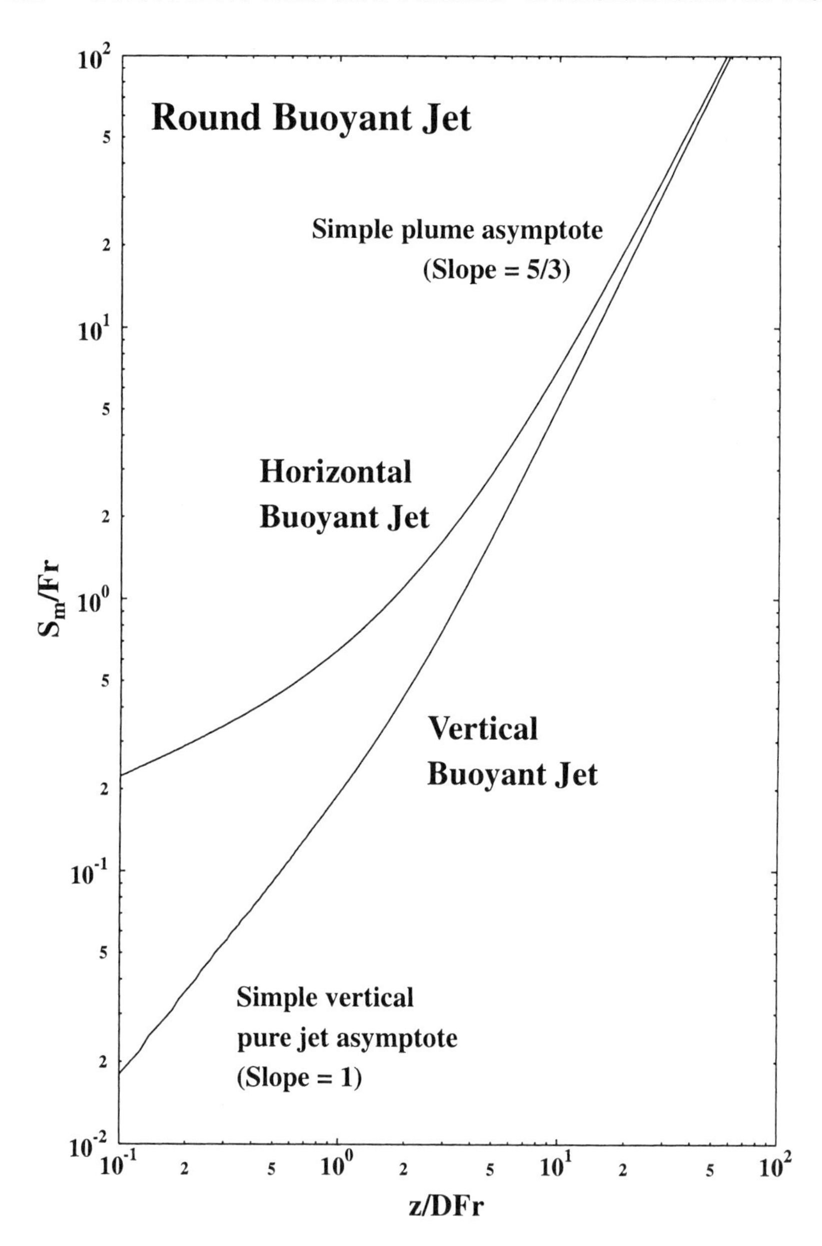

Figure 4.10. Predicted normalized centerline dilution of horizontal and vertical buoyant jet

above the bed (Prob. 4.9). Experiments by Sharp and Vyas (1977) and Sobey *et al.* (1988) have shown the horizontal buoyant jet will be attached if $z_o/l_s \geq 0.1 - 0.15$. Similarly, for high momentum discharges in shallow water, the stratified flow established at the free surface may be hydrodynamically unstable. Recirculating flow cells may be established adjacent to the jet; partially diluted effluent is hence re-entrained into the jet, leading to reduced dilution. A simple criterion for jet stability has been given as $l_s/H \leq 4.3$ or $H/D \geq 0.22Fr$ with apparently little sensitivity to the discharge angle (Lee and Jirka 1981, Jirka 1982). For jet discharges with strong bottom or surface interaction, alternative numerical solution methods may be used (see Chapter 7 and 8).

3. APPLICATION EXAMPLES

The numerical model along with the general numerical solution (Fig.4.9) can be used to illustrate many principles underlying the design of environmental mixing devices. They are illustrated in the following worked examples.

EXAMPLE 4.1 *Consider a single submerged round buoyant jet with* $V_o = 0.5\ m/s,\ D = 0.25\ m$, *and* $\frac{\Delta\rho_o}{\rho_a} = 0.025$. *Estimate the minimum surface dilution in a depth of 25m for* $\phi_o = 0^o$ *and* $\phi_o = 90^o$.

SOLUTION: The source densimetric Froude number for this inclined buoyant jet is

$$Fr = \frac{0.5}{\sqrt{0.025 \times 9.81 \times 0.25}} = 2.0.$$

Compute $z/FrD = 25/(2\times0.25) = 50$. The data in Fig. 4.10 shows that the dimensionless dilution is $S_m/Fr \approx 80$, that is $S_m = 160$ for both the horizontal jet (Case a) and vertical jet (Case b). Alternatively, the Cederwall equation can be used to compute S_m and that gives almost an identical value $S_m = 0.54(2.0)(0.38(50) + 0.66)^{5/3} = \underline{155}$.

For this low value of $Fr = 2.0$, the forced plume trajectory is near vertical over a large portion of its trajectory; thus it is not surprising the dilution is virtually independent of the initial discharge angle.

EXAMPLE 4.2 *Consider now a submerged jet discharging at high velocity in depth of* 10 m, $V_o = 3.1\ m/s,\ D = 0.1\ m$ *and* $\frac{\Delta\rho_o}{\rho_a} = 0.025$. *Estimate the minimum surface dilutions for both vertical and horizontal discharge cases. Consider also the case of a jet discharging downwards at an angle of 30 degrees to the horizontal.*

SOLUTION: The source densimetric Froude namber of this discharge at high velocity is $Fr = 3.1/\sqrt{0.1 \times 0.025 \times 9.81} = 19.8$. Compute

$z/FrD = 10/(19.8 \times 0.1) = 5.1$. From Fig. 4.9, we obain $S_m/Fr \approx 1.8$, that gives a minium dilution $S_m = \underline{34}$ for the vertical jet. For the horizontal jet, we obtain a minimum dilution of $S_m = \underline{55}$, or a similar result of $S_m = 53$ if the Cederwall Equation is used to do the calculation.

For a relatively high momentum discharge, a horizontal discharge is more effective than a vertical one because of its longer trajectory, and hence larger surface area for entrainment. With a downward discharging jet, the momentum flux opposes the buoyancy initially in very much the same way as a dense plume (Chapter 3); however, the much longer trajectory results in higher diluton. Figure 4.11 shows the computed trajectory and average dilution at z=10 m and z=17 m for the three cases. It can be seen the effect of discharge angle is relatively more important at smaller depths. In some cases it may be advantageous to issue the jet downwards to increase initial dilution; however a jet discharging in proximity to a solid boundary may become attached to the bottom due to the *Coanda Effect* (Prob. 4.9). The dilution of such a wall jet can be significantly increased than a free jet, with however potential impacts on the benthos on the sea bed.

EXAMPLE 4.3 *A sewage treatment plant discharges* $0.088\ m^3/s$ *of screened domestic wastewater in the form of horizontal jets into a harbour of mean depth* $12.2\ m$*. As some brackish water is used for toilet flushing, the relative density difference is about* 0.014*. Evaluate the two methods of disposal: a) a single submerged horizontal jet with diameter* $0.37\ m$*; and b) discharge in the form of* $N = 14$ *identical jets (such a design is called a multiport diffuser). The velocity in each case is kept the same. The horizontal jets are adequately spaced apart such that they do not merge before surfacing.*

SOLUTION: For case a), the discharge velocity can be computed to be 0.8 m/s, with $Fr = 3.55$. For case b), the discharge velocity is kept the same at 0.8 m/s, and the corresponding jet diameter for the same total flow is 0.1 m.

Case a			Case b		
	$N = 1$			$N = 14$	
V_o	$=$	$0.8\ m/s$	V_o	$=$	$0.8\ m/s$
D	$=$	$0.37\ m$	D	$=$	$0.1\ m$
Fr	$=$	3.55	Fr	$=$	6.8
$\frac{z}{FrD}$	$=$	9.3	$\frac{z}{FrD}$	$=$	17.9

Using Cederwall equation or the dilution nomogram (Fig.4.10), the computed dilution for z=12.2 m is approximately $\underline{21}$ for the single jet (case a), and 105 for the multiple jets with same velocity (case b). By dividing the given discharge flow into a number of buoyant jets, the total surface area for jet entrainment is increased. A multiple port outfall is therefore a much more powerful mixing device than a single port outfall – in this example a gain in surface dilution by over a factor of six and at no extra pumping cost!

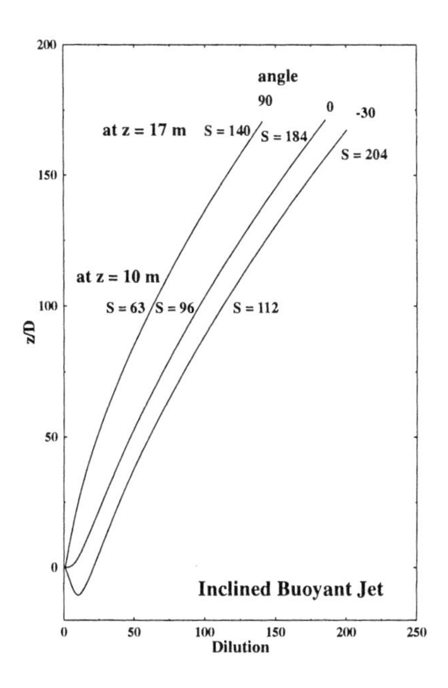

Figure 4.11. Predicted trajectory and dilution for three discharge angles

EXAMPLE 4.4 *In a Scandinavian country, it is proposed to make use of the freshwater used for hydropower to enrich the primary productivity of fjords; the surface layers of these waters are often poor in nutrients. In a fjord of weak currents and relatively stable stratification the design is to discharge freshwater as a submerged buoyant jet. The objective is to lift by jet entrainment the nutrient rich water from depths below 30 m to a depth of about 10 m throughout the long summer, thus extending the growing season of marine organisms for aquaculture (McClimans and Eidnes 2000).*

Figure 4.12 shows the ambient salinity profile for two cases. The jet is discharged at a downwards angle at 30 m depth. Two cases tested in the laboratory are as follows: a) $Q_o = 6\ m^3/s$, $D = 1\ m$, $\phi_o = -35^o$ *and* $Fr = 15.3$; *the ambient salinity at source level is* $S_a = 32.1\ ppt$. *b)*

$Q_o = 6\ m^3/s$, $D = 2\ m$, $\phi_o = -45^o$ and $Fr = 2.65$; the ambient salinity at source level is $S_a = 33.6$ ppt. Compute the unknown jet trajectory and average dilution at the surface or submerged level.

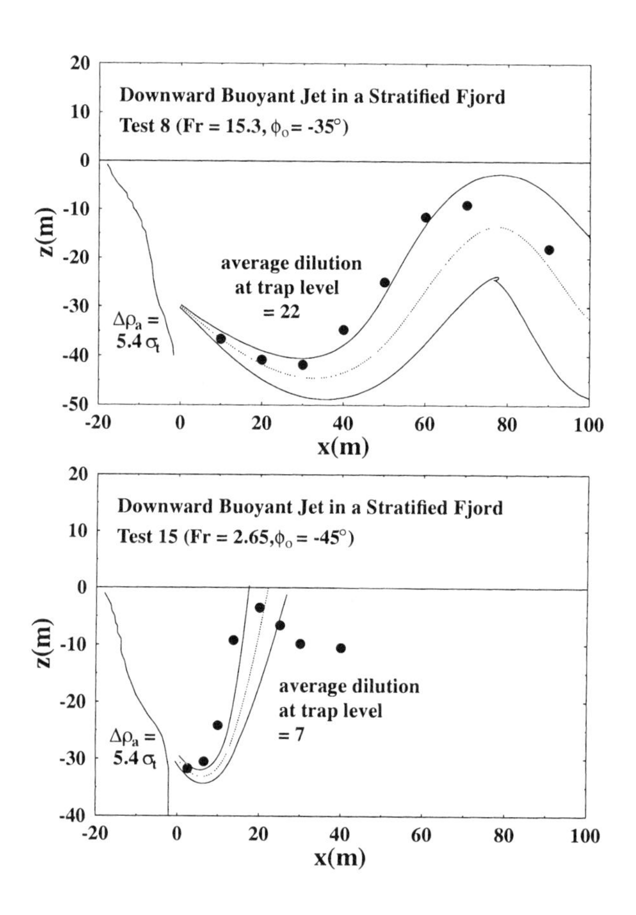

Figure 4.12. Artificial upwelling by buoyant plume (solid circles: observed trajectory)

SOLUTION: The ambient density can be computed as a function of salinity and temperature via the equation of state (Appendix A). Fig. 4.12 shows there is a density difference of $5.4\sigma_t$ units between the surface and bottom waters in the fjord. Natural stratification is typically nonlinear and calculation of initial mixing necessitates the use of a general mathematical model. The source diameter is also not negligible compared with the scale of interest, of the order of 10-20 m. The general integral model (accounting for potential core) is used to compute the trajectory and dilution using the source conditions and ambient density profile (from the equation of state, Appendix A). For the first case the downward jet discharges at a jet densimetric Froude number of $Fr = 15.3$; the jet momentum results in an extended trajectory. The jet mixes with sufficient amounts of the ambient lower layer fluid so that the density of the mixed fluid is less than that of the surface layer; the plume thus stays submerged at an equilibrium trap level of depth around 20 m. The predicted average dilution is about 22. This means that the buoyant plume continually brings a flow of $22 \times 6 = 132$ m^3/s to the euphotic zone. Based on measured nutrient profiles, the supply of nutrient-rich flow from the deep layers can then be computed; the primary production (photosynthesis) of the surface layer is enhanced. On the other hand, case b) is a relatively buoyant discharge, with $Fr = 2.65$. The buoyant plume does not penetrate deep enough into the lower layer, and the plume impinges onto the water surface with a dilution of only 7. It can be seen that the predictions compare well with the laboratory data for both cases. Calculations have shown for this situation that a downward directed jet would give a good mixing in the deeper water below the euphotic zone, and would be less sensitive to changes in stratification. On the other hand, the discharge must be large enough so that the mixed flow has sufficient buoyancy to force it high enough to promote primary production (McClimans and Eidnes 2000).

4. SUMMARY

A general mathematical model is needed for solving environmental mixing problems encountered in practice. In this chapter we have presented a general integral model for an arbitrarily inclined round buoyant jet in stagnant ambient fluid. The model adopts a Lagrangian spreading hypothesis along with the use of top-hat profiles; it predicts the jet characteristics in both the potential core and the region of established flow. Model predictions of dilution and trajectory are well supported by experimental data. The model can readily be extended to treat ambient density stratification. The subject of density stratification is treated in detail in the next chapter.

PROBLEMS

4.1 A discharge of 0.15 m^3/s is released at depth into a fluid of uniform density, with a relative density difference of 0.025 at the point of release. Calculate the initial dilution for a depth of 30 m above the discharge site assuming the release is:

a) a round buoyant plume

b) a round buoyant jet (initial velocity 4 m/s) discharged i) vertically; and ii) horizontally. Compute the momentum length scale l_s, and estimate the jet width at the free surface for case ii).

4.2 The numerical solution of a typical Eulerian integral jet model assumes that the velocity and tracer concentration profiles are Gaussian at the origin (i.e. the flow is fully developed right at source). In reality, it takes a certain distance before turbulent entrainment penetrates into the jet centerline. In dilution prediction, the effect of the zone of flow establishment (ZFE) is sometimes accounted for by a correction factor: Assuming that buoyancy effects are negligible close to the discharge, apply

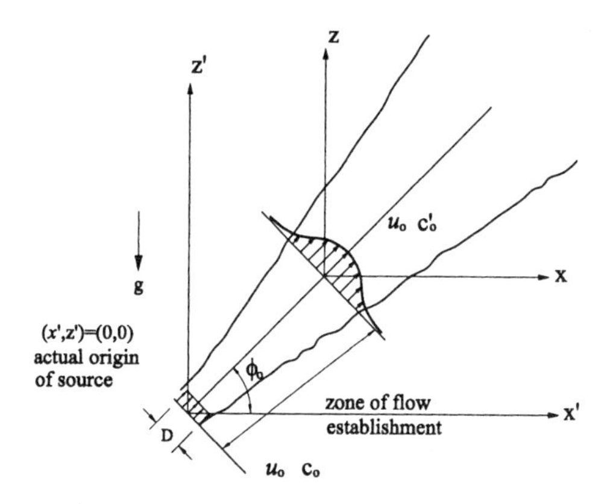

momentum and tracer mass conservation between the discharge and the end of ZFE at $(x, z) = (6.2D\cos\phi_o, 6.2D\sin\phi_o)$ to establish the following relations for a round buoyant jet:

$$b_o = D/\sqrt{2}$$

$$c_o' = c_o\frac{1+\lambda^2}{2\lambda^2}$$

$$S = \frac{2\lambda^2}{1+\lambda^2}S_o$$

where S_o =centerline dilution computed without correction for ZFE.

4.3 Consider a vertical round buoyant jet with the following characteristics:

jet velocity	w_o	=	1 m/s
jet diameter	D	=	0.15 m
density difference	$\Delta\rho_o/\rho_a$	=	0.025
water depth	H	=	10 m

Compute the length scale l_s and the initial dilution in still water. How would the dilution and length scale change if the relative density difference is 0.001 instead? (what situation does this correspond to?) Calculate the power of this jet in kW.

4.4 a) A screening plant discharges a sewage flow of 0.25 m^3/s horizontally through a submerged round outfall of diameter 0.4 m in coastal waters of 15 m depth. Estimate the initial dilution for i) this outfall, and ii) if the outfall is moved to 100 m depth. The mean salinity and temperature of the receiving water are 32 ppt and 20°C respectively. For case i), plot the predicted plume geometry.

b) By keeping the discharge velocity the same as in (a), produce a non-interfering multiport diffuser design that will perform better than the single round outfall in (a) as a water quality control device (for $H = 15$ m). Comment on your results and discuss briefly other advantages, if any, of your design. Practical considerations dictate that the minimum port diameter is 0.1 m.

4.5 A 500 MW nuclear power plant is located on an open coastline. A single port submerged outfall is used to discharge the condenser cooling water into the ocean. The following design data apply:

- Condenser cooling water flow = $10 m^3/s$
- Condenser temperature rise $\Delta T_o = 11\ ^oC$
- Discharge angle $\phi_o = 0^o$
- Discharge velocity $u_o = 5\ m/s$
- Ambient water is stagnant and non-stratified; $T_a = 10\ ^oC$ Salinity = 25 ppt
- Discharge depth $H = 60$ m

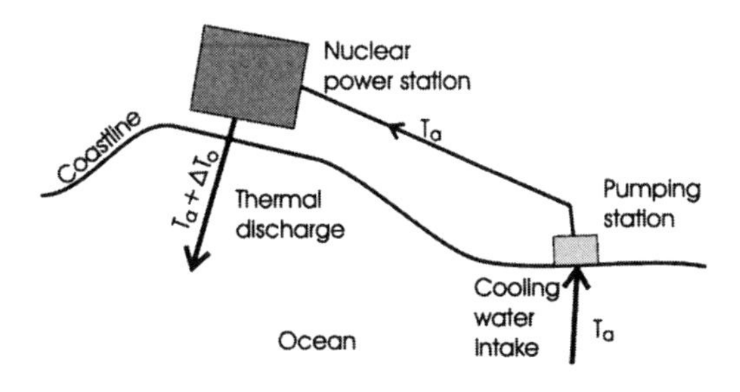

Determine

a) The plume location where the centerline temperature is 1.5 ^oC above the ambient.

b) The centerline temperature and width of the plume when it reaches the surface.

c) The stability of this thermal discharge.

4.6 Consider a vertical buoyant jet discharge in otherwise stagnant ambient fluid, with initial volume flux Q_o, kinematic momentum flux M_o, and specific buoyancy flux F_o; the jet densimetric Froude number is Fr. Explain the significance of the momentum length scale $l_s = {M_o}^{3/4}/{F_o}^{1/2}$. By using the functional form of the jet and plume solutions in Table 3.1 and 2.3, show that the centerlinevelocity w_m, centerlinedilution S_m, and average dilution $\overline{S}$ can be expressed in the following dimensionless forms as a unique function of dimensionless vertical distance above source, z/l_s, over the entire jet-plume range.

$$\frac{{M_o}^{1/2}}{z w_c} \sim f(\frac{z}{l_s})$$

$$\frac{S_m Q_o}{z {M_o}^{1/2}} \sim f(\frac{z}{l_s})$$

$$\frac{\overline{S} Q_o {F_o}^{1/2}}{{M_o}^{5/4}} \sim f(\frac{z}{l_s})$$

$$\frac{S_m}{Fr} \sim f(\frac{z}{l_s})$$

4.7 Analysis of experimental data of the mixing of a vertical buoyant jet suggest that the *average dilution*, $\overline{S}$, can be expressed in terms of the length scales $l_Q = Q_o/M_o^{1/2}$ and $l_s = {M_o}^{3/4}/{F_o}^{1/2}$ by a single formula throughout the jet-plume range:

$$\frac{\overline{S} l_Q}{z} = \frac{\overline{S} Q_o}{z {M_o}^{1/2}} = C_1 \ (1 \ + \ C_2 (z/l_s)^2)^{1/3}$$

where z=vertical distance above the source, and $C_1 = 0.29$ and $C_2 = 0.16$ are empirical constants. Show that this dilution formula is consistent with the limiting cases of a pure jet and plume.

Estimate the initial dilution for a vertical single port sewage discharge with jet velocity $u_o = 1$ m/s, diameter 0.15 m, in depth of 15 m into i) a freshwater lake; and ii) the ocean with relative density difference of 0.025. Assume density of sewage = 1000 kg/m^3.

4.8 A horizontal turbulent heated water jet in the laboratory discharges with jet velocity 1 m/s, jet diameter 0.01 m, and the design jet densimetric Froude number is $Fr = 20$. The jet with the same Froude number can be simulated for two situations, with different ambient temperatures.

For an ambient water temperature of i) $T_a = 15^oC$, and ii) $T_a = 25^oC$, find the corresponding discharge temperature T_o which will give the required Froude number (use the approximate equation of state in Appendix A).

Will the two discharge with the same Fr and submergence result in the same jet trajectory? Please explain.

4.9 <u>*Coanda attachment of buoyant jet to sea bed*</u>
For buoyant jets located close to the bed, the pressure reduction due to the confinement of the entrainment flow demand may cause the plume to cling to the bottom (the so called *Coanda* effect). It is desired to determine the minimum height of the nozzle from the bottom boundary to avoid plume attachment.

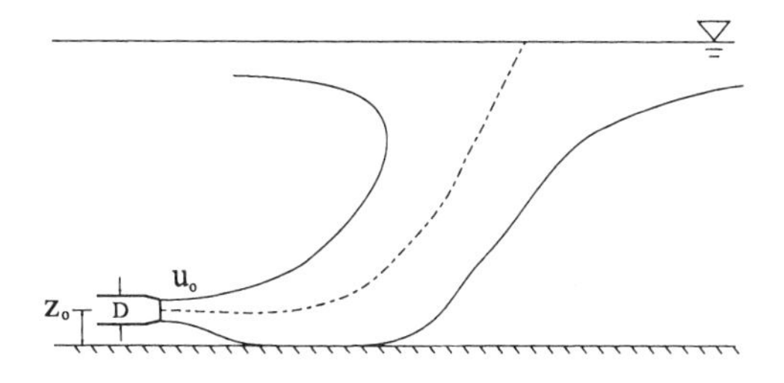

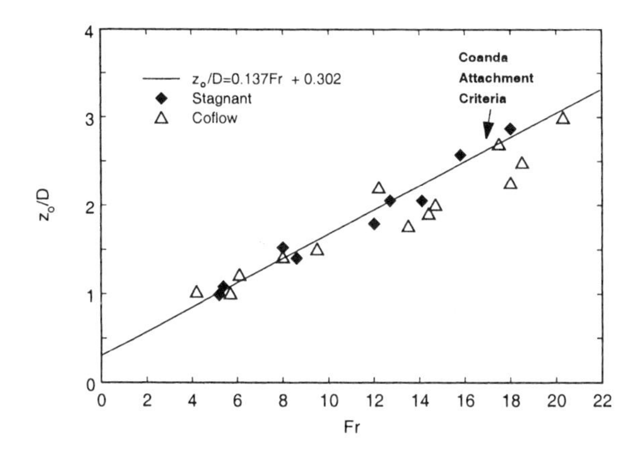

Figure 4.13. Bottom attachment criterion of a horizontal buoyant jet (Wong 1987)

a) Consider a horizontal buoyant jet (jet diameter D) located at a height of z_o above the bottom, discharging with velocity U_o in stagnant fluid. The jet densimetric Froude number is Fr. For a given z_o, the vertical pressure differential increases as the jet velocity (and hence the momentum flux and entrainment velocity). At some point the buoyant force on the plume cannot balance the downward pressure force, and the plume attaches to the bottom. By dimensional reasoning, the limiting height depends on the jet momentum and buoyancy fluxes:

$$z_o = f(M_o, F_o)$$

Hence show that the criterion for bottom attachment can be written as $z_o = Cl_s$ where C = experimental constant. The experiments of Sharp and Vyas (1977) show that $C = 0.15$. Fig. 4.13 shows the data of heated jet experiments by Wong (C.F.Wong, unpublished 1987) who obtained $C = 0.14$. In contrast, Sobey *et al.* (1988) obtained $C \approx 0.1$. Sharp's attachment criterion can alternatively be written as:

$$\frac{z_o}{D} = 0.14\ Fr$$

This means the nozzle must be very close to the bed for the Coanda effect to be noticeable. For $Fr = 10$, the nozzle needs to be located only $1.4D$ from the bottom to avoid plume attachment. In the case of a coflow, the limiting height required is even less.

b) The attachment criterion can also be deduced from kinematic considerations. By assuming a straight horizontal jet over a distance of l_s with a spreading rate of $dB/dx = \beta$, show that the plume will attach if:

$$\beta l_s \quad \geq \quad z_o$$

For a nearly horizontal jet inclined at vertical angle ϕ_o, show that the general attachment criterion is:

$$\tan \phi_o \leq \beta - \frac{z_o}{l_s}$$

where $\beta \approx 0.2$ (Jirka and Doneker 1991).

4.10 *Thermal diffuser discharge in shallow water*:
Thermal power stations located in coastal waters often employ once-through cooling systems. Large flows of condenser cooling water are sometimes discharged from submerged diffusers in the form of high velocity jets. The temperature of the heated effluent can be brought down to acceptable levels within a small mixing zone. After initial mixing, the high momentum flow in the near field is typically vertically mixed, and flows away from the discharge in the form of a 2D jet with bottom friction dissipation (Fig. 4.14).

a) Using boundary layer approximation, the 2D jet beyond the vicinity of the discharge can be described by the following x-momentum equation:

$$x\text{-Momentum:} \qquad u\frac{\partial u}{\partial x} + v\frac{\partial u}{\partial y} = -\frac{\partial \overline{u'v'}}{\partial y} + \frac{f_o}{8H}u^2$$

where H = water depth, f_o= bottom friction coefficient, and the other terms have their usual meaning. Using the integral method, show that the diffuser jet momentum, $M = \rho \int u^2 dy$, is dissipated exponentially downstream.

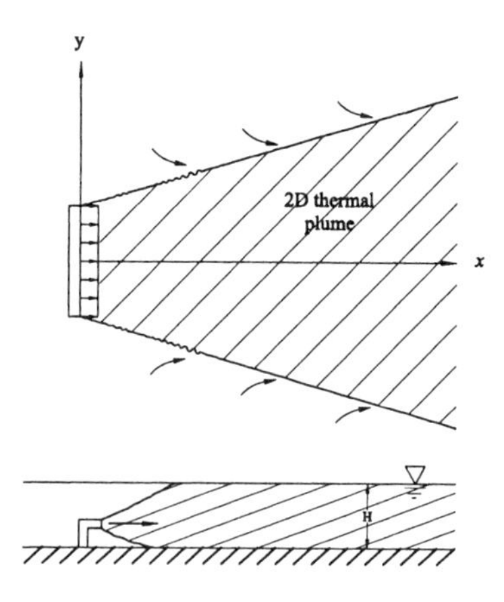

Figure 4.14. Thermal diffuser plume in shallow water

b) Assuming top-hat profiles, obtain an analytical expression for the variation of jet velocity and half-width with distance downstream. Does the jet spread linearly?

4.11 Pollution from sewage outfall damaged by dredger

In a densely populated city, wastewater effluent (after screening to remove the large solids) is discharged into heavily navigated coastal water through a 1.5 km long submarine outfall of diameter 1.5 m. In normal operation the screened flow of $Q_o = 0.75\ m^3/s$ is discharged through a multiport diffuser section fitted with 70 round ports (0.15 m diameter). The jets discharge vertically in a depth of 20 m. On a windy night, a dredger operating in the area accidentally damaged the outfall. In subsequent days the bacteria concentration on a nearby beach 3 km away rose sharply. This caused a public outcry and it was thought that the dredger may be held responsible. You are asked to make an evaluation of this pollution incident.

The concentrations of the key water quality parameters in the screened effluent are: Biochemical Oxygen Demand (BOD) = 300 mg/L; Bacteria concentration = 10^7/100mL.

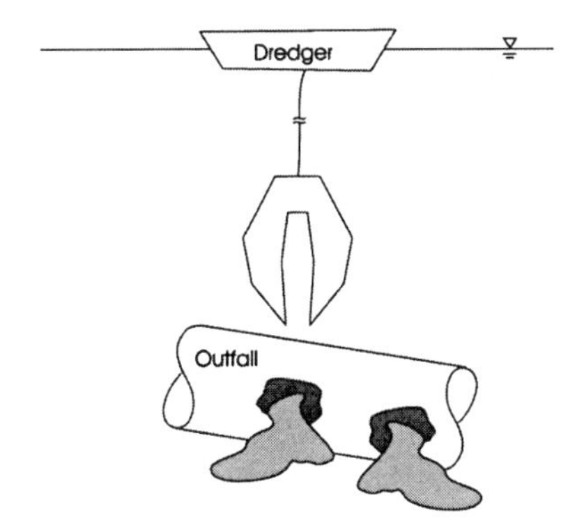

a) Under normal operating conditions, determine the initial dilution in still water achieved by the outfall at the water surface. Assume an initial relative density difference of 0.025 and uniform discharge from all the ports. The jets can be regarded as independent of each other. Estimate the BOD and bacteria concentration at the water surface above the outfall.

From field observations, it is known that if the initial dilution is above 50, sewage slicks are not likely to form; black 'sewage boils' will be absent. In addition, typically it takes 8 hours for the mixed sewage to travel to the nearby beach. If the bacteria concentration further decreases (due to sunlight, sedimentation etc) according to e^{-kt} where k= decay rate, and t=travel time, determine the bacteria concentration on the beach. $k = 0.6$ per hour. The bacteria concentration in the beach water should not exceed 1000/mL.

b) After the outfall was damaged by the dredger, diver inspection revealed two large openings (equivalent diameter 0.35 m each) were punctured on the outfall. An internal hydraulics calculation shows that the sewage flow 'short circuits' and two-thirds of the flow will be discharged through these two holes ($Q = 0.25$ m^3/s per hole) instead. Under these conditions, determine the initial dilution in still water. Estimate the BOD and bacteria concentration at the water surface above the outfall and the bacteria concentration on the beach.

Chapter 5

DENSITY STRATIFICATION

Natural and man-made water bodies are generally density-stratified. The variations in density can be due to temperature, salinity, and particulate matter in suspension. The thermal stratification of lakes and reservoirs is an example (e.g. Harleman 1982). In the presence of the earth's gravity, the fluid of non-uniform density would move towards a state of stable stratification – the heavier fluid tends to stay at the bottom while fluid of smaller density is generally found at a higher elevation. The stable stratification is disturbed when momentum and buoyancy sources, in the form of jets, plumes, puffs and thermals, are present to produce the vertical motions. The rise of buoyant fluid draws fluid from a lower elevation and discharges it into higher elevation. The fall of dense fluid would produce the opposite effect. Small density differences can have a marked effect on the vertical exchange. A variety of significant mixing processes observed in the atmosphere, rivers and lakes, estuaries and the ocean can be modeled as the interaction of jets/plumes/puffs/thermals (JPPT) with a stratified fluid.

Fig. 5.1 shows an example of thermal stratification. At the bottom of a stratified lake, the temperature of the colder water tends to stay constant independent of the seasonal fluctuations but the surface temperature is more variable due to solar heating and inflows such as rivers and industrial and municipal discharges. The stable thermal stratification in the summer is often associated with oxygen depletion in the hypolimnion of eutrophic lakes (Imberger and Patterson 1990). The water quality of these lakes can however be modified by thermal destratification (Zic and Stefan 1988; Schladow and Fisher 1995; Lawrence *et al.* 1997). The temperature stratification of air in the interior space of buildings is another example. Cold dense air tends to stay on the floor as warm air flows towards the ceiling. The placement of heating and cooling equipment and the openings for natural ventilation are important design considerations that can modify the temperature stratification and hence the indoor air quality in modern buildings (Baines and Murphy, 1986; Linden, 1999). In addition to temperature effect, the density of water is affected by turbidity and salinity. The density of the air in the atmosphere is also determined by thermodynamics effects such as compressibility and moisture content (see, e.g., Iribarne and Gobson, 1981).

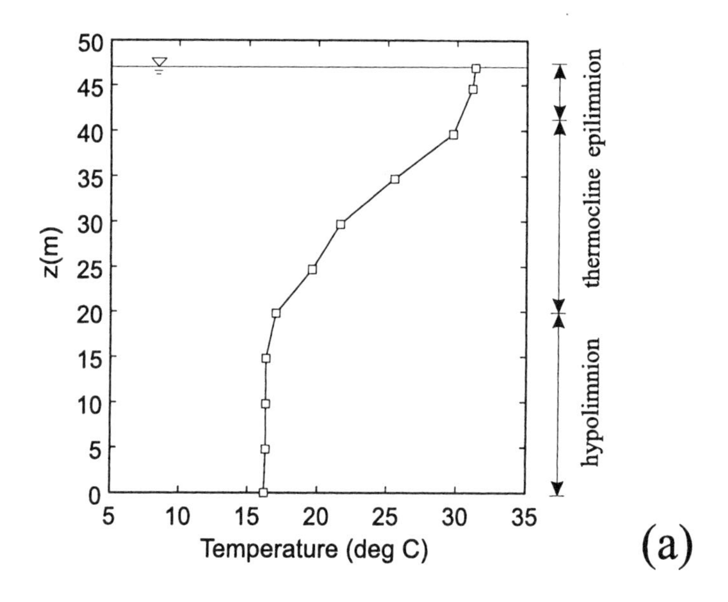

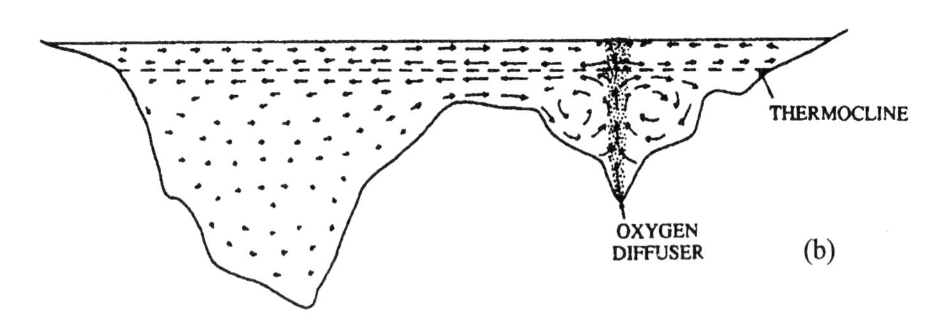

Figure 5.1. (a) Water temperature measured in High Island reservoir, Hong Kong, in July 1990; (b) Thermal de-stratification scheme at Amisk Lake (Lawrence *et al.* 1997)

Ambient		Elemental Volume		Elemental Volume
$\rho_a, \delta(\tilde{\rho}\tilde{V})$	+	$\tilde{\rho}\tilde{V}$	=	$\tilde{\rho}\tilde{V} + \delta(\tilde{\rho}\tilde{V})$
$c_p T_a,\ S_a$		$c_p\tilde{T},\ \tilde{S}$		$c_p\tilde{T} + \delta(c_p\tilde{T})$
				$\tilde{S} + \delta\tilde{S}$

Figure 5.2. Mixing of fluid of density $\tilde{\rho}$ in an elemental volume with a small volume of ambient fluid of density ρ_a. The carrying mass, salt mass and heat in the elemental volume are $\tilde{\rho}\tilde{V}$, $\tilde{S}\tilde{\rho}\tilde{V}$, and $c_p\tilde{T}\tilde{\rho}\tilde{V}$, respectively

1. BUOYANCY VARIATION

The first step in quantifying the effect of density stratification is to evaluate the variation of the buoyancy force with elevation. According to Archimedes Principle, the force of buoyancy in a fluid is the difference in weight between the fluid and it's displaced volume, which per unit *volume* of the fluid, is

$$\mathbf{B} = B\hat{\mathbf{k}} = g(\rho_a - \rho)\hat{\mathbf{k}} \tag{5.1}$$

where $B = g(\rho_a - \rho)$ is the buoyancy, g is the gravitational acceleration, $\hat{\mathbf{k}}$ is the base vector in the $+z$ direction (opposite direction to the force of gravity), and $(\rho - \rho_a)$ is the density of the fluid above that of the ambient. The buoyancy B varies with the elevation z as the fluid moves through and mixes with its surroundings, due to the changes in both ρ and ρ_a. The density of air may vary depending on temperature and moisture content. The density of water is a function of temperature and salinity (see Appendix A). Figure 5.2 shows the mixing of the fluid of density $\tilde{\rho}$ in an elemental volume with its surrounding fluid of density ρ_a. The elemental volume is not a material volume. As puffs and thermals or jets and plumes advance through its surrounding fluid, the mass and associated constituents in the elemental volume increase continuously due to the entrainment of the ambient fluid.

1.1 SALINITY EQUATION

If the density variation is due to dissolved salts, the tracer concentration is the salinity - the mass of dissolved salts per unit mass of the carrying fluid. The salinity is $\tilde{S}$ in the elemental volume and S_a in the ambient. The carrying mass is $\tilde{\rho}\tilde{V}$ in the elemental volume and $\delta(\tilde{\rho}\tilde{V})$ in the small volume from the ambient. Therefore, the mass of the salt is $\tilde{S}\,(\tilde{\rho}\tilde{V})$ in the elemental volume and is $S_a\,\delta(\tilde{\rho}\tilde{V})$ in the small volume from the ambient. After the mixing, the concentration of salt becomes $\tilde{S}+\delta\tilde{S}$ and the mass in the elemental volume becomes $\tilde{\rho}\tilde{V}+\delta(\tilde{\rho}\tilde{V})$. The carrying mass is unchanged. Equating the tracer mass of the system before and after the entrainment (Figure 5.2),

$$S_a\delta(\tilde{\rho}\tilde{V})+\tilde{S}\tilde{\rho}\tilde{V}=(\tilde{S}+\delta\tilde{S})[\tilde{\rho}\tilde{V}+\delta(\tilde{\rho}\tilde{V})], \tag{5.2}$$

which gives the salinity equation

$$\delta[(\tilde{S}-S_a)\tilde{\rho}\tilde{V}]=-\tilde{\rho}\tilde{V}(\delta S_a). \tag{5.3}$$

Once the salinity is calculated using this equation, the density of the water can be computed using the equation of the state. If density is a function of only salinity, we have:

$$\frac{d\rho}{\rho}=\frac{1}{\rho}(\frac{\partial\rho}{\partial S})dS=\gamma_S dS, \tag{5.4}$$

where

$$\gamma_S=\frac{1}{\rho}(\frac{\partial\rho}{\partial S})_T \tag{5.5}$$

is the salinity expansion coefficient. For small salinity differences between the elemental volume and the ambient,

$$(\tilde{\rho}-\rho_a)=\overline{\rho}\,\overline{\gamma}_S(\tilde{S}-S_a), \qquad \delta\rho_a=\rho_a\,\gamma_{Sa}\delta S_a. \tag{5.6}$$

Through these relations, the salinity in Equation 5.3 is eliminated to give the density equation

$$\delta[(\tilde{\rho}-\rho_a)\tilde{\rho}\tilde{V}]\simeq-\tilde{\rho}\tilde{V}(\delta\rho_a), \tag{5.7}$$

since $\overline{\rho}\,\overline{\gamma}_S\simeq\rho_a\gamma_{Sa}$. The density and salinity is linearly related; the proportionality constant $\rho\gamma_S\simeq 0.8$ is practically independent of the salinity and temperature (see Figure A.3 in Appendix A). Multiplying Equation 5.7 by g and dividing by δz, the result is the buoyancy equation

$$\frac{dB}{dz}\simeq\tilde{\rho}\tilde{V}\frac{g}{\tilde{\rho}}\frac{d\rho_a}{dz} \tag{5.8}$$

where $B=g(\rho_a-\tilde{\rho})\tilde{V}$ is the buoyant force acting on the elemental volume. The buoyancy force B is positive if $\rho<\rho_a$. The gradient of this force dB/dz is positive if $d\rho_a/dz>0$.

If density variation is due to suspended solid, such as sediments, or other dissovled solids, the density change in the fluid is expected to be proportional to tracer mass concentration, and Equation 5.8 would be equally applicable under those conditions. See method of excess in chapter 9.

1.2 TEMPERATURE EQUATION

If density variation is due to temperature, the tracer concentration is heat per unit mass of carrying fluid. The heat in the elemental volume is $c_p\tilde{T}\,(\tilde{\rho}\tilde{V})$. The heat in the small volume from the ambient is $c_pT_a\delta(\tilde{\rho}\tilde{V})$, where c_p = specific heat at constant pressure. Equating the heat before and after the entrainment (Figure 5.2),

$$c_p\tilde{T}\,(\tilde{\rho}\tilde{V}) + c_pT_a\,\delta(\tilde{\rho}\tilde{V}) = [\tilde{\rho}\tilde{V} + \delta(\tilde{\rho}\tilde{V})][c_p\tilde{T} + \delta(c_p\tilde{T})] \tag{5.9}$$

which gives the heat equation

$$\delta[c_p(\tilde{T} - T_a)\tilde{\rho}\tilde{V}] = -\tilde{\rho}\tilde{V}\,\delta(c_pT_a). \tag{5.10}$$

The temperature variation in this equation is related to the density variation:

$$(\tilde{\rho} - \rho_a) = -\overline{\rho}\,\overline{\gamma}_T(\tilde{T} - T_a), \qquad \delta\rho_a = -\rho_a\,\gamma_{Ta}\delta T_a \tag{5.11}$$

where

$$\gamma_T = -\frac{1}{\rho}(\frac{\partial\rho}{\partial T})_S \tag{5.12}$$

is a thermal expansion coefficient; $\overline{\gamma}_T$ is the average of the coefficient over the temperature range ($\tilde{T}$, T_a) and γ_{Ta} is the coefficient at temperature T_a. The thermal expansion coefficient is temperature dependent. For ideal gas, $\gamma = T^{-1}$ (where T is the absolute temperature). For water, the coefficient can be determined by the equation of the state given in Appendix A. Eliminating the temperature in Equation 5.10 using Equation 5.11, gives

$$\delta[(\frac{c_p}{\overline{\gamma}_T\overline{\rho}})(\rho_a - \tilde{\rho})\tilde{\rho}\tilde{V}] = -\tilde{\rho}\tilde{V}(\frac{c_p}{\gamma_{Ta}\rho_a})\,\delta\rho_a. \tag{5.13}$$

Multiplying both sides by g and let $\tilde{B} = (\rho_a - \tilde{\rho})g\tilde{V}$, gives the buoyancy equation

$$\delta[(\frac{c_p}{\overline{\gamma}_T\overline{\rho}})\,\tilde{\rho}\tilde{B}] = -\tilde{\rho}\tilde{V}\,(\frac{c_p}{\gamma_{Ta}\rho_a})\,g\delta\rho_a. \tag{5.14}$$

According to this equation, the buoyancy in the elemental volume $\tilde{B}$ is changed on the one hand due to the variation of the thermal expansion coefficient and on the other hand by the variation of the density in the ambient with elevation. If the thermal expansion coefficient and specific heat are assumed to be a constant and if $\overline{\rho}/\tilde{\rho} \simeq 1$ over a small range of temperature, Equation 5.14 becomes

$$\frac{dB}{dz} \simeq \tilde{\rho}\tilde{V}\ \frac{g}{\rho_a}\frac{d\rho_a}{dz}. \tag{5.15}$$

Either Equation 5.8 or 5.15 can be written as:

$$\frac{dB}{dz} \simeq -\tilde{\rho}\tilde{V}\,N^2 \tag{5.16}$$

in terms of the stratification frequency

$$N = \sqrt{-\frac{g}{\rho_a}\frac{d\rho_a}{dz}}. \tag{5.17}$$

Displacement of a parcel of fluid in a stably stratified environment is expected to oscillate about its equilibrium position close to this frequency (see Figure 5.3).

EXAMPLE 5.1 *The water temperature in High Island reservoir (see Figure 5.1a) can be represented by a three-layer profile. The temperature near the surface of the reservoir, in the epilimnion from $z = 42$ m to 47 m, is essentially constant at 31 ^oC. The bottom layer, in the hypolimnion from $z = 0$ to 20 m, has a lower temperature of 17^oC. Find the stratification frequency N in the thermocline assuming that the density variation in the middle layer is linear from z=20 m to 42 m.*

SOLUTION: The density of fresh water can be calculated by the formula in Appendix A. In the surface layer (epilimnion), the temperature is $T_e \simeq 31$ ^{o}C and the density is $\rho_{31} = 995.35$ kg/m^3. The density in the bottom layer (hypolimnion), where the temperature $T_h \simeq 17$ ^{o}C, is $\rho_{17} =$ 998.78 kg/m^3. The density changes rapidly over the 22 m depth of the thermocline. According to Equation 5.17, the stratification frequency in the thermocline is

$$N = \sqrt{-\frac{g}{\rho_a}\frac{d\rho_a}{dz}} = \sqrt{\frac{9.81(998.78 - 995.35)}{\frac{1}{2}(998.78 + 995.35) \times 22}} = \underline{0.0392}\ \text{rad/s}.$$

The corresponding period of oscillation is

$$T = \frac{2\pi}{N} = \underline{160}\ \text{s}.$$

The water mass in this upper layer of the lake is expected to oscillate with a dominant internal wave period of 160 s.

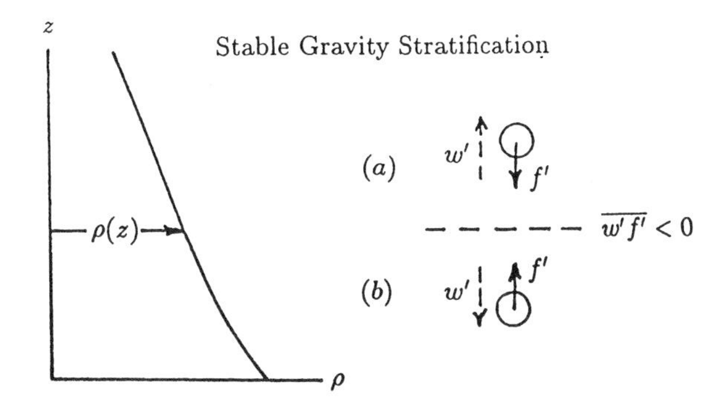

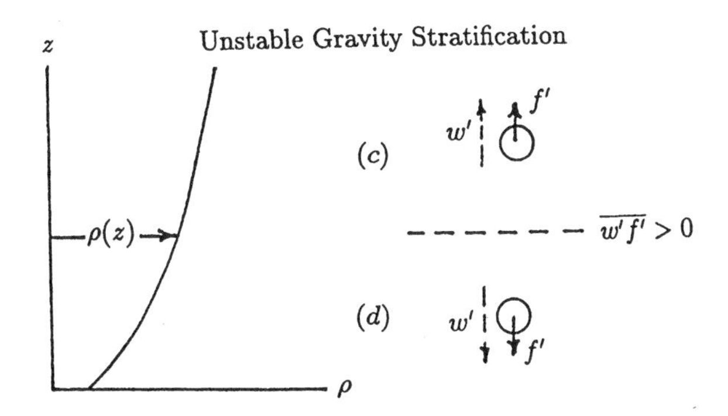

Figure 5.3. Displacement of a fluid parcel in a stratified fluid: (a) and (b) show oscillation about the static equilibium level in stable stratification, while (c) and (d) show the tendency for the displaced fluid to move further from the equilibrium in unstable stratification; w' and f' are the velocity and buoyancy fluctuations associated with the motion.

1.3 STRATIFICATION FREQUENCY

Density stratification, or gravity stratification, is said to be stable if $d(\rho_a g)/dz < 0$ and unstable if $d(\rho_a g)/dz > 0$. Fluid in a stable stratification tends to approach the static equilibrium. As shown in Figure 5.3, the displacement of a fluid in stable stratification upward would make the fluid heavier than its surrounding and hence tends to fall back to its equilibrium position. Conversely, a fluid element moved downward would become lighter than its surrounding and tends to return to its equilibrium position. The oscillation of a displaced fluid about its equilibrium level is defined by the stratification frequency, or the so-called Brunt-Väisälä frequency

$$N = \sqrt{-\frac{g}{\rho_a}\frac{d\rho_a}{dz}}. \tag{5.18}$$

In the atmosphere, the density gradient can be determined by detecting the oscillation frequency of the weather balloon, according to the above equation. In surface waters and the sea, the density structure is measured by a conductivity-temperature-depth (CTD) probe, from which the salinity and density can be determined.

The stratification frequency, N, is a parameter for the relative effect between gravity stratification and the inertia. We speak of gravity stratification because the effect comes from the variation of the fluid's specific weight, $\delta(\rho_a g)$, not the variation in the fluid's density. The distinction between specific weight and density is significant in dimensional analysis. However, 'density stratification' is the common usage to describe the gravity effect.

In marine and coastal waters buoyancy variation is often a combination of the temperature and salinity changes over the depth. The so-call sigma unit ($\sigma_t = \rho$ (kg/m^3) $- 1000$) is defined as the density excess relative to the density of pure water at 4^oC. The density of sea water can be determined from the equation of state (Appendix A), from which the Brunt-Väisälä frequency can be obtained. Figure 5.4 shows a vertical profile of salinity and temperature measured in the Pearl River estuary in the Spring season. The measurements were made in the northwestern waters of Hong Kong as part of a post-operation monitoring study at the Urmston Road Sewage Outfall (see chapter 10). The salinity is 24 ppt near the free surface and 32 ppt at 16 m depth. The surface and bottom temperatures are 16° and 15.8^o respectively. As the temperature is practically constant over the depth, the density stratification is primarily due to variation of salinity. It has essentially a two-layer structure: the sigma value varies rapidly from $\sigma_t = 17$ to $\sigma_t = 22$ over the top 5 m surface layer and then changes from $\sigma_t = 22$ to $\sigma_t = 23.5$ over the 11 m of bottom layer. The Brunt-Väisälä frequency in the upper 5-m layer is

$$N = \sqrt{-\frac{g}{\rho_a}\frac{d\rho_a}{dz}} \simeq \sqrt{\frac{9.81 \times (22 - 17)}{1000 \times 5}} = 0.099 \text{ rad/s}. \qquad (5.19)$$

The 11-m of water below the surface layer has an average Brunt-Väisälä frequency approximately equal to

$$N = \sqrt{-\frac{g}{\rho_a}\frac{d\rho_a}{dz}} \simeq \sqrt{\frac{9.81 \times (23.5 - 22)}{1000 \times 11}} \simeq 0.036 \text{ rad/s}. \qquad (5.20)$$

The two-layer density stratification in this case is primarily influenced by the fresh water outflow from the Pearl River, which changes with season. The order of magnitude of buoyancy frequency in this coastal water is about 10^{-1} rad/s to 10^{-2} rad/s. These values obtained in the

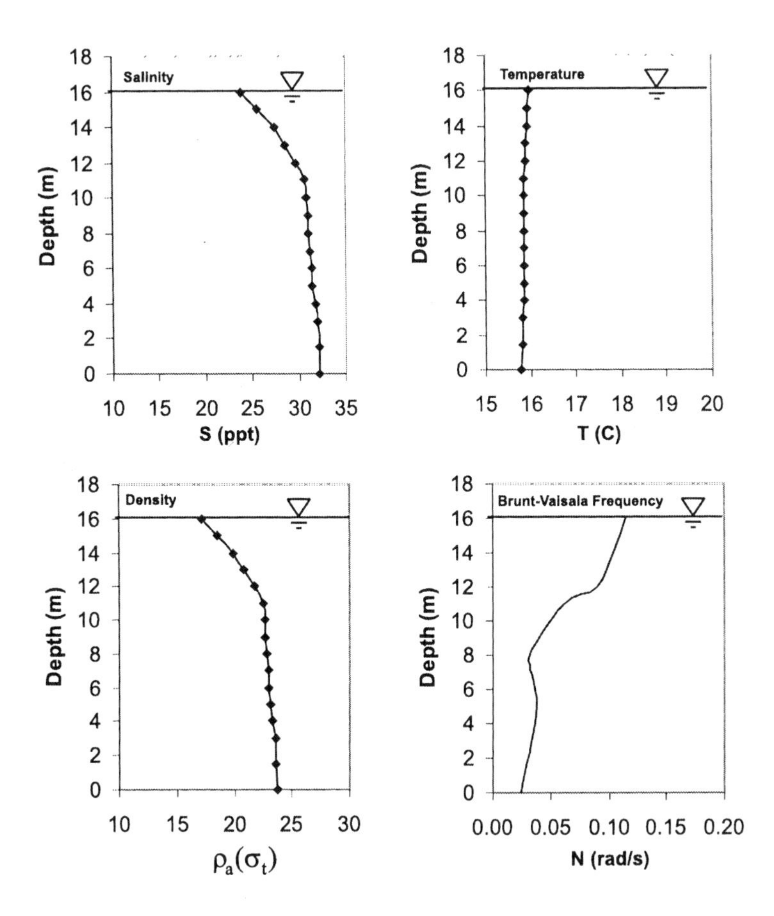

Figure 5.4. Profile of salinity, temperature, density and Brunt-Väisälä frequency in the Pearl River Estuary, China, obtained in March 1995. Measurements were made in the northwestern waters of Hong Kong.

river estuary are large compared with data obtained in the open sea. As will be seen later, the existence of even much weaker density variations can have a significant impact on environmental mixing and transport (Fig. 5.11 and Example 5.5).

2. THERMALS IN STRATIFIED FLUID

Thermals are isolated volumes of turbulent fluid produced by instantaneous sources of buoyancy. These are classified as round thermals if the source of buoyancy is derived essentially from an instantaneous point source and as line thermals if the buoyancy is from a line source. As the thermal moves through a stratified environment, the buoyancy force associated with the thermal varies with the changing densities of both the thermal and its surrounding fluid following Equation 5.8.

The rise and fall of the thermals in stratified fluid are determined by two basic equations: the buoyancy equation and momentum equation. Multiplying Equation 5.8 by dz/dt gives the buoyancy equation

$$\frac{d\tilde{B}}{dt} = -(-\frac{g}{\rho_a}\frac{d\rho_a}{dz})\rho_a\tilde{V}\frac{dz}{dt}, \tag{5.21}$$

which in compact form is

$$\frac{d\tilde{B}}{dt} = -N^2\frac{\rho_a}{\tilde{\rho}}\tilde{M} \tag{5.22}$$

where $\tilde{M} = \tilde{\mathcal{M}}\tilde{w}$ is the momentum, $\tilde{\mathcal{M}} = \tilde{\rho}\tilde{V}$ the mass, $\tilde{w} = dz/dt$ the velocity of the thermal and z elevation of the thermal above its source. The rate of momentum change is equal to the buoyancy force acting on the thermal (Newton's second law). The rate of the change is determined by the momentum equation

$$(1+k)\frac{d\tilde{M}}{dt} = \tilde{B} \tag{5.23}$$

where k is an added mass coefficient for the fraction of the momentum in the non-turbulent (irrotational) fluid surrounding the thermal (see also Prob. 8.3 for an explanation of added mass). Given a distribution of stratification profile, $N(z)$, and initial values for the momentum and buoyancy at the source, M_o and B_o, respectively, the values of $\tilde{B}(t)$ and $\tilde{M}(t)$ at subsequent time are determined by numerical integration of the buoyancy and momentum equations, Equations 5.22 and 5.23.

The space-time relation is obtained by integration of the equation $\tilde{M}(t) = \tilde{\mathcal{M}}\tilde{w} = \tilde{\mathcal{M}}dz/dt$ where $\tilde{\mathcal{M}} = 4\rho\pi R^3/3$ for a round thermal and $\tilde{\mathcal{M}} = \rho\pi R^2$ for a line thermal, where R is the radius of the thermal. Hence,

$$\tilde{M}(t) = \rho\frac{4\pi R^3}{3}\frac{dz}{dt} \quad \text{for round thermals}$$

$$\tilde{M}(t) = \rho\pi R^2\frac{dz}{dt} \quad \text{for line thermals} \tag{5.24}$$

Using a linear spreading hypothesis, the radius R is related to the height of the thermal:

$$dR = \beta|dz|. \tag{5.25}$$

Given the spreading coefficient β and the added-mass coefficient k, Equations 5.22, 5.23, 5.24, and 5.25 form a complete set of equations to find $R(t)$ and $z(t)$ by numerical integration.

2.1 ROUND THERMALS

The round thermals in a linear density-stratified ambient are first examined as an example. For a linear density profile, $d\rho_a/dz$ = constant. The density structure of the environment is defined by a single parameter - the Brunt-Väisälä or stratification frequency. Under this condition, the asymptotic solutions of the thermals can be obtained by dimensional considerations to within a constant.

The maximum rise, z_m, of the round thermal is a function of fluid density, ρ, the initial buoyancy, B_o, and the buoyancy frequency, N. The dimensions of these variables are

$$[\frac{B_o}{\rho}] = [g'_o][\mathcal{V}] = \frac{L^4}{T^2}, [N] = \frac{1}{T} \tag{5.26}$$

The dimensionless parameter of this problem is

$$\frac{z_m N^{\frac{1}{2}}}{(B_o/\rho)^{\frac{1}{4}}}. \tag{5.27}$$

Since this is the only dimensionless parameter, it must be equal to a constant; that means

$$z_m = \text{constant } (\frac{B_o}{\rho})^{\frac{1}{4}} N^{-\frac{1}{2}} \tag{5.28}$$

For the case of linear ambient stratification, an analytical solution for an instantaneous point source of buoyancy can be obtained. Differentiating the momentum equation (Equation 5.23) once with time, and then making use of the buoyancy equation (Equation 5.22) to eliminate $\tilde{B}$, gives the equation of the simple harmonic oscillator

$$\frac{d^2\tilde{M}}{dt^2} + N'^2\tilde{M} = 0 \tag{5.29}$$

where

$$N' = N\sqrt{\frac{\rho_a}{\rho(1+k)}} \tag{5.30}$$

is a modified frequency. A similar equation for B,

$$\frac{d^2\tilde{B}}{dt^2} + N'^2\tilde{B} = 0, \tag{5.31}$$

can be obtained. The solution for both the momentum $\tilde{M}$ and the buoyancy force $\tilde{B}$ associated with the thermal are simple harmonic with an oscillation frequency equal to N'.

In the stable stratification, the buoyancy is not a constant. If momentum is initially zero, M and B would vary sinusoidally as:

$$\tilde{B} = B_o \cos(N't), \tag{5.32}$$

$$\tilde{M} = -\frac{\rho}{\rho_a N^2}\frac{d\tilde{B}}{dt} = \frac{B_o}{N'(1+k)}\sin(N't) \tag{5.33}$$

In the presence of a stable density stratification, the thermal buoyancy reduces with the depth and becomes zero at a depth when $N't = \pi/2$. At this depth the momentum of the thermal is maximum as $\tilde{M}$ is 90 degree out of phase with $\tilde{B}$. Due to inertia, the thermal continues to move downward until $N't = \pi$. The momentum becomes zero when $N't = \pi$ as the maximum depth of the thermal penetration is attained.

For a point source, since $R = \beta z$ and $\tilde{M} = 4\pi\rho R^3 \tilde{w}/3$, Equation 5.33 becomes

$$\frac{4\pi}{3}\rho\beta^3 z^3 \frac{dz}{dt} = \frac{B_o}{N'(1+k)}\sin(N't). \tag{5.34}$$

Integrating once with time, gives

$$\frac{\pi}{3}\rho\beta^3 z^4 = \frac{B_o}{N'^2(1+k)}[1-\cos(N't)]; \tag{5.35}$$

that is

$$z = [\frac{3B_o(1-\cos(N't))}{\rho\pi\beta^3 N'^2(1+k)}]^{\frac{1}{4}} = [\frac{3B_o(1-\cos(N't))}{\rho_a\pi\beta^3 N^2}]^{\frac{1}{4}} \tag{5.36}$$

where the constant of integration has been selected so that $z = 0$ at $t = 0$. The maximum depth of penetration, z_m, where $\tilde{M} = 0$, is obtained by setting $N't = \pi$:

$$z_m = [\frac{6B_o}{\rho_a\pi\beta^3 N^2}]^{\frac{1}{4}} = 2.64\,(\frac{B_o}{\rho_a})^{\frac{1}{4}} N^{-\frac{1}{2}} \tag{5.37}$$

The equilibrium level, z_e, where $\tilde{B} = 0$, is obtained by setting $N't = \frac{1}{2}\pi$:

$$z_e = [\frac{6B_o}{\rho_a\pi\beta^3 N^2}]^{\frac{1}{4}} = 2.22\,(\frac{B_o}{\rho_a})^{\frac{1}{4}} N^{-\frac{1}{2}} \tag{5.38}$$

where $\beta = 0.34$ is adopted (see also Prob. 9.20). Compared to a spreading angle of 0.17 for a round plume, round thermals have been observed to spread at a much wider angle, and thus they dilute more with height (cf Prob. 1.5). The added mass coefficient is assumed to be the same as that for a sphere, with $k = 0.5$. Note that the added mass coefficient affects the frequency but not the maximum level of the plume.

Experiments on discrete buoyant clouds in a stably stratified fluid (Morton *et al.*1956) have resulted in a maximum rise coefficient of 2.66 for z_m, in close agreement with the prediction. It should be noted that the Lagrangian approach outlined above is not limited to small density differences (Escudier and Maxworthy 1973; see Prob. 5.5).

EXAMPLE 5.2 *Thermal in Linear Density Stratification*
A 1.0 m^3 of dredge spoil contaminated with 10.0 kg of cyanide is accidentally released into a stably-stratifed lake with a linear temperature stratification. The temperature varies linearly with the depth - 20 °C at the surface and 10 °C at a depth of 20 m. The Brunt-Väisälä period for such a temperature profile is 232 seconds. Find the depth of penetration of the toxic spill and estimate the cyanide concentration as the cloud begins to spread out into a horizontal layer in the stratified ambient. Assume that the cloud spreads at a rate $\beta = d\tilde{R}/dz = 0.34$. The density of the dredge spoil is 1100 kg/m^3 (relative density difference 0.1).

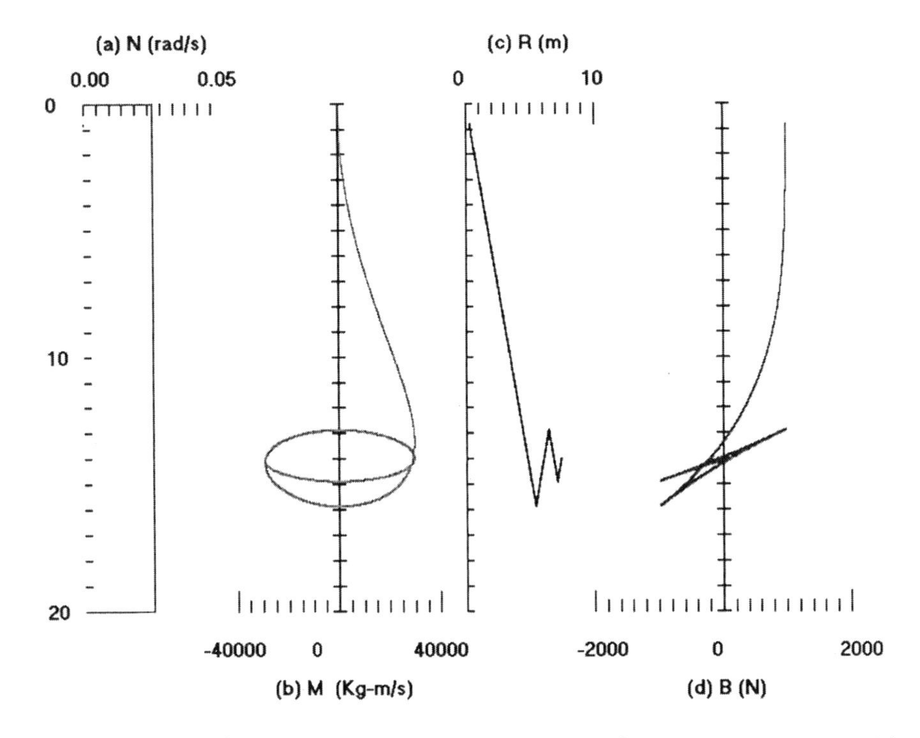

Figure 5.5. Round thermal in a uniform density stratification; N = 0.0271 rad/s. Momentum (M), radius (R) and buoyancy (B) of the thermal vs depth of advance.

SOLUTION: The density of the dredge spoil ρ_o = 1100 kg/m^3. The density of the surrounding fluid is $\rho_a = \rho_w \simeq$1000 kg/m^3. The initial negatively buoyancy force associated with the contaminated waste cloud is $B_o = g(\rho_o - \rho_w)V_o = 9.81 \times (1100 - 1000) \times 1.0 = 981$ N. The buoyancy frequency is $N = 2\pi/232 = 0.0271$ rad/s. The maximum depth of penetration is then given by Eq. 5.37.

$$z_m = 2.64 \times [\frac{981}{1000}]^{\frac{1}{4}}/\sqrt{0.0271} = \underline{16.0} \text{ m.} \tag{5.39}$$

As the cloud entrains fluid from its surrounding, the radius of the cloud is assumed to spread at a rate $\beta = dR/dz = 0.34$. At this maximum depth the radius of the thermal is $R = 0.34 \times 16.0 = 5.44$ m. The volume of the thermal is V $= 4\pi R^3/3 = 674$ m^3. The average cyanide concentration in the thermal is $\tilde{c} = 10.0/674 = \underline{0.0148}$ kg/m^3 (i.e. 14.8 mg/L).

Numerical integration of Equations 5.22, 5.23, 5.24, and 5.25 is conducted for an initial momentum $M_o = 0$, an initial buoyancy force $B_o = 9.81 \times (1100 - 1000)$ N, and a uniform Brunt-Väisälä frequency $N = 0.0271$ rad/s. Figure 5.5 (a)-(d) shows the results of the numerical integration: the stratification frequency profile $N(z)$, thermal momentum $\tilde{M}(z)$, radius R, and buoyancy force $\tilde{B}(z)$, respectively. The cyanide cloud penetrates the thermocline to a maximum depth of $z = 16$ m and then starts to oscillate with frequency equal to N' at an equilibrium level of $z = 14$ m. The oscillation is manifested in the momentum and buoyancy of the cloud (Figure 5.5b and 5.5d) which varies respectively from $\tilde{M}$ =+3000 Kg-m/s to -3000 Kg-m/s and $\tilde{B}$ = -1000 N to +1000 N; $\tilde{M}$ and $\tilde{B}$ are 90 degrees out of phase. The buoyancy force at maximum depth is upward. Subsequently, after reaching the maximum depth, the thermal reverses its direction and rises back to its equilibrium level. Further mixing will take place before the intrusion of the thermal fluid into the stratified ambient to form a horizontal layer but the concentration would probably not be further reduced by more than a factor of 2. Internal waves are radiated from the oscillating cloud. The amplitude of the oscillation is quickly diminished due to radiation damping, which is however ignored in the present analysis.

EXAMPLE 5.3 *Thermal in Nonlinear Density Stratification*

Consider the accidental release of 1.0 m^3 of toxic dredge spoil into the 20 m deep lake, in this example for a more realistic temperature profile. The temperature is constant at T_s=20 $^o C$ in the surface layer, and T_b=10 $^o C$ in the bottom layer. The temperature varies linearly from T_s to T_b over a 5 m thick thermocline. Study the thermal behaviour if the surface layer thickness is (a) 5 m and (b) 4.5 m. The density of the dredge spoil is 1100 kg/m^3 as in Example 5.2.

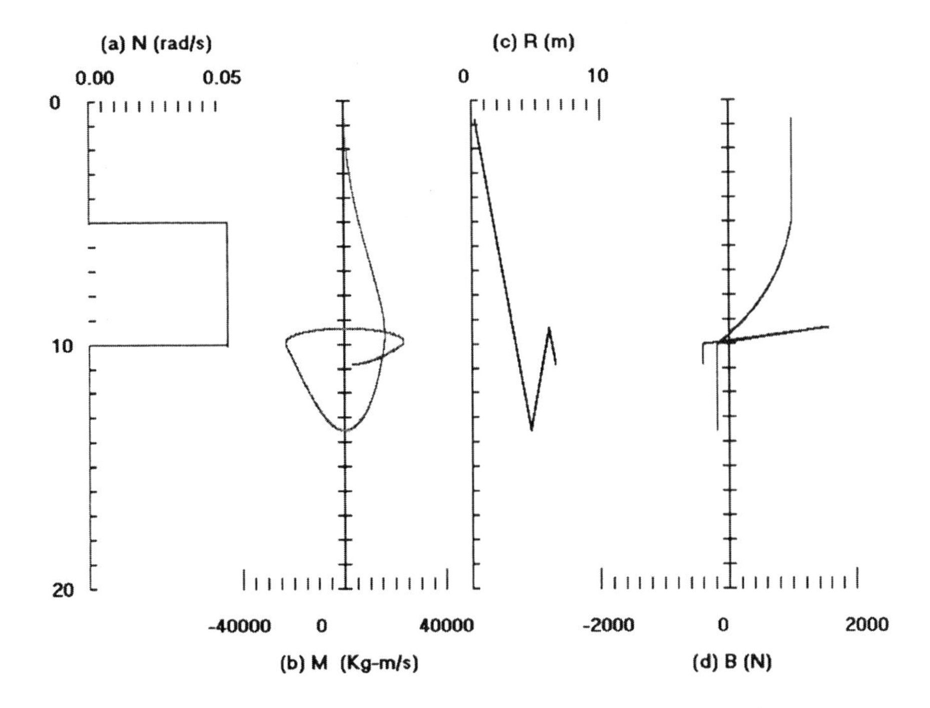

Figure 5.6. Round thermal in non-uniform stratification, Case A. Thermal advance arrested by a thermocline with $N = 0.0542$ rad/s for $z = 5 - 10$ m, The thermal continues its advance to maximum depth before it reverses and oscillates about an equilibrium depth

SOLUTION: Numerical integration of Equations 5.22, 5.23, 5.24, and 5.25 is conducted for an initial momentum $M_o = 0$, an initial buoyancy force $B_o = 9.81 \times (1100 - 1000)$ N, and a uniform Brunt-Väisälä frequency $N = 0.0271$ rad/s are conducted for two cases with nonlinear density stratification. The results for two case of slightly different thermocline thickness, to be referred to as case A and case B, are shown in Figures 5.6 and 5.7, respectively.

Case A: Thermal Trapped by the Thermocline

For the first case, the upper edge of the thermocline is located at a depth of 5 m below the surface - i.e. $T(z = 5) = 20^oC$ and $T(z = 10) = 10^oC$. Fig. 5.6 shows the numerical solution of the thermal behaviour in the non-linearly stratified lake. The change of temperature from 20^oC to 10^oC over a 5 m thick thermocline gives a stratification frequency $N = 0.0542$ rad/s within the thermocline - twice that in the previous example due to a four-fold increase in ambient density gradient. As the

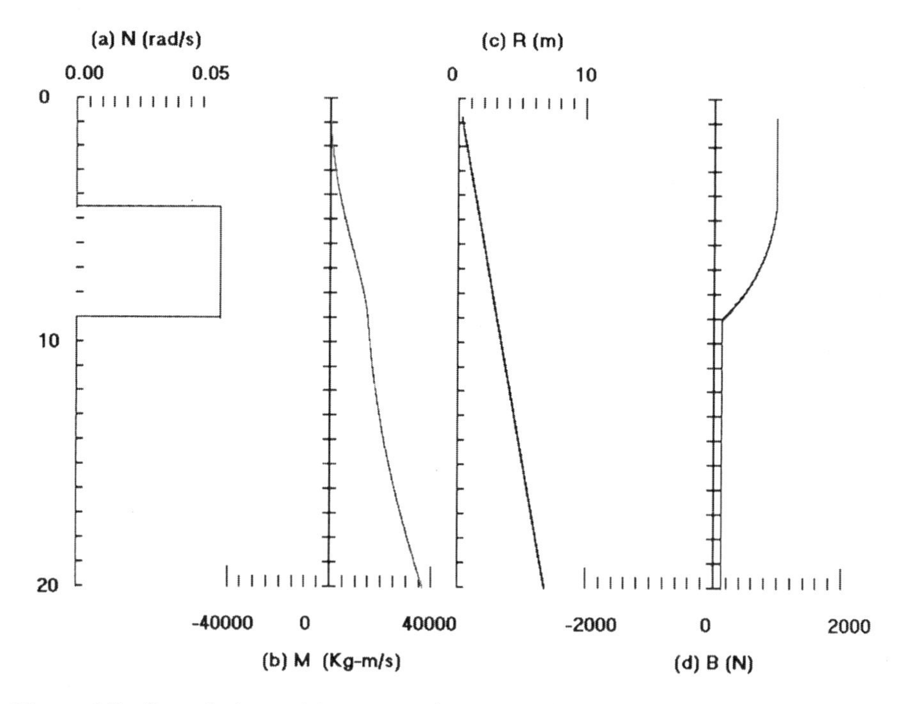

Figure 5.7. Round thermal in non-uniform stratification, Case B. Thermal penetrates a thermocline and continues into the hypolimnion; $N = 0.0571$ rad/s for $z = 4.5$ m to 9.0 m.

thermal advances through the surrounding fluid, the radius R is assumed to increase linearly with depth of advance z following a linear relation as shown in Figure 5.6c.

In this case, the buoyancy force acting on the thermal stays constant in the upper layer but begins to reduce in the thermocline and becomes negative as the thermal pass through the lower interface at a depth of 10 m. Below the 10 m depth, the buoyancy force is upward. Despite the upward force of buoyancy, due to the inertia the thermal continues the downward movement until the maximum depth of penetration is reached at $z = 13.4$ m. The upward force finally brings the thermal back to the thermocline. After a few oscillations, the thermal finally settles at an equilibrium depth of 9.8 m. In this case, the cyanide cloud is trapped in an area near the lower interface but within the thermocline.

Case B: Thermal Penetration Through the Thermocline

The situation is quite different if the thermocline is located at a slightly shallower depth with slightly smaller thickness (4.5 m to 9 m). The buoyancy frequency is slightly increased to $N = 0.0542 \times \sqrt{(10-5)/(9-4.5)}$ $= 0.0571$ rad/s. In this case, as shown in Figure 5.7, the buoyancy force is still downward after passing through the lower interface of the thermocline at 9 m depth. With the downward force the cyanide cloud is not trapped as in case A but continues its advance to the hypolimnion of the lake.

Whether a thermal penetrates the thermocline is dependent on the volume of the thermal $\tilde{V}$ and the change in density across the thermocline $\delta\rho_a$, according to the basic formula given by Equation 5.8. The thermal volume, $\tilde{V}$, increases with the distance from its source. The reduction in buoyancy is directly proportional to the density difference and volume. The deeper the depth of the thermocline, the greater would be the volume of the thermal and consequently harder for the thermal to penetrate the thermocline. Large volume leads to high rate of reduction in buoyancy force. The rebound of the thermal by the deep thermocline is the consequence of the high reduction rate that reverses the direction of the buoyancy force. In other words, by the time the thermal hits the thermocline, it would have mixed sufficiently with the top layer fluid to have a density smaller than that of its surroundings.

2.2 LINE THERMAL

As for the round thermal, a simple dimensional analysis can be made for the line thermal. The dimensions of the variables for the line thermal are

$$[\frac{B_o}{\rho}] = [g_o'][\mathcal{V}] = \frac{L^3}{T^2}, [N] = \frac{1}{T} \tag{5.40}$$

In this case, B_o is the buoyancy of the instantaneous source per unit length of the thermal. The dimensionless parameter of this problem is

$$\frac{z_m(B_o/\rho)^{-\frac{1}{3}}}{N^{\frac{2}{3}}}. \tag{5.41}$$

This is the only dimensionless parameter of the problem and it must be equal to a constant. Hence,

$$z_m = \text{constant } (\frac{B_o}{\rho})^{\frac{1}{3}} N^{-\frac{2}{3}} \tag{5.42}$$

The constant was found from the experimental observation by Wright (1984) to be 2.30.

The momentum of the line thermal is

$$\pi\rho\beta^2 z^2 \frac{dz}{dt} = \frac{B_o}{N'(1+k)} \sin(N't). \tag{5.43}$$

Integrating once with respect to the time, gives

$$\frac{\pi}{3}\rho\beta^2 z^3 = \frac{B_o}{N'^2(1+k)}[1 - \cos(N't)]; \tag{5.44}$$

that is

$$z = [\frac{3B_o(1 - \cos(N't))}{\rho\pi\beta^2 N'^2(1+k)}]^{\frac{1}{3}} = [\frac{3B_o(1 - \cos(N't))}{\rho_a\beta^2\pi N^2}]^{\frac{1}{3}} \tag{5.45}$$

the constant of integration has been selected so that $z = 0$ at $t = 0$. The maximum rise occurs when $N't = \pi$; i.e.,

$$z_m = [\frac{6B_o}{\rho_a\beta^2\pi N^2}]^{\frac{1}{3}} \tag{5.46}$$

Assuming $\beta = 0.34$ and the added mass coefficient of a line thermal is similar to that of a cylinder, $k = 1$ (see numerical determination of k in Chapter 8), we have

$$z_m = 2.54\,(\frac{B_o}{\rho})^{\frac{1}{3}} N^{-\frac{2}{3}} \tag{5.47}$$

The equilibrium level is determined by setting $N't = \pi/2$:

$$z_e = [\frac{3B_o}{\rho_a\beta^2\pi N^2}]^{\frac{1}{3}} = 2.02\,(\frac{B_o}{\rho})^{\frac{1}{3}} N^{-\frac{2}{3}} \tag{5.48}$$

These levels may be compared with formula for the advected line thermal (buoyancy-dominated far field , BDFF, of a plume in crossflow, see Chapter 8):

$$z_m = 2.30\,(\frac{B_o}{\rho})^{\frac{1}{3}} N^{-\frac{2}{3}} \tag{5.49}$$

$$z_e = 1.85\,(\frac{B_o}{\rho})^{\frac{1}{3}} N^{-\frac{2}{3}} \tag{5.50}$$

obtained from the experimental observation by Wright (1984). The formula derived from atmospheric plume data in stratified cross wind by Briggs (1965),

$$z_e = 1.78\,(\frac{B_o}{\rho})^{\frac{1}{3}} N^{-\frac{2}{3}}$$

is also in agreement with the equation (5.48) obtained by Lagrangian method. The buoyancy per unit length of the line thermal is equal to the buoyancy flux F_o divided by the crossflow velocity U_a, i.e., $B_o = F_o/U_a$ (see Chapter 8). The source size in the laboratory experiment is finite, but the formulae in Equations 5.49 and 5.48 are derived for the line source of infinitesimally small size. As pointed out by Wright (1984), the coefficients for the maximum rise and the equilibrium level generally should be lower for the laboratory sources of finite size.

3. PLUMES IN STRATIFIED FLUID

The continuous release of buoyancy produces a plume. In stable density stratification, the buoyancy flux of the plume is not a constant but reduces with height. The sketch in Figure 5.8 shows the rise and leveling off of the plume at the equilibrium level. Initially, the plume fluid at the source is lighter than its surrounding environment. The reduction of the density in the environment also causes the buoyancy flux of the plume to decrease with elevation. The buoyancy flux becomes zero at the neutral buoyancy level as the density difference between the plume and its surrounding is reduced to zero. Due to inertia the plume overshoots the neutral buoyancy level to form a dome. The fluid in the dome is heavier than its surrounding. Further mixing is expected as the plume fluid from the dome falls back to spread out laterally to form a horizontal layer (equilibrium level). Figure 5.9 shows a negatively buoyant two-dimensional plume (inclined at 45° to vertical) in a linearly stratified fluid in a laboratory experiment. The spreading layer in this case is quite thick compared with the depth of penetration. The maximum depth of the dome, z_m, and the depth of the horizontal layer, h_e, can be determined from such a laboratory experiment.

3.1 ROUND PLUME IN LINEARLY STRATIFIED ENVIRONMENT

The Brunt-Väisälä frequency is constant in a linearly stratified fluid. Asymptotic formulae of plumes in uniform density stratification can be derived using dimensional analysis. The dimensions of the *specific* buoyancy flux of the round plume F_o is,

$$[F_o] = [g_o'][Q] = \frac{L^4}{T^3} \tag{5.51}$$

It can then be shown that the height of rise is proportional to a length scale based on this buoyancy flux, F_o, and the buoyancy frequency N (Prob. 5.3):

$$z_e \sim F_o^{\frac{1}{4}} N^{-\frac{3}{4}} \tag{5.52}$$

Briggs (1969) has correlated field and laboratory data of plume rise and proposed the following formula for the rise of the plume:

$$z_e = 3.76\, F_o^{\frac{1}{4}} N^{-\frac{3}{4}}. \tag{5.53}$$

Figure 5.10 shows Briggs' correlation of the data with this formula. The laboratory experiments of Wong and Wright (1988) gave a similar formula for the maximum height of the plume

$$z_m = 4.5\, F_o^{\frac{1}{4}} N^{-\frac{3}{4}}. \tag{5.54}$$

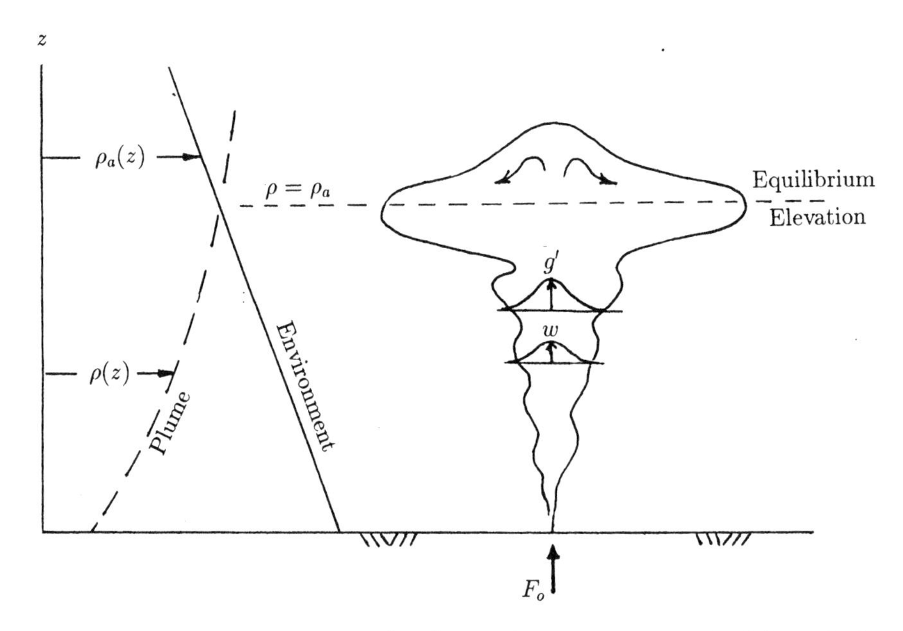

Figure 5.8. Plume rise in a stratified fluid. The plume entrains fluid from the lower levels. Buoyancy flux becomes negative as the plume overshoots the neutral buoyancy level. The mixed fluid settles back and spreads at the equilibrium level.

Wong and Wright (1988) also provided a formula for the thickness of the horizontal spreading layer at the equilibrium level as follows:

$$h_e = 1.7\, F_o^{\frac{1}{4}}\, N^{-\frac{3}{4}}. \tag{5.55}$$

This thickness of the spreading layer is quite large. It accounts for 38% of the maximum rise of the plume.

Lagrangian method may be generally employed in the formulation for linear and nonlinear stratification effects. The elemental volume is defined as the volume due to release of buoyancy at the source of the plume over a period of one time unit. Substituting volume flux, Q, for $\tilde{V}$, and the specific buoyancy flux, $F = g'Q$, for $\tilde{B}$, the mass balance equation (Equation 5.8) becomes

$$\frac{dF}{dz} = Qg\frac{d\rho_a}{dz}, \tag{5.56}$$

which can be re-written, in terms of the buoyancy frequency as

$$\frac{dF}{dt} = -N^2 M \tag{5.57}$$

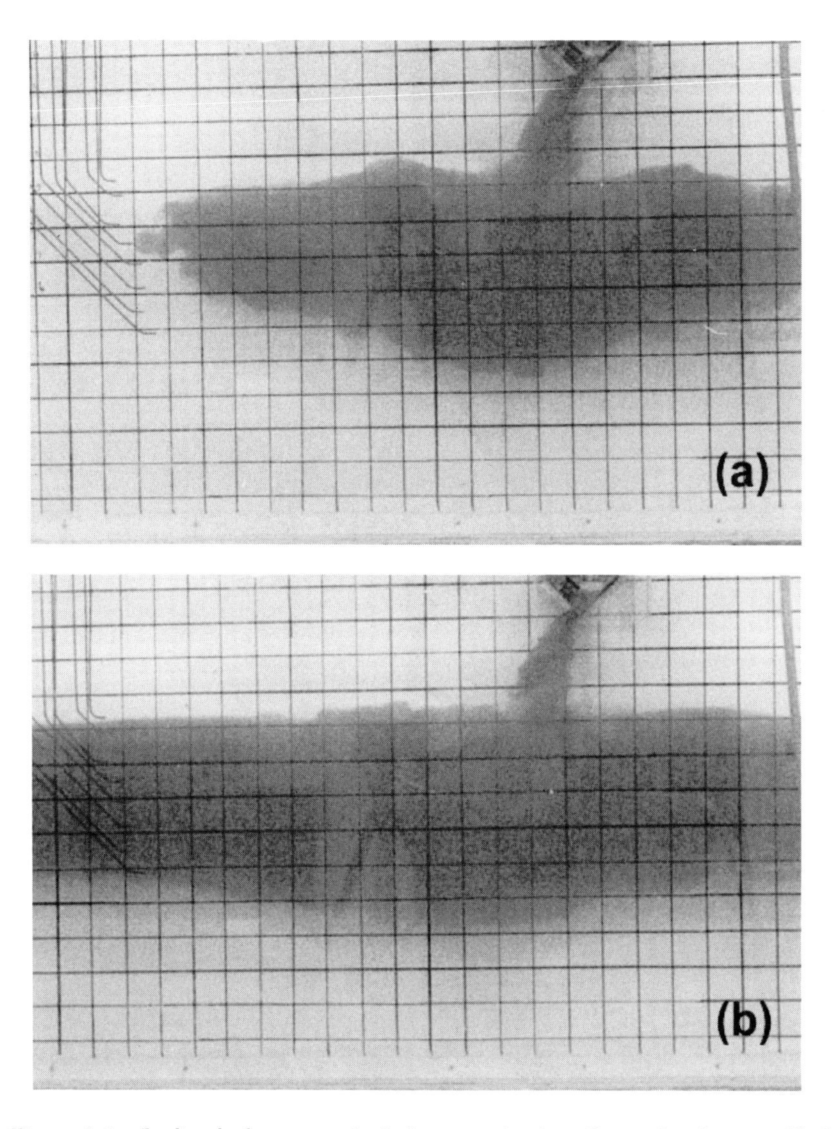

Figure 5.9. Inclined plane negatively buoyant jet in a linear density stratified fluid: (a) shortly after discharge; (b) when spreading layer is established ($\phi_o = 45^o$, $Q = 5.84$ cm^2/s, $g'_o = 20.31$ cm/s^2, $N = 0.485$ s^{-1})

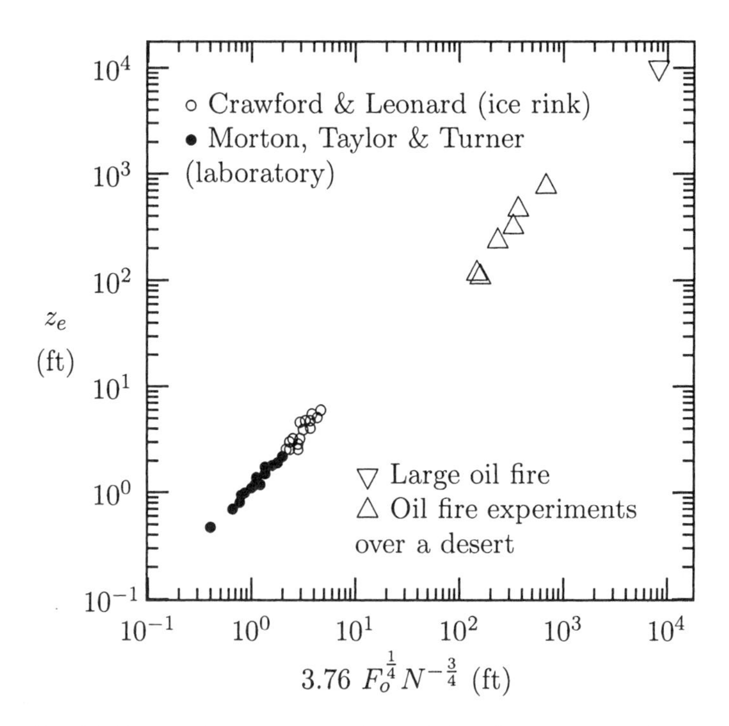

Figure 5.10. Field and laboratory experimental data for the rise of plumes in linearly stratified environment; data reproduced from Briggs (1969) and Turner (1973).

where $M = \tilde{w}Q$ is the *specific* momentum flux. This momentum in the material volume, $\tilde{\rho}M$, increases with time at a rate equal to the buoyancy force, ρF:

$$\frac{dM}{dt} = F. \tag{5.58}$$

Equations 5.57 and 5.58 combine to give

$$\frac{d^2M}{dt^2} + N^2M = 0. \tag{5.59}$$

If the frequency N is a constant, the equation is simple harmonic and the solution for a source with an initial momentum M_o is

$$M = M_o \cos(Nt) + \frac{F_o}{N} \sin(Nt), \tag{5.60}$$

An asymptotic solution is obtained by ignoring the initial momentum flux at the source. Equation 5.60 with $M_o = 0$ is

$$\pi\tilde{b}^2\tilde{w}^2 = \frac{F_o}{N} \sin Nt \tag{5.61}$$

The square root of the above is

$$\sqrt{\pi}\tilde{b}\tilde{w} = \sqrt{\pi}\beta z \frac{dz}{dt} = \sqrt{\frac{F_o}{N}} \sin Nt. \tag{5.62}$$

Integrating with time once, gives

$$\frac{1}{2}\beta\sqrt{\pi}z^2 = \sqrt{\frac{F_o}{N}} \int_0^t \sqrt{\sin Nt} dt \tag{5.63}$$

This leads to the equation for the penetration distance $z(t)$

$$z = \sqrt{\frac{2}{\beta\pi^{\frac{1}{2}}}}\; I(\tau)^{1/2}\; F_o{}^{\frac{1}{4}} N^{-\frac{3}{4}} \tag{5.64}$$

where $\tau = Nt$ and $I(\tau) = \int_o^\tau \sqrt{\sin\tau} d\tau$. The maximum height of the plume rise occurs when $M = 0$. Equation 5.63, integrating from $t = 0$ to $t = \pi/N$, gives the maximum height

$$z_m = \sqrt{\frac{2}{\beta}} \pi^{-\frac{1}{4}} [I(\pi)]^{\frac{1}{2}} F_o{}^{\frac{1}{4}} N^{-\frac{3}{4}}. \tag{5.65}$$

where the integral $I(\pi) = \int_0^\pi \sqrt{\sin\tau} d\tau = 2.396$. For $\beta = 0.17$ we have

$$z_m = 4.0\; F_o{}^{\frac{1}{4}} N^{-\frac{3}{4}}. \tag{5.66}$$

Strictly speaking, the integral formulation is invalid shortly after the plume passes the neutral buoyancy level, when the plume starts to re-entrain mixed fluid from the spreading layer. Nevertheless, the linear spreading hypothesis gives a reasonable solution, which is essentially the same as the following solution obtained by Morton *et al.*(1956) using the entrainment hypothesis:

$$z_m = 1.15\alpha_G^{-\frac{1}{2}}\; F_o{}^{\frac{1}{4}} N^{-\frac{3}{4}} = 4.02\; F_o{}^{\frac{1}{4}} N^{-\frac{3}{4}} \tag{5.67}$$

where the plume entrainment coefficient of $\alpha_G = 0.082$ is adopted. The predicted plume rise compares also favorably with the observed rise coefficient of 4.5 by Wong and Wright (1988) (Eq. 5.54).

The volume flux, $Q = \pi\tilde{b}^2\tilde{w}$, in the plume increases with distance from the source. The formula for the round plume in stratified fluid is derived using Equations 5.62 and 5.64 as follows:

$$Q = \pi\tilde{b}^2\tilde{w} = \beta z\; \pi^{\frac{1}{2}} F_o{}^{\frac{1}{2}} N^{-\frac{1}{2}} \sqrt{\sin Nt} \tag{5.68}$$

Hence,

$$Q = \sqrt{2\beta\pi^{\frac{1}{2}}}\; [I'(\tau)]^{\frac{1}{2}}\; F_o{}^{\frac{3}{4}} N^{-\frac{5}{4}} \tag{5.69}$$

where $\tau = Nt$ and $I'(\tau) = \sin\tau \int_o^\tau \sqrt{\sin\tau'}d\tau'$. The maximum volume flux (for $\beta = 0.17$) is

$$Q_m = 0.94\ F_o^{\frac{3}{4}} N^{-\frac{5}{4}}, \tag{5.70}$$

which occurs at $\tau = Nt = 0.66\ \pi$ with $I' = 1.481$. According to this analysis, the average dilution of the round plume as it impinges onto the equilibrium layer is

$$S = \frac{Q_m}{Q_o} = 0.94\ Q_o^{-1} F_o^{\frac{3}{4}} N^{-\frac{5}{4}}. \tag{5.71}$$

The above average dilution coefficient is consistent with the laboratory observations. Wong and Wright (1988) obtained the formula $S_m = 0.80\ Q_o^{-1} F_o^{\frac{3}{4}} N^{-\frac{5}{4}}$ based on their laboratory data; S_m is the *minimum* dilution obtained in the horizontal spreading layer, which is expected to be less than the average dilution. Similar to previous investigations (e.g. Fan 1967), their experiments show that the entrainment of ambient fluid up to the maximum height of rise is not significantly affected by the presence of a stable ambient density gradient.

3.2 PLANE PLUME IN LINEARLY STRATIFIED ENVIRONMENT

The plane plume is also called the line plume because it is produced by a line source of buoyancy. Similar derivation can be carried out to find the asymptotic solution of the plane plume in an environment of uniform density stratification. The *specific* buoyancy flux per unit length of the line source has a dimension

$$[F_o] = [g_o'][Q] = \frac{L^3}{T^3} \tag{5.72}$$

By dimensional analysis the maximum height of rise can be shown to correlate with a length scale based on F_o and the buoyancy frequency

$$z_m \sim F_o^{\frac{1}{3}} N^{-1} \tag{5.73}$$

The Lagrangian method can similarly be applied for the plane plume. The momentum per unit length of line plume is

$$M = \frac{F_o}{N} \sin(Nt); \tag{5.74}$$

which is,

$$2\tilde{b}\tilde{w}^2 = \frac{F_o}{N} \sin Nt. \tag{5.75}$$

The square root of the above is

$$\sqrt{2\beta} z^{\frac{1}{2}} \frac{dz}{dt} = \sqrt{\frac{F_o}{N} \sin Nt} \tag{5.76}$$

Integrating with time once gives

$$\frac{2}{3}\sqrt{2\beta} z^{\frac{3}{2}} = \sqrt{\frac{F_o}{N}} \int_0^t \sqrt{\sin Nt}\; dt \tag{5.77}$$

that leads to the depth of the plume penetration

$$z = (\frac{9}{8\beta})^{\frac{1}{3}} F_o^{\frac{1}{3}} N^{-1} [I(\tau)]^{\frac{2}{3}} \tag{5.78}$$

where $\tau = Nt$ and $I(\tau) = \int_o^\tau \sqrt{\sin \tau'} d\tau'$. The maximum height of the plume rise occurs when $M = 0$. The integration of the above, from $Nt = 0$ to $Nt = \pi$, gives $I(\pi) = 2.396$ and the formula for the maximum height

$$z_m = 3.36\; F_o^{\frac{1}{3}} N^{-1}. \tag{5.79}$$

Wright and Wallace (1979) and Wallace and Wright (1984), based on their laboratory data, have proposed a coefficient of 3.6 for the maximum rise and a coefficient 1.84 for the horizontal spreading layer thickness. The thickness of the layer is 51% of the maximum height (Figure 5.9). The laboratory data give the height of the horizontal spreading layer

$$z_{et} = 2.99\; F_o^{\frac{1}{3}} N^{-1} \tag{5.80}$$

at the top of the layer, and

$$z_{eb} = 1.15\; F_o^{\frac{1}{3}} N^{-1}. \tag{5.81}$$

at the bottom of the horizontal spreading layer. Similar values for a plume have been obtained for an inclined buoyant jet (Lee and Cheung 1986).

The volume flux per unit length of the line plume is equal to the width of the top-hat profile, $2\tilde{b}$, times the velocity of the top-hat profile, $\tilde{w}$; i.e.,

$$Q = 2\tilde{b}\tilde{w} = 2\beta z \frac{dz}{dt}. \tag{5.82}$$

Given Equation 5.78 for the variation of z as a function of time t,

$$Q = (3\beta)^{\frac{1}{3}} F_o^{\frac{1}{3}} N^{-1} [I(\tau)]^{\frac{1}{3}} \sqrt{\sin \tau} \tag{5.83}$$

If $\beta = 0.17$, the *average* dilution at the plume impingement of the equilibrium layer would be

$$S = \frac{Q_{\max}}{Q_o} = 0.76\; Q_o^{-1} F_o^{\frac{2}{3}} N^{-1}. \tag{5.84}$$

The maximum volume flux $Q_{\max}$ occurs shortly after the plume pass the equilibrium level at time $Nt = 0.62\pi$. The *minimum* dilution formula proposed by Wallace and Wright (1984) based on laboratory data obtained in the horizontal spreading layer is

$$S_m = 0.88\, Q_o^{-1} F_o^{\frac{2}{3}} N^{-1} = 0.24 \frac{F_o^{\frac{1}{3}} z_m}{Q_o} \tag{5.85}$$

After impingement the plume undergoes further mixing; the dilution in the spreading layer is expected to be somewhat higher than the dilution at the point of plume impingement.

In addition to the above two-dimensional plume experiments, Roberts *et al.*(1989) studied the dilution of the waste field formed with a line plume of finite length in a stratified crossflow. Three-dimensional effects are considered in his experiments which generally revealed higher dilutions and hence lower trap levels. For near stagnant conditions, the rise height and dilution coefficients are 2.6 and 0.37 respectively (cf 3.36 and 0.24 in Eq. 5.79 and 5.85).

EXAMPLE 5.4 *The diffuser pipe of the Orange County wastewater outfall is 1829 m (6000 ft) long and is located at 61 m (200 ft) below the sea surface. The monthly average density profiles in the vicinity of the outfall in the coastal waters of Southern California are shown in Fig. 5.11 (data given in Koh and Brooks, 1975). The total flow from the outfall is 13.2 m^3/s. Neglecting the effect of coastal current, estimate the dilution of effluent plume in February when the density of the coastal water varies from σ_t = 25.5 to 24.5 over the 60 m depth of water. Assume $\sigma_t = 0$ for the effluent.*

SOLUTION: The water in February is least stratified. Over the entire depth of 60 m, σ_t varies from 25.5 to 24.9. Assuming that the variation $\Delta\sigma_t = 0.6$ is linear over the 60 m depth, the Brunt-Väisälä frequency is

$$N = \sqrt{-\frac{g}{\rho_a}\frac{d\rho_a}{dz}} \simeq \sqrt{\frac{9.81 \times (25.5 - 24.9)}{1000 \times 60}} = 0.0099 \text{ rad/s.}$$

The buoyant effluent from the diffuser rises essentially as a line plume. The discharge per unit length of the 1829 m diffuser diffuser is Q_o = 13.15/1829 = 7.19×10^{-3} m^2/s. The specific buoyancy flux is

$$F_o = g_o' Q_o = 9.81 \times \frac{25.5}{1000} \times \frac{13.15}{1829} = 1.799 \times 10^{-3} \;\; \text{m}^3/\text{s}^3$$

The length scale associated with the rise of this line plume is $l_s = F_o^{\frac{1}{3}} N^{-1} = 12.28$ m. The maximum rise of the effluent plume is (Eq. 5.79)

$$z_m = 3.36\, F_o^{\frac{1}{3}} N^{-1} = \underline{41.3} \text{ m.}$$

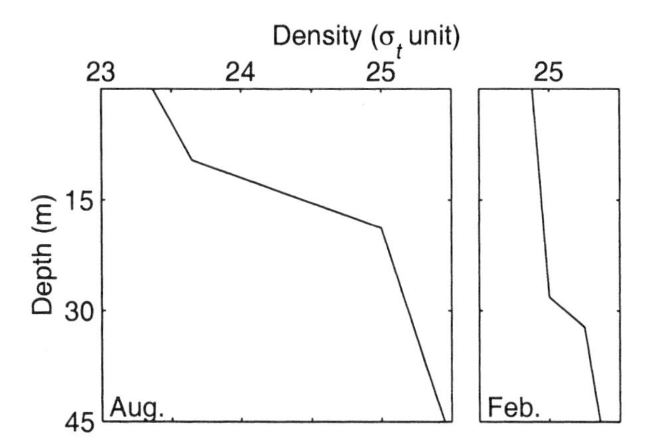

Figure 5.11. Monthly-averaged density profiles (in σ_t units) used for design of Orange County Outfall, California. Reproduced from Fischer *et al.* (1979).

The top and bottom of the horizontal spreading layer are $z_t = 2.99\ l_s = 38.2$ m and $z_b = 1.15\ l_s = 14.1$ m above the source at the diffuser, respectively (Equations 5.80 and 5.81). The average dilution at the plume impingement is $S = 0.76\ Q_o^{-1} F_o^{\frac{2}{3}} N^{-1} = \underline{157}$ (Equation 5.84). The minimum dilution ratio in the horizontal spreading layer is estimated to be $S_m = 0.88\ Q_o^{-1} F_o^{\frac{2}{3}} N^{-1} = \underline{182}$ using Equation 5.85.

It is remarkable how such a small density stratification of the ocean in February can arrest the rise of the plume and cause the spreading of the plume along a horizontal layer. The dilution of the plume is significantly reduced by the presence of the density stratification. In an un-stratified ocean, the same line plume at the base of the spreading layer would have a volume flux per unit length of the diffuser equal to

$$Q = 2\beta(1-\beta)z(\frac{F_o}{2\beta})^{\frac{1}{3}}$$

$$= 2 \times 0.17 \times (1 - 0.17) \times 60\ (\frac{1.799 \times 10^{-3}}{2 \times 0.17})^{\frac{1}{3}} = 2.95\ \mathrm{m}^3/\mathrm{s}$$

and an average dilution equal to $S = Q/Q_o = 410$, which is 2.6 times greater than the same plume trapped at depth 21.8 m to 45.9 m (i.e., from $z = 38.2$ m to 14.1 m) by the density stratification. In the above, the plume is assumed to reach the ocean surface when its elevation is within one half-width from the surface, that is when the elevation $z = 60(1-\beta)$ m.

The above calculations of the plume in linear density stratification using the one-dimensional model has produced results that are generally in agreement with formulae derived from laboratory experimental data. The results are also consistent with the numerical solution of the 2D Navier-Stokes equations using the Large Eddy Simulation (LES) method (Altai and Chu, 2001; Chu and Altai, 2001). The LES simulation results in Figure 5.12 shows how the turbulent line plume develops in the stratified environment. The dynamics of plume penetration and the spreading of the horizontal layer in stratification are shown as movie no.1 in the CD attached with this monograph.

3.3 PLUMES IN ARBITRARY DENSITY STRATIFICATION

The Lagrangian method has been illustrated for the case of linear density profile, for which basic laboratory data is available to validate the theory. In practice, the ambient density profile is typically nonlinear. Many practical problems cannot be described by the asymptotic formulae. Nevertheless, they can be readily analyzed by the Lagrangian method which is generally applicable for arbitrary density stratification profile. Discharge below a thermocline may or may not be trapped by the thermocline. The dilution characteristics of the un-trapped plume above the thermocline are very different from the trapped plume below the thermocline.

EXAMPLE 5.5 *The diffuser pipe of the Orange County wastewater outfall is 1829 m (6000 ft) long and is located at 61 m (200 ft) below the sea surface. The total flow from the outfall is 13.2 m^3/s. Neglecting the effect of coastal current, estimate the dilution of effluent plume in August when the density structure of the coastal water is essentially two-layer. Ambient density is approximately constant in the upper and lower layer. The density changes rapidly across the thermocline from $\sigma_t = 23.5$ to 25.5 from 10m to 20m depth (see Figure 5.11).*

SOLUTION: Figure 5.11 shows the ambient density profile near the Orange County outfall in southern California. The density stratification is greatest in August. Across the thermocline in August, σ_t varies from 25.5 to 23.5 over a depth of 10 m with a Brunt-Väisälä frequency at the thermocline equal to

$$N = \sqrt{-\frac{g}{\rho_a}\frac{d\rho_a}{dz}} \simeq \sqrt{\frac{9.81 \times (25.5 - 23.5)}{1000 \times 10}} = 0.044 \text{ rad/s.} \tag{5.86}$$

A simple analysis of the line plume in August may be made by assuming the density to be constant and neglecting the effect of the stratification

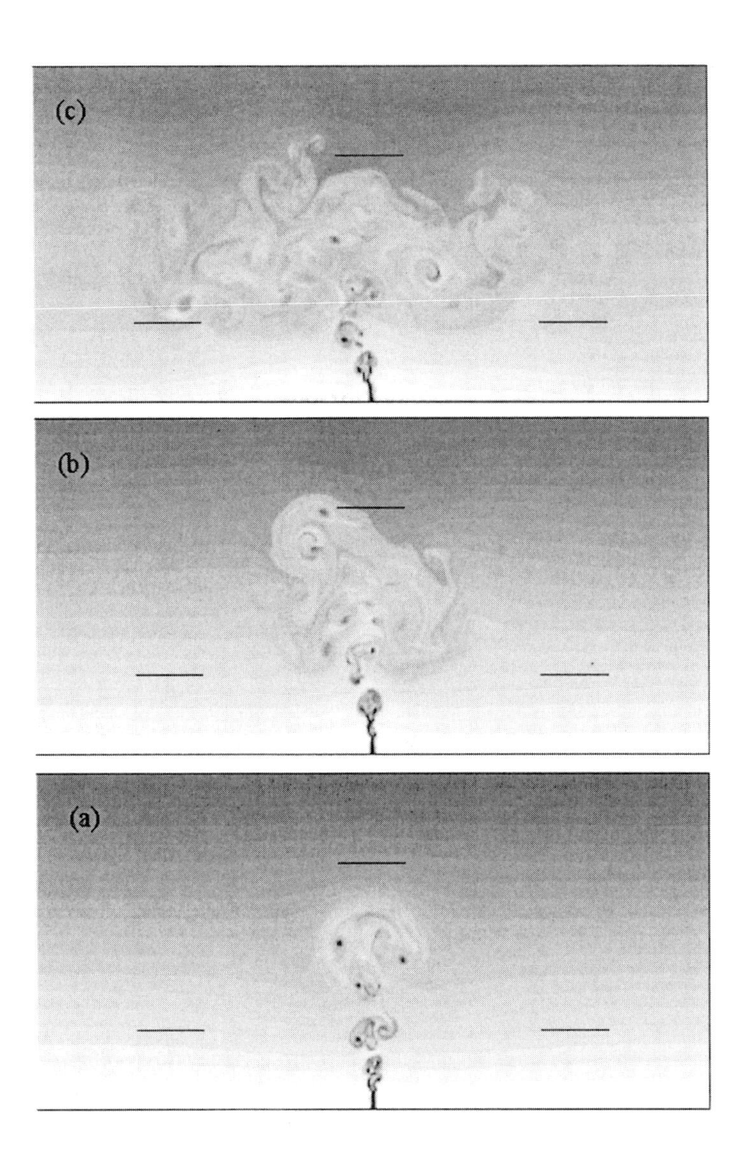

Figure 5.12. LES (Large Eddy Simulation) of plane plume in stagnant linearly stratified fluid. LES based on the data of Orange County California Outfall in February; σ_t varies linearly from 24.9 to 25.5 over the 60m depth. Computed plume at: (a) $t =$ 352 s, (b) $t =$ 701 s, (c) $t =$ 1430 s. The horizontal lines are the predicted maximum rise ($z_m = 44$ m) at the dome, the top ($z_{et} = 38.2$ m) and bottom ($z_{eb} = 14.1$ m) elevation of the horizontal spreading layer from Equations 5.79, 5.80 and 5.81, respectively.

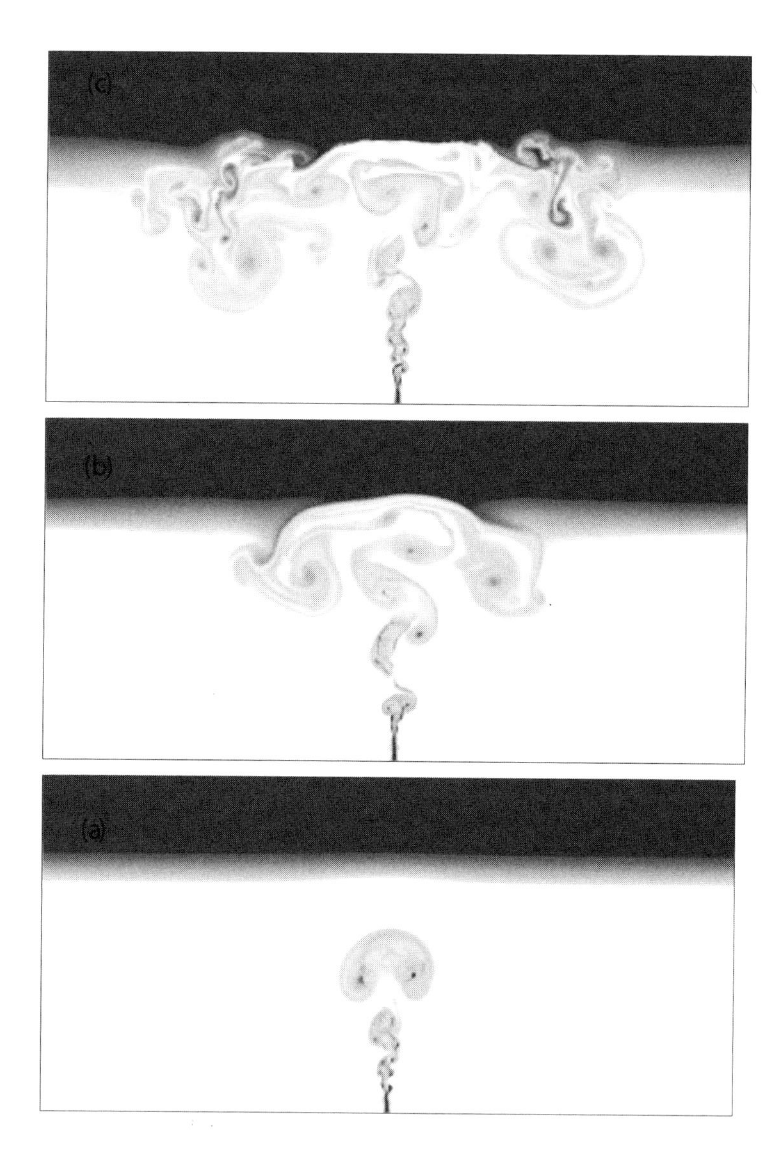

Figure 5.13. LES of plane plume impingement at a thermocline. LES using the data of the Orange County California Outfall and August stratification; σ_t varies from 23.5 to 25.5 from $z = 10$ to $z = 20$ m. Predicted concentration field at (bottom to top) time (a) $t = 251$ s, (b) $t = 499$ s, (c) $t = 752$ s.

for the first 40 m rise of the line plume. The dilution of the plume below the thermocline is calculated by the plume formula in un-stratified fluid

$$Q = 2\beta z(\frac{F_o}{2\beta})^{\frac{1}{3}}$$

The plume impinges on the thermocline as the plume rises a distance of $z = 40(1 - \beta)$ m. The volume flux at the plume impingement is

$$Q = 2 \times 0.17 \times (1 - 0.17) \times 40 \ (\frac{1.799 \times 10^{-3}}{2 \times 0.17})^{\frac{1}{3}} = 1.97 \text{ m}^2/\text{s}$$

The corresponding dilution ratio is $S = Q/Q_o = \underline{294}$. The effluent density from the diffuser is practically the same as the fresh water, which has a sigma unit $\sigma_t \simeq 0$. The sigma unit of the surrounding sea is 25.5. The sigma unit of the effluent is initially below its surrounding by a deficit of $\Delta\sigma_t = 25.5$. Dilution of the deficit gives $\Delta\sigma_t = 25.5/294 = 0.0867$ at the plume impingement to the thermocline. Therefore, $\sigma_t = 25.5 - 0.0867 \simeq 25.4$ at the impingement. The plume is so diluted at the thermocline by the surrounding sea, such that its density is now heavier than the water above the thermocline. The LES (Large Eddy Simulation) of the line plume in Figure 5.13 shows how the plume is trapped because the diluted plume fluid is heavier than the surrounding fluid. The dynamics of the trapped plume is shown as movie no.2 in the CD.

The plume may not be trapped if the diluted plume fluid is lighter than the fluid at the thermocline. Figure 5.14 shows how the plume advances through a thermocline of weak density stratification. The dynamics of plume penetration through a thermocline is shown as movie no.3 in the CD. For this 2D situation, as the plume fluid is added to the upper layer above the thermocline, the interface between the layers moves down at a rate equal to the rate the volume of the plume fluid enters the upper layer. The following section decribes how the stratification in the ambient is developed as buoyant fluid is fed to the environment by the plume.

4. PLUME IN A CONTAINER

The supply of buoyant fluid by a plume into a container introduces a stratified environment, which affects the rise of the plume and, at the same time, depends on the rate of buoyancy supplied by the plume. The mutual interaction of the plume with its environment can lead to very complex phenomena. The heating and cooling of a room is one example. In the absence of a fan to mix the air, a stove is known to be able to produce strong temperature stratification. As the plume

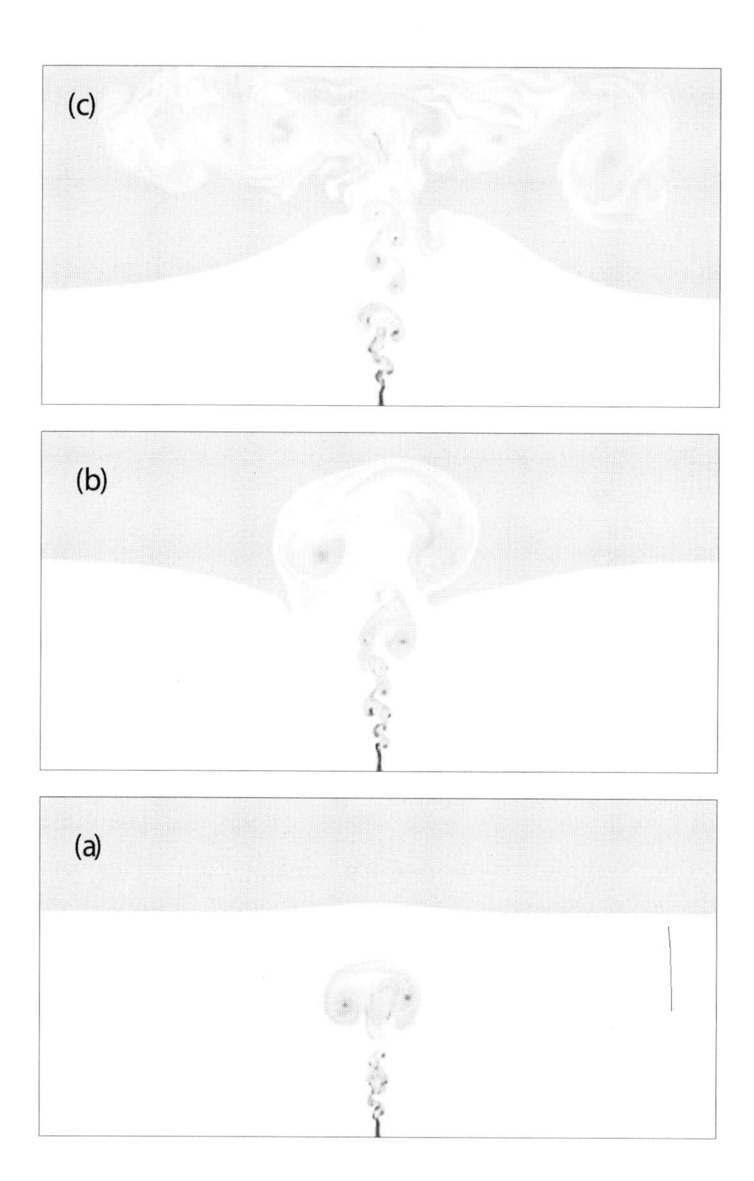

Figure 5.14. LES of plane plume penetration at a weak thermocline. In this example, σ_t varies only slightly from 25.4 to 25.5 across the thermocline. The depth of water and the buoyancy flux of the plume is the same as the case shown in figure 5.13. Sequence of images are predicted plume concentration fields at (from bottom to top) $t = 251$ s, 503 s, and 1004 s.

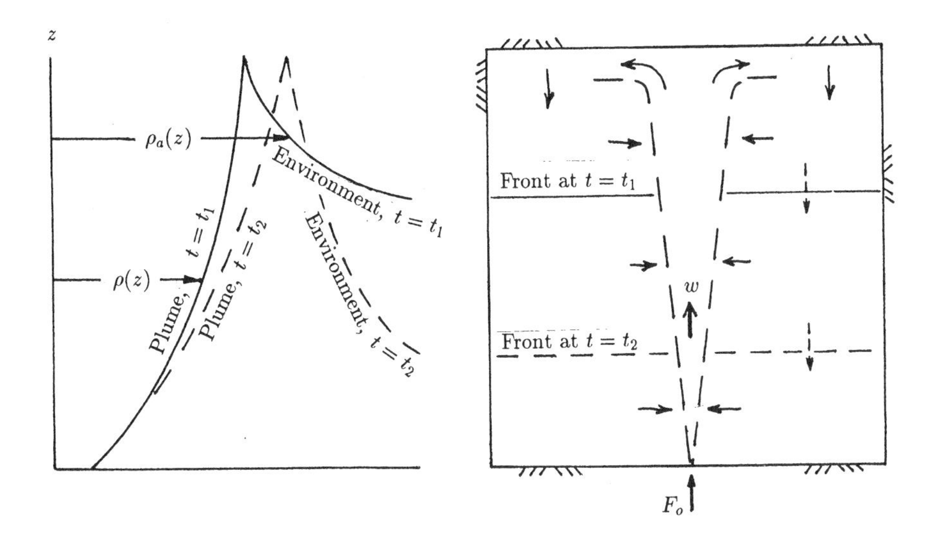

Figure 5.15. The development of density stratification in a container by a round plume, after Baines and Chu (1996)

produced by the stove impinges on the ceiling, it spreads out laterally to form a warm and stratified layer. Heat waves move down from the ceiling as plume fluid accumulates in the stratified layer below the ceiling. A similar development of stratification is observed as lateral inflows of cold and turbid waters enter a lake. The vertical progression of density stratification in the lake, and in the room, is closely related to the fluid entrainment into the plume.

The development of density stratification in a container of constant cross-sectional area by a round plume was examined by Baines and Turner (1969). Figure 5.15 shows the problem examined. The density stratification is produced by a steady inflow of buoyant fluid located at the bottom of the container. This sketch shows how the density profile in the environment develops from density fronts which originate at the top where the plume spreads laterally. The location of the initial front from starting the flow is shown in the sketch on the right. The density profiles at these times are shown on the left. The downward velocity of the stratified interfaces in the environment is induced by the upward flow of the plume. The volume flux of the plume, Q, is equal to the downward flow of the environment:

$$Q = -AW \tag{5.87}$$

in which A is the horizontal cross section of the container and W the vertical velocity in the stratified environment. Neglecting mixing across the stratified interfaces, the density is constant following the fluid motion in the environment; i.e.,

$$\frac{\partial \rho_a}{\partial t} + W \frac{\partial \rho_a}{\partial z} = 0. \tag{5.88}$$

The solution by Baines and Turner (1969) is based on the quasi-steady assumption by which the density in the environment was assumed to increase at the same rate at all points in the plume and in the environment. For this quasi-steady state, the rate of buoyancy change in the container is equal to the influx of buoyancy by the plume:

$$-V_R g \frac{\partial \rho_a}{\partial t} = F_o \tag{5.89}$$

where V_R is the volume of the container. First, eliminating $\partial \rho_a / \partial t$ from Equations 5.88 and 5.89,

$$V_R W g \frac{\partial \rho_a}{\partial z} = F_o. \tag{5.90}$$

Then, eliminating W from Equations 5.87 and 5.90,

$$Q g \frac{\partial \rho_a}{\partial z} = -\frac{A}{V_R} F_o = -\frac{F_o}{H} \tag{5.91}$$

where $H = A/V_R$ is the depth of the container. With this density stratification, and applying to Equation 5.56,

$$\frac{dF}{dz} = -\frac{F_o}{H}. \tag{5.92}$$

Hence, the buoyancy flux in the plume decreases linearly with height above the source in the quasi-steady state. The buoyancy flux is F_o at the bottom of the container and it decreases to zero at the top where the plume impinges the ceiling of the container; i.e.,

$$F = F_o(1 - \frac{z}{H}). \tag{5.93}$$

The momentum equation for the motion of the plume is

$$\frac{dM}{dt} = F_o(1 - \frac{z}{H}) \tag{5.94}$$

(see Equation 5.58). The momentum flux of the round plume is

$$M = \pi \tilde{b}^2 \tilde{w}^2. \tag{5.95}$$

If $\tilde{b} = \beta z$,

$$M = \pi \beta^2 z^2 (\frac{dz}{dt})^2 \tag{5.96}$$

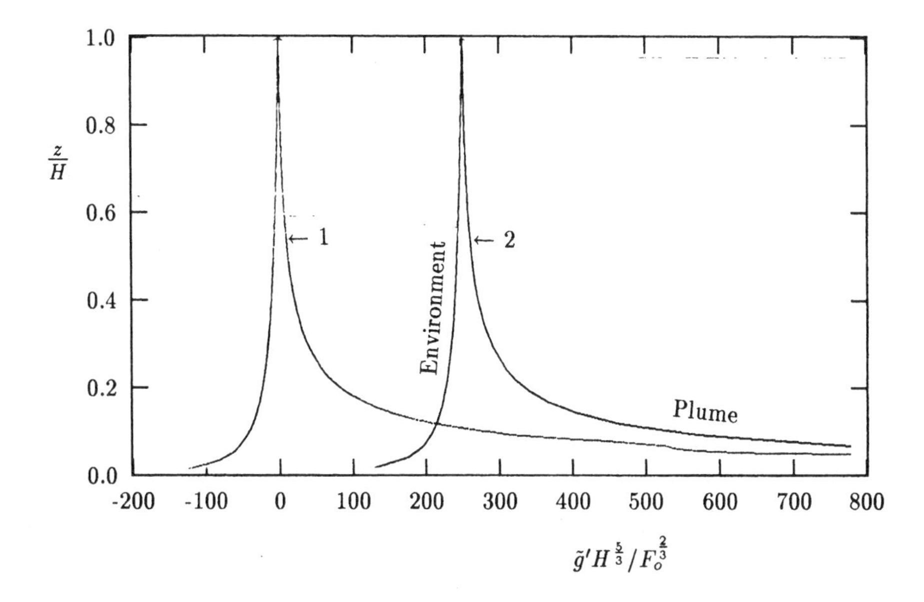

Figure 5.16. Buoyancy profiles in plume and environment, after Baines and Chu (1996). Curve 1 at time t. Curve 2 at time $(t + 200\, A/F_o^{\frac{1}{3}} H^{\frac{4}{3}})$.

which can be re-written as

$$\frac{dz^2}{dt} = \frac{2}{\beta}\sqrt{\frac{M}{\pi}}. \tag{5.97}$$

The pair of equations, Equations 5.94 and 5.97, is integrated numerically to give M and z^2. The volume flux of the plume is

$$Q = \beta z \sqrt{\pi M}. \tag{5.98}$$

The ambient density stratification, according to Equation 5.91, is

$$\frac{d\rho_a}{dz} = -\frac{F_o}{gHQ}. \tag{5.99}$$

Figure 5.16 shows the two density profiles produced by the plume with curve 2 identifying a later time. The initial density front approaches the floor asymptotically with time while the density in the ceiling of the container increases steadily toward the source density.

The mutual interaction between plumes and its stratified environment has many applications. Baines and Murphy (1986) analyzed and verified the density distribution produced by the failure of a steam line in the interior of a power plant. Germeles (1975) extended it to the unsteady case and applied it to the filling of tanks of liquid natural gas. Baines *et al.* (1993) examines the effect of fountain and its application to clean room technology. The method has been applied to line plumes, periodic thermals, unsteady source discharge (Killworth and Turner, 1982) and laminar plumes (Worster and Leith, 1985). The equations differ in each case but are derived from the same method of separating the flow into plume and stratified regions and then connecting the regions by entrainment relation. The quasi-steady solution is presented here as an example. The general solution for the unsteady flow involves partial differential equations (Equation 5.88) and can be obtained numerically, for example, by method of characteristics.

5. SUMMARY

Thermals and plumes in linear stratification are analyzed in this chapter using the Lagragian method. The predicted rise heights are in very close agreement with experimental and field observations in all cases. Table 5.1 summarizes the asymptotic formulae obtained using the Lagrangian method and the formulae obtained from experimental and field data.

In practice, the stratification is often not linear, the initial momentum may not be small, and the size of source is not generally negligible. Solution to these more complex problems of finite momentum and buoyancy sources in non-unform stratification are readily obtained by numerical integration of the governing equations. We have carried out the numerical calculations in this chapter of the thermal penetration problem in a nonlinear stratification using the equation obtained by the Lagrangian method and compared successfully with the results of LES (Large Eddy Simulation). Further development of the Lagrangian method for jets and plumes in coflows and crossflows will be considered in subsequent chapters.

Table 5.1. Asymptotic formulae for buoyancy sources in linear density stratification obtained using the Lagrangian method and the formulae derived from experimental and field data

Flows	Lagrangian Model	Experimental & Field Data
round thermal	$z_m = 2.64\,(\frac{B_o}{\rho})^{\frac{1}{4}} N^{-\frac{1}{2}}$ $z_e = 2.22\,(\frac{B_o}{\rho})^{\frac{1}{4}} N^{-\frac{1}{2}}$ $(\beta = 0.34, k = 0.5)$	$z_m = 2.66\,(\frac{B_o}{\rho})^{\frac{1}{4}} N^{-\frac{1}{2}}$ (Morton *etal.* 1956)
line thermal	$z_m = 2.54\,(\frac{B_o}{\rho})^{\frac{1}{3}} N^{-\frac{2}{3}}$ $z_e = 2.02\,(\frac{B_o}{\rho})^{\frac{1}{3}} N^{-\frac{2}{3}}$ $(\beta = 0.34, k = 1.0)$	$z_m = 2.30\,(\frac{B_o}{\rho})^{\frac{1}{3}} N^{-\frac{2}{3}}$ (Wright, 1984) $z_e = 1.85\,(\frac{B_o}{\rho})^{\frac{1}{3}} N^{-\frac{2}{3}}$ (Wright, 1984) $z_e = 1.78\,(\frac{B_o}{\rho})^{\frac{1}{3}} N^{-\frac{2}{3}}$ (Briggs, 1965)
round plume	$z_m = 4.00\,F_o^{\frac{1}{4}} N^{-\frac{3}{4}}$ $(\beta = 0.17)$	$z_m = 4.50\,F_o^{\frac{1}{4}} N^{-\frac{3}{4}}$ (Wong & Wright, 1988) $z_e = 3.76\,F_o^{\frac{1}{4}} N^{-\frac{3}{4}}$ (Briggs, 1969)
plane plume	$z_m = 3.36\,F_o^{\frac{1}{3}} N^{-1}$ $(\beta = 0.17)$	$z_m = 3.60\,F_o^{\frac{1}{3}} N^{-1}$ $z_{et} = 2.99\,F_o^{\frac{1}{3}} N^{-1}$ $z_{eb} = 1.15\,F_o^{\frac{1}{3}} N^{-1}$ (Wright & Wallace, 1979)

B_o = buoyancy force; F_o = specific buoyancy flux.

PROBLEMS

5.1 The integral jet model formulation outlined in Chapters 3 and 4 (for the case of constant ambient density) can be extended to the case of a stratified ambient fluid. Consider a stable ambient density distribution $\rho_a(z)$, with the density decreasing vertically upwards. For this case, it is also reasonable to assume that the velocity profile and the concentration excess or density deficit with respect to the **local** ambient value are also Gaussian:

$$
\begin{aligned}
c(s,r) - c_a[z(s)] &= (c_m - c_a)e^{-\frac{r^2}{\lambda^2 b^2}} \\
\rho_a[z(s)] - \rho(s,r) &= (\rho_a - \rho_m)e^{-\frac{r^2}{\lambda^2 b^2}}
\end{aligned}
$$

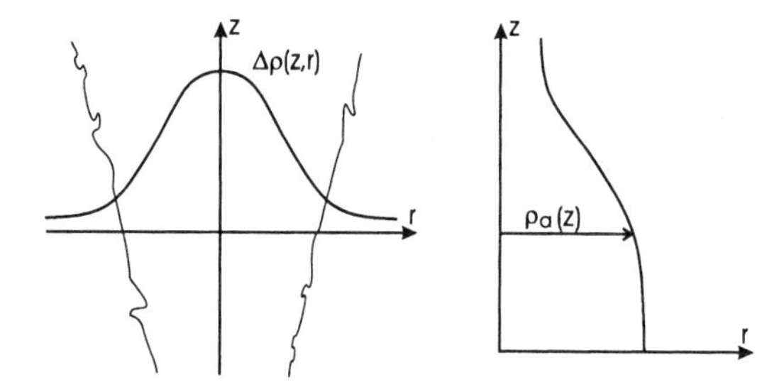

Show that the governing equations remain the same as that for the case of ρ_a = constant, except with the equations for conservation of tracer mass and density deficit flux replaced by:

$$
\begin{aligned}
\frac{d}{ds}\int u\Delta c\, dA &= Q\frac{dc_a}{dz}\sin\phi \\
\frac{d}{ds}\int u\Delta \rho\, dA &= Q\frac{d\rho_a}{dz}\sin\phi
\end{aligned}
$$

where $c_a(z)$ is the ambient tracer concentration, Q is the volume flux, and s=natural streamwise co-ordinate. (Hint: use both the conservation of tracer mass and the continuity equations).

5.2 Consider a total discharge flow of 2.2 m^3/s from a multiport diffuser at 50 m depth when there is ambient stratification in the ocean. The 50 ports are spaced apart to avoid interference between adjacent jets, and the density of the discharge is ρ_o = 1000 kg/m^3.

A typical vertical ambient density profile in the summer is shown in the accompanying graph (Fig. 5.17). Determine if the sewage field will stay submerged beneath the surface under these conditions.

Use VISJET or any reasonable approximate method (e.g. assume linear density stratification over height of rise).

5.3 A solid sphere of diameter d (volume V) is released from the free surface of a linearly stratified liquid with density variation $\rho_a(z) = \rho_o + (d\rho_a/dz)\, z$, where $d\rho_a/dz$

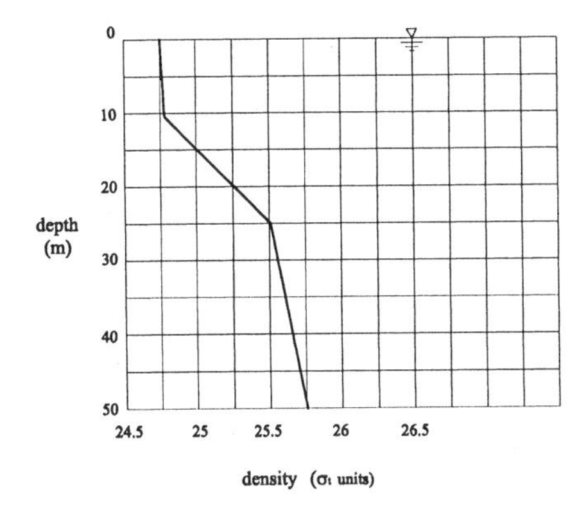

Figure 5.17. Ambient density profile in summer

= constant, and z = vertical distance below free surface. The density of the sphere is $\rho_d > \rho_o$. The sphere sinks to an equilibrium position z_t within the fluid.

a) At the equilibrium position, show using hydrostatics that regardless of the size of the sphere, the buoyant force acting on the sphere (which balances the weight of the sphere $W = \rho_d g V$) depends only on the density at z_t (= vertical distance from free surface to center of sphere):

$$F_b = \rho_a(z_t)gV = (\rho_o + \frac{d\rho_a}{dz}z_t)gV$$

Show that the equilibrium level of the sphere is

$$z_t = g'_o(\frac{g}{\rho_d}\frac{d\rho_a}{dz})^{-1}$$

where z = positive downwards (same direction as gravity), and $g'_o = (\rho_d - \rho_o)g/\rho_d$.

b) The above experiment corresponds to a negatively buoyant release in linearly stratified fluid. Thus for a general positively buoyant plume released at $z = 0$ in a linearly stratified environment, it is reasonable to postulate that the maximum height of rise depends on:

$$z_m = fn(Q_o, g'_o, -\frac{g}{\rho_d}\frac{d\rho_a}{dz})$$

where z=vertical co-ordinate (positive upwards) above source. Since for small density differences, Q_o and g'_o are both represented in the specific buoyant flux, $F_o = Q_o g'_o$, we can write:

$$z_m = fn(F_o, -\frac{g}{\rho_d}\frac{d\rho_a}{dz})$$

Hence z_m can be obtained by dimensional reasoning (cf Eq. 5.52 and Prob. 5.6).

5.4 *Generation of linear stratification*

In laboratory studies it is often desired to generate a linear vertical stratification, i.e. $d\rho_a/dz$ = constant. This can be efficiently done using the following method (Oster, G 1965, *Scientific American*, 213(2), 70-79).

Assume a linear stratification is to be prepared using a salt solution in a rectangular flume of plan cross-section area A_f and volume V_f. Prepare in a separate container a salt solution with density equal to the desired ambient density at the bottom layer. The container has an initial volume of salt solution equal to about $V_o = V_f/2$. The salt solution (constantly well-stirred by mixing) is then pumped into the flume at a certain flow rate Q. Simultaneously freshwater (with zero salt concentration) is fed into the container at a flow rate of $Q/2$. It can be shown the salt concentration of the solution in the container (which is pumped meticulously into the flume with minimal disturbance) decreases linearly with time.

a) The time required to fill the flume is $t_f = V_f/Q = 2V_o/Q$; this is the same as the time required to empty the container, $t_d = V_o/(Q/2)$.
If $V(t)$ and $c(t)$ are the volume and concentration of the salt solution in the container, By performing a mass and salt balance (assuming full mixing in the container) show that:

$$\frac{dV}{dt} = -\frac{Q}{2}$$
$$\frac{d(Vc)}{dt} = -Qc$$

b) Show that the concentration of the salt solution pumped into the flume will be linearly decreaing with time:

$$\frac{c(t)}{c_o} = 1 - \frac{t}{t_d}$$

where c_o = initial salt concentration (corresponding to the lowest layer in the flume), and $t_d = V_o/(Q/2)$ is the "detention time" in the container.

As the flume is filling up with the salt solution, $A_f dh/dt = Q$, and $t = (A_f/Q)h$, show that the salt concentration (hence density) in the flume varies linearly with depth as:

$$\frac{c(h)}{c_o} = 1 - \frac{h}{h_f}$$

where $h(t)$ = depth of salt solution in the flume at time t, and h_f = final depth of linear stratification in the flume.

5.5 *Model for turbulent round thermal*

Consider an isolated mass of buoyant fluid (initial density ρ_o, volume V_o) released at time $t = 0$ in ambient fluid of uniform density ρ_a, with $\rho_o < \rho_a$. The flow is driven by the initial buoyant force, $B_o = (\rho_a - \rho_o)gV_o$, which stays constant by virtue of conservation of tracer mass. The thermal mixes with the ambient fluid by turbulent entrainment. Both experiments and numerical calculations show that the round thermal is a vortex-ring like flow, which is self-similar and can be described by its characteristic radius $R(t)$, average velocity $W(t)$, and density $\rho(t)$.

a) Applying the equation of motion for the round thermal (assumed to be a sphere), show that the momentum equation is:

$$\frac{d}{dt}[\frac{4}{3}\pi R^3(\rho + k\rho_a)W] = \frac{4}{3}\pi R^3(\rho_a - \rho)g$$

where k accounts for the virtual mass of the unsteadily moving thermal. The virtual mass coefficient is assumed to be the same as a solid sphere, $k = \frac{1}{2}$.

Show that the thermal velocity is given by:

$$W = \frac{\rho_a - \rho}{\rho + k\rho_a} gt$$

Obtain limiting expressions for (i) $\Delta\rho/\rho_a \ll 1$ and (ii) $\rho_a \gg \rho$ $(\Delta\rho/\rho \approx 1)$.

b) Entrainment hypothesis - If it is assumed that the ambient fluid is entrained into the thermal at a rate proportional to its surface area and velocity:

$$\frac{d}{dt}(\frac{4}{3}\pi R^3) = (4\pi R^2)\alpha W$$

where α is the entrainment coefficient (assumed constant). Noting that $W = dz/dt$ show that a linear spread is predicted, with $R = \beta z$, and $\beta = \alpha$ for this case.

5.6 a) Consider a round buoyancy-dominated wastewater plume in a linearly stratified ocean. Deduce to within a constant the maximum height of rise of the mixed effluent, z_m, in terms of the specific buoyancy flux F_o and the stratification parameter $\epsilon = -g/\rho_o(d\rho_a/dz) = N^2$, where N = buoyancy frequency, and the other terms have their usual meaning. Explain briefly your physical reasoning.

b) In the deep ocean, hot sulfide ore-bearing solutions ($\sim$ 350°C) are sometimes released from local spots on the sea bed. The fine sulfide particles (typical settling velocity = 1/500 cm/s) are carried with the hot, grey-black water from these hydrothermal 'submarine vents'. When the buoyant plume ('black smoker') reaches a certain level, it spreads sideways and the suspended sulfide particles may settle back to the sea floor to form sulfide deposits. A typical mean horizontal velocity and temperature at 1-2 km depth are 1 cm/s and 10°C respectively.

i) Assuming the flowrate and temperature of the source to be 10^4 cm^3/s and 350°C ($\Delta\rho/\rho = 0.01$) respectively, estimate the maximum height of rise of the black smoker. The stratification frequency can be taken as N = 3×10^{-4} s^{-1}. Neglect the effect of ambient current.

ii) Massive sulfide deposits of horizontal dimensions 100 to 1000 m are sometimes found on the sea floor. Comment on the possibility that these confined ore deposits are due to the settling out of particles from the rising effluent or from the spreading layer of the black smoker plume.

5.7 a) Consider a vertical round jet in a linearly stratified ocean. Deduce to within a constant the maximum height of rise of the mixed effluent, z_m, in terms of the specific momentum flux M_o and the stratification parameter $\epsilon = -g/\rho_o(d\rho_a/dz) = N^2$, where N = stratification frequency. Show that:

$$z_m = C\,(\frac{M_o}{\epsilon})^{1/4} = C{M_o}^{1/4}N^{-1/2}$$

Experiments show that $C = 3.8$.

b) Wastewater from a small town next to a popular beach is discharged in the form of a single round jet in coastal water of 12 m depth. The jet diameter is $D = 0.036$ m and the discharge flow is $Q_o = 0.02$ m^3/s. In the Spring month of April the receiving water is weakly stratified with a salinity gradient of 0.1 ppt/m. The bottom salinity and temperature are 32.5 ppt and 26°C respectively, and the effluent density is $\rho_o = 998$ kg/m^3.

Compute the jet densimetric Froude number Fr, the jet momentum and buoyancy fluxes, and the momentum length scale L_s. If the jet is discharged vertically, assuming a momentum jet, estimate the maximum height of rise z_m and the dilution. Will the beach be protected?

Determine z_m for a corresponding plume.

5.8 For the wastewater discharge described in Prob. 5.7, suppose the jet is discharged horizontally in the offshore direction with the same velocity and other parameters unchanged. Assess if the wastewater jet will be trapped beneath the free surface. a) If $\overline{S}$ is the average dilution up to the trapping level z_t, then the density of mixed sewage at the trapping level ρ_m can be estimated by assuming the entrained ambient flow has an average density between source level and z_t. Show that:

$$\rho_m = \frac{\rho_o Q_o + \frac{1}{2}(\rho_a(0) + \rho_a(z_t))(\overline{S} - 1)Q_o}{\overline{S}Q_o}$$

Since $\rho_m = \rho_a(z_t)$, show that:

$$z_t = \frac{2(\rho_a(0) - \rho_o)}{(-d\rho_a/dz)\overline{S}}$$

b) The above equation shows that if $\overline{S}$ is sufficient, trapping can occur, with $z_t \leq H$. Experiments show that the mixing up to the trapping level is not much affected by the stratification. Use the Cederwall equation to estimate the dilution of the horizontal buoyant jet and assess the likelihood of trapping. (Hint: first assume $z_t = H$, determine horizontal jet dilution $\overline{S}$, then compute z_t and iterate).

5.9 For a plane buoyant jet in a linearly stratified fluid, with source momentum flux Q_o, specific buoyancy flux F_o, and stratification frequency $N = \sqrt{-g/\rho_a(d\rho_a/dz)}$, show that the following length scales can be obtained:

$$\begin{aligned} l_s &= M_o/F_o^{2/3} \\ l'_b &= F_o^{1/3}/N \\ l'_m &= M_o^{1/3}N^{-2/3} \end{aligned}$$

l_s/l'_b measures the relative importance of the initial momentum, initial buoyancy, and the ambient density stratification. For a buoyant jet that behaves like plumes, $l_s/l'_b \ll 1$, show by dimensional analysis that maximum height of rise is $z_m \sim l'_b = C_1 F_o^{1/3} N^{-1}$ (Eq. 5.79). For a buoyant jet that behaves like momentum jets, $l_s/l'_b \gg 1$, show that $z_m = C_2 M_o^{1/3} N^{-2/3}$. Experiments show that $C_1 = 3.6$ and $C_2 = 2.3$ for a vertical plane buoyant jet, and $C_1 = 3.6$ and $C_2 = 1.9$ for a jet inclined at 45° to horizontal.

5.10 A 20 port submerged diffuser for a marine terminal discharges treated ship ballast water at a flow of $Q_s = 1.06$ m^3/s at 45° to the horizontal. The ambient water (70 m depth) is near stagnant and is weakly stratified, with a variation of 0.58 σ_t units over the bottom 50 m. The waste effluent density is $\rho_o = 1020.7$ kg/m^3, while the bottom ambient density is $\rho_a(0) = 1024.1$ kg/m^3. If the diffuser length is $L = 60$ m, and the jet velocity is $U_o = 4.2$ m/s, determine the volume flux $Q_o = Q_s/L$ of the equivalent slot (2D) buoyant jet, and the momentum and buoyancy fluxes (M_o and F_o). Estimate the maximum height of rise and the initial dilution.

For a 45° discharge, the minimum diluton in the spreading layer is given by $S_m \sim F_o^{1/3} l'_b/Q_o$ for plume like conditions (dilution coefficient=0.81), and $S_m \sim (M_o l'_m)^{1/2}/Q_o$ for jet like conditions (dilution coefficient=0.6).

Chapter 6

TURBULENT ROUND JET IN COFLOW

The jet in coflowing current is examined in this chapter as a prelude to the general Lagrangian formulation of a jet in non-uniform ambient current, with a three-dimensional trajectory. Like simple jets and plumes, the axisymmetric coflowing jet is an important limiting case. It is a shear flow which has been extensively studied (e.g. Maczynski 1962; Antonia and Bilger 1973, 1974; Shirazi *et al.*1974; Knudsen 1988; Wood 1993; Nickels and Perry 1996). The detailed velocity field has been measured by many investigators and more recently the tracer concentration has been studied using Laser-induced fluorescence (LIF) techniques (Chu, Lee and Chu 1999; Wang and Davidson 2001). It has a number of turbulent spreading characteristics that are critical to understanding the Lagrangian method. In the following, the key observed features of a coflowing jet will be first summarized. The subtleties and difficulties of an integral formulation or turbulent closure for a jet in a current are explained and discussed. Using a general Lagrangian integral model based on a spreading hypothesis, the coflowing jet properties are then calculated. The predictions of jet width, velocity excess, and centerline dilution are compared with experimental data and the implications on jet modelling are discussed.

1. SUMMARY OF EXPERIMENTAL OBSERVATIONS

Figure 6.1 shows a schematic diagram of the round jet in a coflow. The jet is issued from a circular nozzle of diameter D at a velocity of U_o. The velocity of the coflow is U_a. The jet velocity is the coflow velocity, U_a, plus the excess velocity, ΔU_e, as shown in Figure 6.1. For a coflowing jet,

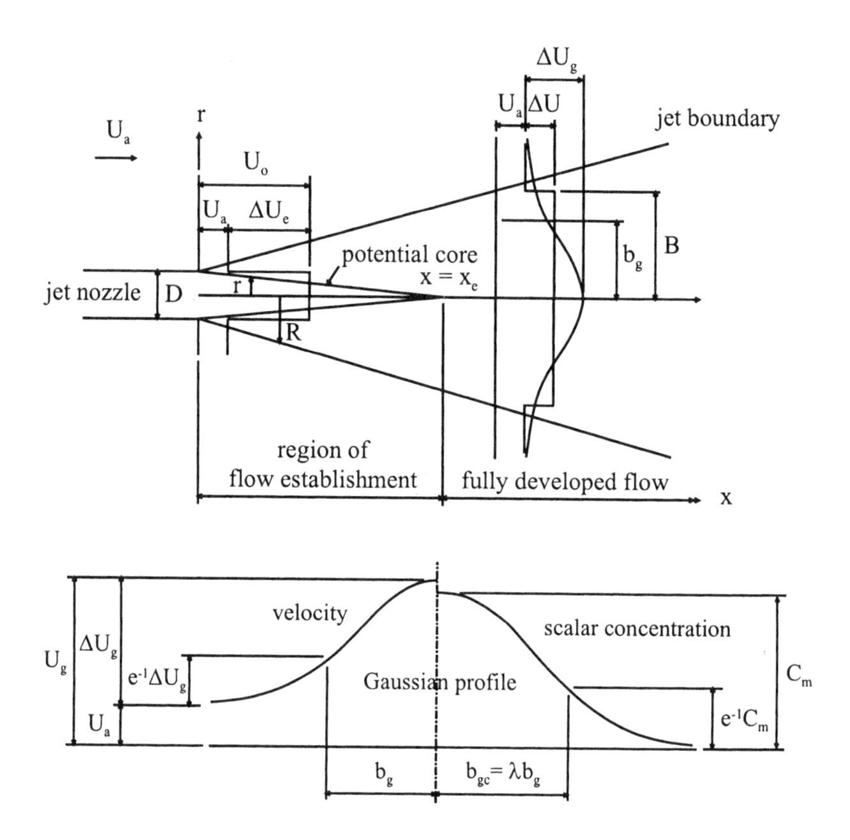

Figure 6.1. a) Schematic diagram of round jet in coflow; b) representation of characteristic jet properties

using a similar derivation as in Chapter 2, it can be readily shown that the specific *jet excess momentum*, $M_e = \int uu_e dA = \int u(u - u_a)dA$, is conserved (Prob. 6.1). Two parameters are significant for the coflowing jet. These are the coflow velocity, U_a, and the excess momentum flux, M_{eo}. Similar to previous chapters, we will use M_{eo} to denote the specific (kinematic) excess momentum flux unless otherwise stated. Since the dimension of $[M_{eo}]$ is $\mathrm{L}^4/\mathrm{T}^2$ and the dimension of $[U_a]$ is L/T, the excess momentum length scale

$$l_m^* = \frac{M_{eo}^{1/2}}{U_a} \tag{6.1}$$

and the coflow velocity U_a are the length scale and velocity scale of the problem, respectively.

1.1 CROSS-SECTIONAL IMAGES

The jet mixing can be studied by measuring the concentration of a tracer at different x-sections from the source using a technique known as Laser-induced fluorescence (LIF). The source fluid is a fluorescent dye solution of known concentration. A sheet of laser light can be made to illuminate a given section against a completely dark background; the resulting laser-induced fluorescent image of the jet cross-section can be video-recorded. The fluorescence intensity at each point is proportional to the tracer concentration. With careful calibration, the dye concentration at each point can be determined from the measured light intensity. This technique provides a powerful means to visualize the phenomenon, and to measure the instantaneous tracer concentration field not achievable with probe-based techniques.

Fig. 6.2 shows the instantaneous images and the time-averaged concentration of the jet cross-section at $x = 40D$ downstream of the nozzle in an experiment with $U_o = 0.796\ m/s$, $U_a = 0.05\ m/s$, and $D = 1.02\ cm$ (Chu 1996; Chu *et al.*1999). The time-averaged image is obtained from a series of 250 video frames obtained over a one minute period of time. The cross-sectional shape of the jet is constantly varying with time as a core of turbulent eddies that moves around the jet centerline. The time-averaged profile partly reflects the variation of concentration within this dominant turbulent eddy structure, and perhaps also the movement of the dominant eddy around the centerline. This behaviour is not due to the coflow; very similar turbulent structure was observed in the cross-sectional images of the jet in stagnant environment.

1.2 GAUSSIAN PROFILES

Figure 6.3 shows time-averaged concentration (C) contours at two sections of the coflowing jets. The dimension of the jet is expressed in pixel coordinates ($\sim$ 23 pixels/cm). Despite the irregular shape of the instantaneous images, the time-averaged contours are circular in shape with very good axisymmetry. It is not surprising then that the radial profile of time-averaged concentration $C(x, r)$ is self-similar and Gaussian. In Figure 6.4, the measured concentration C normalized by the centerline maximum $C_m = C(x, 0)$ is plotted against the dimensionless radial distance r/b_{gc}, where b_{gc} is the concentration half-width defined by the radial location at which $C = e^{-1}C_m$. It can be seen the radial concentration variation is self-similar; the data at different sections collapse

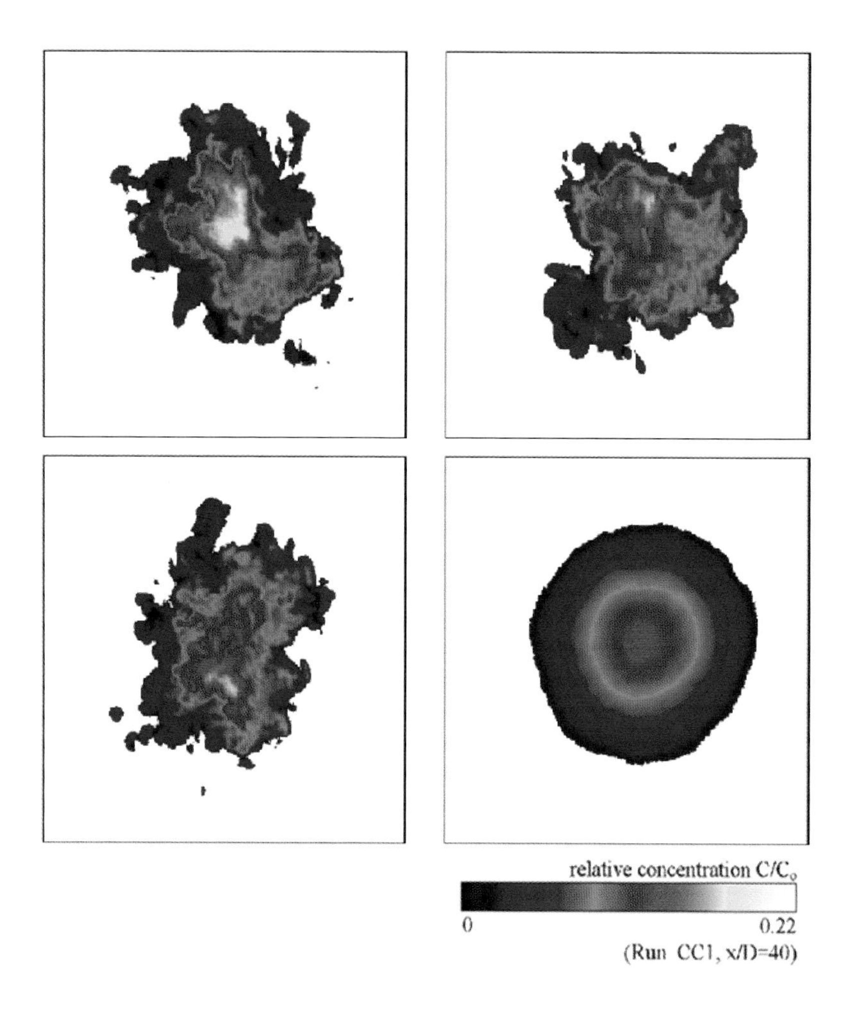

Figure 6.2. Instantaneous images and time-averaged concentration field (lower right corner) in the cross-section of coflowing jet (U_o = 0.796 m/s, U_a = 0.05 m/s, and D = 1.02 cm). Concentration lower than the visual threshold not shown.

nicely onto one curve and can be well-approximated by the Gaussian distribution:

$$\frac{C}{C_m} = \exp(-\frac{r^2}{b_{gc}^2}) \tag{6.2}$$

(see definitions of C_m and b_{gc} in Figure 6.1.)

For a coflowing jet, it is the velocity excess above the ambient that determines the entrainment. Figure 6.5 shows the excess velocity measured across a coflowing jet by Laser Doppler Anemometry (LDA). The excess velocity $u_e(x,r) = u(x,r) - U_a$ normalized by the maximum ($u_{em} = \Delta U_g$) is plotted against r/b_g, where b_g is a similarly defined velocity half-width. The self-similar velocity excess can also be described by:

$$\frac{u_e}{\Delta U_g} = \exp(-\frac{r^2}{b_g^2}) \tag{6.3}$$

(see definitions of ΔU_g and b_g in Figure 6.1.)

The width of the Gaussian concentration profile, b_{gc}, is generally wider than that of the velocity profile, b_g. Unlike the case of a jet in stagnant ambient, however, the width of a coflowing jet varies nonlinearly with the distance from the source.

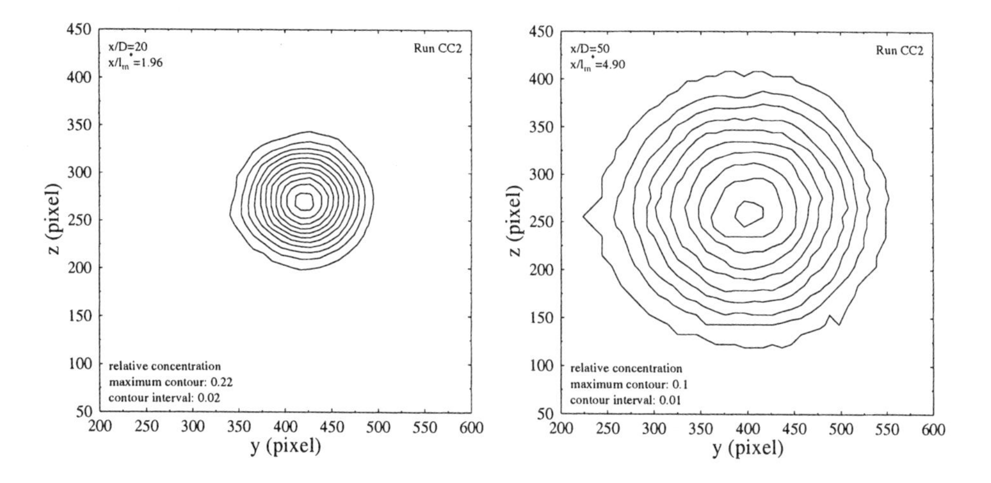

Figure 6.3. Cross-sectional time-averaged concentration contours of a coflow jet

2. INTEGRAL MODEL

Figure 6.1 shows the schematic definition of an axisymmetric turbulent jet with source diameter D and exit velocity U_o in a coflow U_a. Near the source, a jet mixing layer develops within the zone of flow establishment (ZFE) which extends from the jet nozzle up to the end of the potential core at $x = x_e$; the velocity within the potential core is constant, with an excess velocity equal to $U_o - U_a$. In the zone of established flow (ZEF), $x \geq x_e$, the turbulent-mean velocity profile of the fully turbulent flow is approximately Gaussian, with a centerline excess velocity $\Delta U_g = U_g - U_a$ and nominal half-width b_g. Our objective is to develop an integral model that predicts the variation of the jet

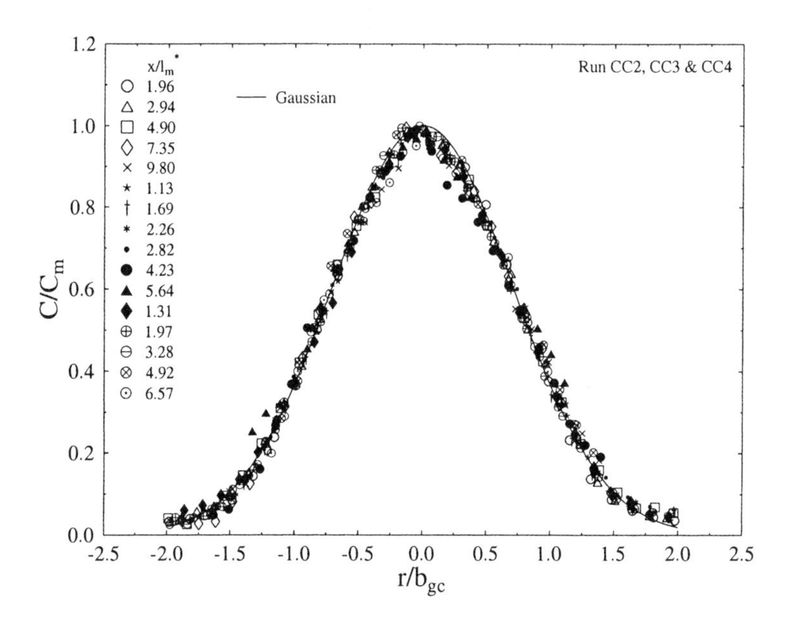

Figure 6.4. Radial profile of time-averaged concentration in a coflow jet

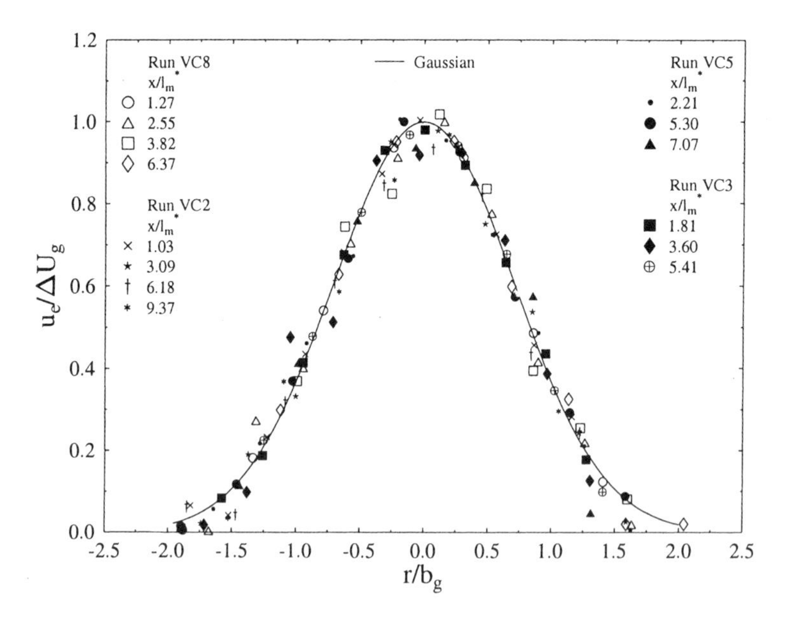

Figure 6.5. Measured radial profile of time-averaged excess velocity in a coflow jet

width b_g and the excess velocity ΔU_g, so that the dilution or the tracer concentration in the coflowing jet can be determined.

2.1 THE NATURAL BUT INCORRECT FORMULATION

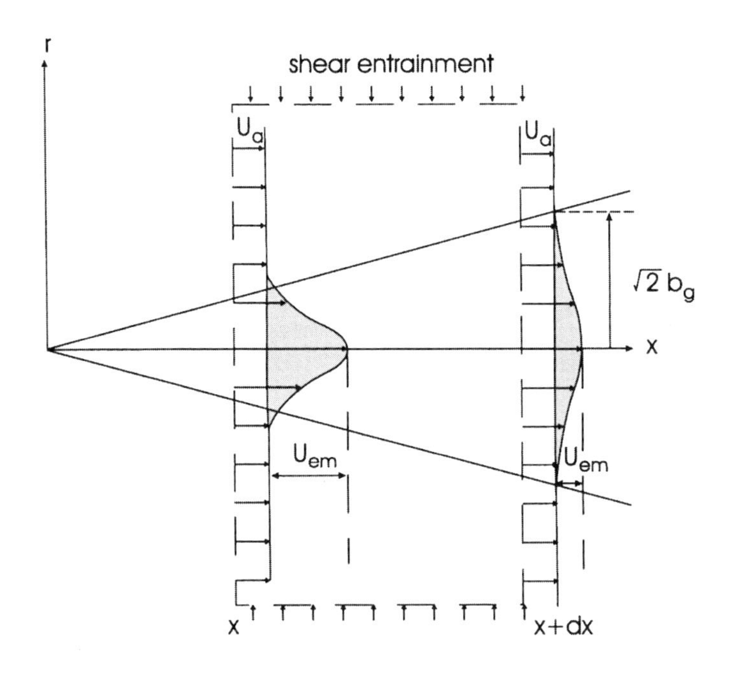

Figure 6.6. Control volume formulation of a coflow jet

The modelling of a jet in a current is considerably more complicated than the stagnant case. The presence of the ambient current introduces a number of subtleties and difficulties that can be illustrated using the coflowing jet. Building on the theory of jets and plumes, one can formulate the mass and momentum balances for a control volume bounded between x and $x + dx$, as shown in Figure 6.6. It seems reasonable to write:

$$\frac{dQ}{dx} = \frac{d}{dx}\int u dA = Q_e \tag{6.4}$$

$$\frac{dM}{dx} = \frac{d}{dx}\int u^2 dA = Q_e U_a \tag{6.5}$$

where Q, M are the jet volume and momentum flux, and Q_e is the entrainment flow. The above simply relates the increase in jet volume flux due to turbulent entrainment, and the change in jet momentum flux resulting from the addition of momentum due to the entrained ambient flow. However, we note that the jet volume flux Q is not a bounded integral; hence some definition of jet edge is required. It is customary to define $r = \sqrt{2}b_g = B$ (where B is the top-hat half-width) as the edge of the jet. It is also reasonable to assume in accordance with the entrainment hypothesis that the turbulent entrainment is proportional to the centerline jet excess velocity $u_{em} = \Delta U_g$. After substituting the volume and momentum flux in terms of centerline variables of the Gaussian distribution, we then have (Prob.6.2):

$$\frac{d}{dx}\int_0^{\sqrt{2}b_g} u dA = \frac{d}{dx}(\pi b_g^2(u_{em} + 2U_a)) = 2\pi\alpha b_g u_{em} \tag{6.6}$$

$$\frac{d}{dx}\int_0^{\sqrt{2}b_g} u^2 dA = \frac{d}{dx}(\frac{\pi}{2}b_g^2(u_{em} + 2U_a)^2) = 2\pi\alpha b_g u_{em} U_a \tag{6.7}$$

In a coflow of uniform velocity, it can be shown that the excess momentum flux is conserved (Prob. 3.14 or Chu and Lee 1996).

$$M_e = \int_0^{\infty} u_e(u_e + U_a)2\pi r dr = \frac{\pi}{2}b_g^2 u_{em}(u_{em} + 2U_a) \tag{6.8}$$

is constant and equal to the source excess momentum flux $M_{eo} = (U_o - U_a)U_o A_o$, where $u_{em} = U_g - U_a = \Delta U_g$ is the centerline excess velocity, and $A_o = \pi D^2/4$ the cross-sectional area of the nozzle.

The plausibility of the above commonly used integral model can be examined. First, the entrainment hypothesis as formulated includes the advective flux due to the ambient flow U_a as part of the turbulent entrainment, and is hence physically unreasonable. As shown in Figure 6.6, it seems clear the entrainment due to velocity shear should affect only the excess volume flux $\Delta Q = \int u_e dA$. That is, the entrainment hypothesis should be modified to:

$$\frac{d}{dx}\int_0^{\sqrt{2}b_g} u_e dA \approx \frac{d}{dx}\int_0^{\infty} u_e dA = \frac{d}{dx}(\pi b_g^2 u_{em}) = 2\pi\alpha b_g u_{em} \tag{6.9}$$

Adopting this entrainment hypothesis then, by expanding Equation 6.7 and making use of Eq. 6.9, it can be shown that

$$\frac{d}{dx}M_e = -4U_a^2\frac{db_g}{dx} \tag{6.10}$$

Since $\frac{db_g}{dx}$ is in general not zero, the above formulation clearly does not conserve jet excess momentum. What is at fault here? The terms on the RHS reflects the increase in jet momentum flux due to the fact the jet is expanding - i.e. the ambient advection into the conical surface of the differential control volume has been neglected in Equation 6.7

(Schatzmann 1979). Only with the integration limits properly taken into consideration is the excess momentum conservation recovered. The correct equations for the coflow jet are then Equations 6.9 and 6.8 respectively.

We see that an apparently innocent formulation has resulted in many problems for the coflowing jet. Many integral models (e.g. Fan 1967; Abraham 1970; and Hirst 1972 upon which the USEPA model DKHDEN is based) were formulated heuristically using a control volume approach as in the above Equation 6.6. Consequently they do not model shear entrainment correctly nor satisfy excess momentum conservation; they cannot be expected to predict the coflowing jet well. In addition, there is the key question of the entrainment coefficient. Is it a constant as in the simple jet? It turns out the entrainment assumption with a constant shear entrainment coefficient fails miserably for the coflowing jet.

2.2 JET SPREADING HYPOTHESIS

We adopt a general spreading hypothesis (Chu and Lee 1996): the change in width of the shear layer, *in a Lagrangian frame of reference (moving with the eddies),* is assumed to be proportional to the relative velocity between the jet element and its surroundings.

Let $\tilde{u}$ be the characteristic jet velocity and U_a be the ambient velocity. For the steady jet in coflow, the spreading hypothesis reduces to a particularly simple form:

$$\frac{D\tilde{b}}{Dt} = |\tilde{u}|\frac{d\tilde{b}}{dx} = \beta|\tilde{u}_e| \tag{6.11}$$

where t and x represent time and distance along jet trajectory, $\tilde{b}$ is the characteristic half-width, $\tilde{u_e} = \tilde{u} - U_a$ is the characteristic jet excess velocity, and β is the proportionality constant.

If β is a universal constant, this relation must hold for $U_a = 0$. This implies that $\beta = \beta_s$, where β_s is the jet spreading rate in stagnant ambient (e.g. Fischer *et al.*1979, Wood *et al.*1993); the exact value of β_s, however, depends on the choice of velocity profile in the integral model. At the outset, it is not clear what characteristic velocity should be adopted in Equation 6.11. The centerline excess jet velocity has been used in some studies (Antonia and Bilger 1974, Knudsen 1988, Wood *et al.*1993), while recent works (Wright 1994, Chu 1994) suggest that the average excess velocity may be more appropriate. In our notation, and supported by the observations, the physics of the flow, such as transport of mass and momentum, can be characterized by the ensemble average properties of a coherent dominant eddy structure consisting of a range of turbulent eddies which govern the mixing with the irrotational ambient fluid. To facilitate this concept, the 'top-hat' profile is employed to represent the flow phenomenon and the average velocity is then considered as the characteristic velocity. This choice of characteristic velocity must ultimately be tested against experiments.

The excess velocity can be expressed in term of a 'top-hat' profile:

$$u_e = \begin{cases} \Delta U & \text{if } r \leq B \\ 0 & \text{otherwise} \end{cases}$$

where ΔU, and B are the excess velocity and half-width of an equivalent jet with a sharp boundary and uniform velocity, $\Delta U + U_a$, carrying the same mass flow and excess momentum flux as the actual jet. By equivalence of excess volume flux (the part of the volume flux due to turbulent entrainment), $\int u_e dA$, and *excess* momentum flux $M_e = \int u u_e dA$, the following relations between the two profiles can be obtained (Problem 6.3):

$$\pi \Delta U B^2 = \pi \Delta U_g \, b_g^2 \tag{6.12}$$

$$\pi(\Delta U + U_a)\Delta U B^2 = \frac{\pi}{2}\Delta U_g^2 \, b_g^2 + \pi U_a \, \Delta U_g \, b_g^2 \tag{6.13}$$

Thus, we have,

$$\Delta U = \frac{\Delta U_g}{2} \tag{6.14}$$

$$B = \sqrt{2}\, b_g \tag{6.15}$$

For total tracer mass flux, we have

$$\Gamma = \int_0^\infty C(u_e + U_a) 2\pi r dr$$

$$= \int_0^\infty (\Delta U_g \, e^{-(r/b_g)^2} + U_a) C_g \, e^{-(\frac{r}{\lambda b_g})^2} 2\pi r dr$$

$$= \pi \lambda^2 \, b_g^2 \, C_g (U_a + \frac{1}{1+\lambda^2}\Delta U_g) \tag{6.16}$$

By conservation of tracer mass flux, $\Gamma = \Gamma_o = C_o U_o A_o$, where C_o is concentration at the source. The centerline dilution S_c can hence be determined from Equation 6.16,

$$S_c = \frac{C_o}{C_m} = \frac{\lambda^2 \, \pi B^2 \, (U_a + \frac{2}{1+\lambda^2}\Delta U)}{2 \; Q_o} \tag{6.17}$$

where $Q_o = U_o A_o$ is the source volume flux. Note that the above equation only requires knowledge of the average flow properties (ΔU and B). In Eq. 6.16 the turbulent mass transport $\int u'c'dA$ (of the order of 10 percent of the total mass flux) has tacitly been ignored. This assumption is justified primarily by comparison of the predicted mean flow quantities with data.

2.3 GOVERNING EQUATIONS

Adopting the top-hat half-width $B = \sqrt{2}\, b_g$ and excess velocity $\Delta U = U - U_a = \Delta U_g/2$ as the characteristic length and velocity scales (Figure 2a), the jet spreading hypothesis (Equation 6.11) can be rewritten as,

$$\frac{DB}{Dt} = |U_a + \Delta U| \frac{dB}{dx} = \beta_s |\Delta U| \tag{6.18}$$

With the well-known spreading rate $db_g/dx \simeq 0.109$ of pure jet in stagnant ambient, the value of β_s can be taken as $\sqrt{2} \times 0.109 \simeq 0.154$ in our definition of jet edge $B = \sqrt{2}\, b_g$.

The conservation of excess momentum flux M_e can be expressed as (for ZEF):

$$\begin{aligned} M_{eo} = M_e &= \int u(u - U_a)\, dA \\ &= \pi B^2 U \Delta U = \pi B^2 (\Delta U + U_a) \Delta U \end{aligned} \tag{6.19}$$

Equation 6.19 can be rewritten as

$$(\frac{\Delta U}{U_a})^2 + (\frac{\Delta U}{U_a}) - \frac{l_m^{*2}}{\pi B^2} = 0 \tag{6.20}$$

where $l_m^* = M_e^{1/2}/U_a$ is the excess momentum length scale.

Introducing the dimensionless variables $U^* = \Delta U/U_a$, $B^* = B/l_m^*$ and $x^* = x/l_m^*$, the momentum equation (Eq. 6.20) and the spreading hypothesis (Eq. 6.18) become

$$\left. \begin{aligned} U^{*2} + U^* - \frac{1}{\pi B^{*2}} &= 0 \\ \frac{dB^*}{dx^*} = \beta_s \frac{\Delta U}{\Delta U + U_a} &= \beta_s \frac{U^*}{1 + U^*} \end{aligned} \right\} \tag{6.21}$$

This pair of equations, subjected to initial conditions, can then be numerically integrated using a 5th order Runge-Kutta method to obtain the solution for $\Delta U(x)$ and $B(x)$ at different downstream locations from the source. The centerline dilution S_c is then determined from Equation 6.17.

2.4 APPROXIMATE INITIAL CONDITIONS

For strong jets, $x/l_m^* \leq O(10)$, the initial spreading rate of the jet at exit can be expressed as,

$$\frac{dB}{dx} = \beta_s \frac{\Delta U}{\Delta U + U_a} = \beta_s (1 - U_a/U_o) \simeq \frac{D}{2x_o} \tag{6.22}$$

where x_o is the virtual origin of the jet. As a first approximation the initial conditions at the jet nozzle exit relative to the virtual origin can then be expressed as,

$$\left.\begin{aligned} \Delta U_o^* &= \tfrac{\Delta U_o}{U_a} \\ B_o^* &= \tfrac{D}{2l_m^*} \\ x_o^* &= \frac{-D}{2\beta_s\left(1 - U_a/U_o\right)} \end{aligned}\right\} \tag{6.23}$$

Strictly speaking, the above governing equations are only applicable to the region after the jet flow is fully developed and the velocity and tracer concentration profiles become self-similar, i.e. $x > x_e$. Alternatively, the initial conditions of the zone of established flow can be obtained as a solution of a model for the zone of flow establishment. Within the ZFE ($x \leq x_e$), the potential core with constant velocity $U = U_o$ is bounded by the free shear layer developed between the jet and the ambient. The spreading of this mixing layer and the location where the mixing layer intersects the jet axis, x_e, can also be predicted using the same spreading hypothesis. After determining the length of potential core x_e, the integration of the governing equations for the ZEF can then be performed starting from this location.

2.5 PREDICTION OF POTENTIAL CORE LENGTH

Referring to Figure 6.1, let r and R be the half-width of the potential core and location of the edge of the shear layer in the ZFE respectively. The characteristic half-width $\tilde{b}$ of the mixing layer can then be expressed as $R - r$. In the ZFE, the average velocity $\tilde{u}$ of the mixing layer can be taken as the average of the initial jet velocity U_o and ambient velocity U_a, i.e. $\tilde{u} = (U_o + U_a)/2$, and the average excess velocity is chosen as the characteristic excess velocity of the mixing layer $\tilde{u}_e$, i.e. $\tilde{u}_e = \tilde{u} - U_a = (U_o - U_a)/2$.

Applying the spreading hypothesis (Equation 6.11) and integrating from $x = 0$ to x_e, we have

$$\int_{x=0}^{x=x_e} d(R - r) = \int_0^{x_e} \beta_s \frac{U_o - U_a}{U_o + U_a}\, dx \tag{6.24}$$

By conservation of excess momentum, we have

$$M_e = \pi \frac{D^2}{4} U_o (U_o - U_a)$$

$$= \pi r^2 U_o (U_o - U_a) + \pi (R^2 - r^2) \frac{U_o + U_a}{2} \frac{U_o - U_a}{2} \tag{6.25}$$

Imposing the boundary condition at $x = x_e$, i.e. at the end of the potential core, $r = 0$, we have,

$$\frac{R}{D} = \frac{1}{\sqrt{1 + U_a/U_o}} \tag{6.26}$$

Substitute into Equations 6.24, we have

$$\frac{x_e}{D} = \frac{\sqrt{1 + U_a/U_o}}{\beta_s (1 - U_a/U_o)} \tag{6.27}$$

It can be seen that the length of the potential core increases with the ambient current to jet velocity ratio; as $U_a/U_o \to 1$, $x_e/D \to \infty$, i.e. spreading rate of the mixing layer is very slow. For the limiting case of stagnant ambient, $U_a/U_o = 0$, using $\beta_s = 0.154$, the length of the potential core is determined to be $x_e = (1/\beta_s)D \simeq 6.5D$ - which agrees with the generally accepted value of $\simeq 6.2D$ (Fischer *et al.*1979).

2.6 ALTERNATIVE DEFINITION OF CHARACTERISTIC VELOCITY

Alternatively, one is often tempted to choose the centerline excess velocity ΔU_g of the Gaussian profile as the characteristic excess velocity (Antonia and Bilger 1974, Knudsen 1988). The jet spread hypothesis, Equation 6.11, can then be written as,

$$\frac{dB}{dx} = \beta_s \frac{\Delta U_g}{U_a + \Delta U_g} \tag{6.28}$$

Note that when the excess velocity is much less than the ambient velocity, i.e. in the weak jet regime with $\Delta U \ll U_a$, this equation implies a spreading rate,

$$\frac{dB}{dx} \simeq \beta_s \frac{\Delta U_g}{U_a} = \beta_s \frac{2\Delta U}{U_a} \tag{6.29}$$

that is **twice** that given by Equation 6.18 or 6.21.

3. ASYMPTOTIC SOLUTIONS: STRONG AND WEAK-JET

The asymptotic properties of two limiting cases may be considered. The effect of the coflow velocity can be ignored in the near field of a strong jet where $\Delta U >> U_a$. The initial development of the strong jet is expected to be close to the case of the jet in stagnant fluid. The excess velocity reduces rapidly with the distance from the source; both the jet spread and excess velocity decay are governed by the pure jet solution if the jet momentum flux is replaced by the excess momentum flux M_{eo}. The width and velocity of the coflow jet in the near field (strong jet) region, $x/l_m^* \leq 10$, are then given by the following in terms of dimensionless scales (see Equations 2.2):

$$\frac{b_g}{l_m^*} = 0.114(\frac{x}{l_m^*}) \tag{6.30}$$

$$\frac{\Delta U_g}{U_a} = 7.0(\frac{x}{l_m^*})^{-1} \tag{6.31}$$

For a strong jet, the solution is assumed to be the same as those for the jet in stagnant environment.

However, far from the source (the strongly-advected region) where ΔU $<< U_a$, the effect of the excess velocity ΔU may be ignored except its contribution to the spreading process. The asymptotic properties of the weak jet are as follows. Let $(U_a + \Delta U) \simeq U_a$, the momentum equation becomes

$$U_a \Delta U \pi B^2 = M_{eo}, \tag{6.32}$$

The spreading equation is

$$U_a \frac{dB}{dx} = \beta \Delta U. \tag{6.33}$$

Multiplying both side of the spreading equation by $U_a \pi B^2$,

$$U_a^2 \pi B^2 \frac{dB}{dx} = \beta U_a \Delta U \pi B^2 = \beta M_{eo} \tag{6.34}$$

The equation can be integrated once with respect to x to give a one-third power law

$$B = (\frac{3\beta}{\pi})^{\frac{1}{3}} [\frac{M_{eo}}{U_a^2}]^{\frac{1}{3}} x^{\frac{1}{3}} \tag{6.35}$$

for the spreading of the weak jet in the far field. The corresponding excess velocity follows the minus two-third power law

$$\frac{\Delta U}{U_a} = [9\beta^2\pi]^{-\frac{1}{3}}[\frac{M_{eo}}{U_a^2}]^{\frac{1}{3}}x^{-\frac{2}{3}} \qquad (6.36)$$

The use of $\beta = \sqrt{2}db_g/dx = \sqrt{2}(0.114) = 0.16$ for the spreading rate and using the relation between Gaussian and top-hat variables results in:

$$b_g = 0.38[\frac{M_{eo}}{U_a^2}]^{\frac{1}{3}}x^{\frac{1}{3}} \qquad (6.37)$$

$$\frac{\Delta U_g}{U_a} = 2.23[\frac{M_{eo}}{U_a^2}]^{\frac{1}{3}}x^{-\frac{2}{3}} \qquad (6.38)$$

Alternatively, the adoption of $\beta = 0.17$ based on velocity intermittency data (Chu 1994) would have resulted in almost the same coefficients for the above equations (0.39 and 2.14 respectively). The one-third power law for the width and the minus-two-third power law for the excess velocity are in good agreement with experimental data (see later discussion and Figures 6.7 and 6.8). When normalized by the length and velocity scales, the coflow jet solution in the far field then becomes:

$$\frac{b_g}{l_m^*} = 0.385(\frac{x}{l_m^*})^{\frac{1}{3}} \qquad (6.39)$$

$$\frac{\Delta U_g}{U_a} = 2.14(\frac{x}{l_m^*})^{-\frac{2}{3}} \qquad (6.40)$$

4. COMPARISON OF THEORY WITH EXPERIMENTAL DATA

4.1 JET SPREADING RATE

Although the shape of the concentration contour and the mixing characteristics are similar to those of a pure jet, the jet spread is, however, not linearly related to the distance x. For the same discharge, the half-width of coflowing jet is less than that of the pure jet and depends on the ambient to jet velocity ratio $R' = U_a/U_o$. Figure 6.7 shows the comparison of theory (Equation 6.21) with the measured concentration half-width b_{gc} in the LIF experiments of Chu (1996) and previous studies. The normalized results of different R' generally collapse on the same curve suggesting that l_m^* is the correct length scale. The present results using LIF techniques are comparable with the experimental results of Knudsen (1988) and McQuivey *et al.*(1971) using probe-based techniques. It can also be seen that the jet in coflow can be well-predicted by the present spreading hypothesis using the top-hat velocity as the characteristic velocity (Equation 6.18), especially in the strong jet region with $x/l_m^* \leq 10$. Inclusion of the potential core does not result in significant difference in the results; although the initial phase is better predicted. On the other hand, using the centerline maximum velocity as the characteristic velocity (Equation 6.28) results in over-prediction of the jet half-width. The data give support to a suggestion of Wright (1994) on using the top-hat velocity ΔU as the velocity scale.

Using $\lambda = b_{gc}/b_g$, the velocity half-width b_g obtained from the detailed point velocity measurements (Section 5.3) can be converted into concentration half-width $b_{gc} = \lambda b_g$, with $\lambda = 1.19$. Figure 6.8 shows the inferred b_{gc} against the numerical prediction. The predicted b_{gc} compares well with the data, as well as those results determined directly using LIF technique (Figure 6.7). This gives further support to the integral model and the use of $\lambda \simeq 1.2$.

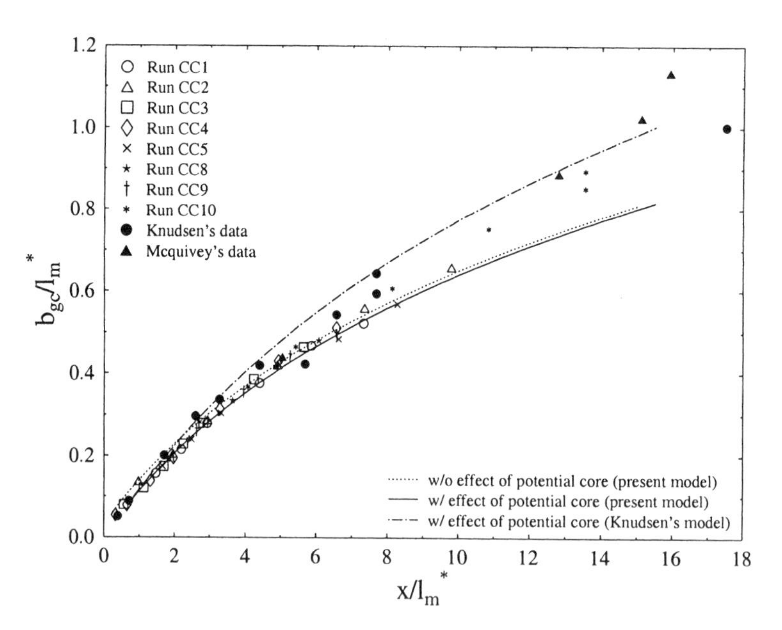

Figure 6.7. Predicted concentration half-width of coflow jet versus horizontal distance (empty symbols: data of Chu 1996)

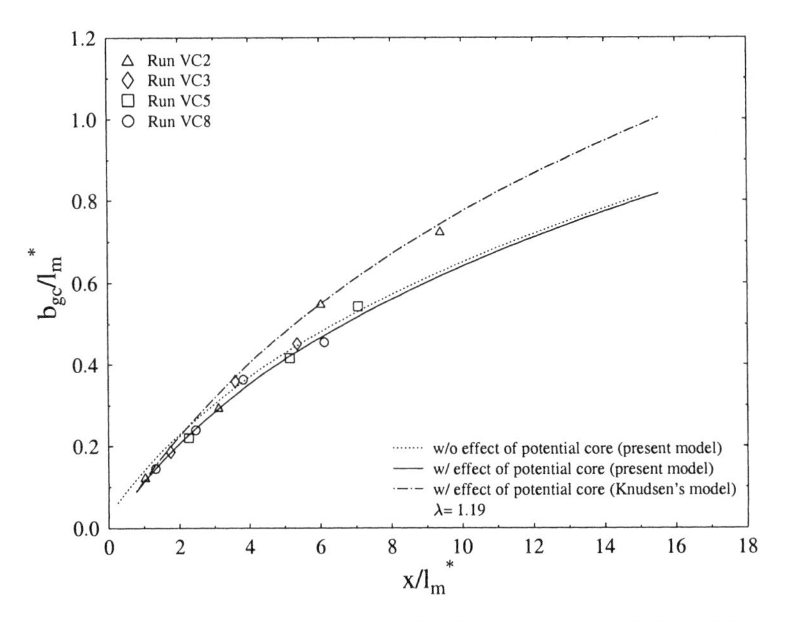

Figure 6.8. Predicted concentration half-width of coflow jet (inferred from velocity) versus horizontal distance

4.2 CENTERLINE DILUTION

In the strong jet region, by dimensional analysis, the centerline dilution can be written as $S_c(x) = C_o/C_m(x) \sim (M_e^{1/2}/Q_o)\,x$. Or equivalently,

$$S_c(x)\,\frac{R'}{1-R'} = k\frac{x}{l_m^*} \tag{6.41}$$

where k is the proportionality constant. The normalized measured centerline dilution $S_cR'/(1-R')$ is plotted against the normalized distance x/l_m^* in Figure 6.9. The results are comparable with the data of Knudsen (1988) and McQuivey *et al.*(1971). In the strong jet region, say $x/l_m^* \leq 5$, the normalized dilution is found to vary linearly with x/l_m^*, with $k = 0.174$; this is almost the same as the predicted value of 0.170 and the value inferred from the data of Ayoub (1971) by Lee (1989). The predicted centerline dilution using the present hypothesis (Equation 6.41) is well supported by the experimental results. However, using the alternative spreading hypothesis (Equation 6.28) results in overprediction of the centerline dilution - giving $k = 0.221$. It is interesting to note that despite the narrower jet spread in a coflow, the centerline dilution is only slightly less than the linear dilution predicted by the straight jet in stagnant ambient; this is apparently related to the slower velocity decay in a coflow.

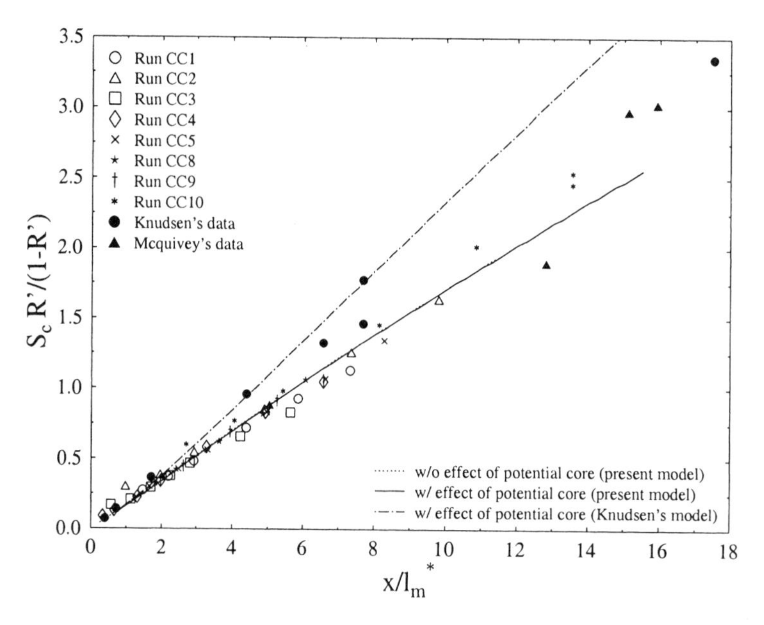

Figure 6.9. Centerline dilution of coflow jet versus downstream distance

4.3 CENTERLINE EXCESS VELOCITY DECAY

The ratio of the centerline excess velocity over ambient velocity $\Delta U_g/U_a$ can be determined directly from LDA velocity measurements. The normalized excess velocity is plotted against x/l_m^* in Figure 6.10 along with the data of McQuivey (1971) and the data of Nickels and Perry (1996). Alternatively, by conservation of excess momentum, $\Delta U_g/U_a$ can be determined indirectly from the concentration half-width (and using $\lambda = 1.2$). The centerline velocity excess inferred from the LIF data and previous concentration data are also shown as a comparison. It can be seen the present data are consistent with previous studies, and the theory is well-supported by the experiments. The prediction follows closely the asymptotic relation in the strong jet region (cf Table 2.3), $\Delta U_g/U_a = 7.0\ M_{eo}^{1/2} x^{*-1}$, as well as the -2/3 law in the weak jet region; the best fit of Nickel and Perry's data (x/l_m^* up to 60), $\Delta U_g/U_a = 2.67\ M_{eo}^{1/2} x^{*-2/3}$, is in reasonable agreement with the present spreading hypothesis. Inclusion of the potential core also does not introduce any significant difference in the result. On the whole, the predicted excess velocity using $\beta \approx 0.16 - 0.17$ is about 20 percent lower than the observed values in Nickel and Perry's experiments. At present, there is no concentration data in the weak-jet region; the experiments have to be done by a towed jet, as the weak-jet behaviour is expected to be influenced by the ambient turbulence in the coflow. The present results seem to suggest a slightly lower spread rate in the weak-jet or wake region, consistent with related studies.

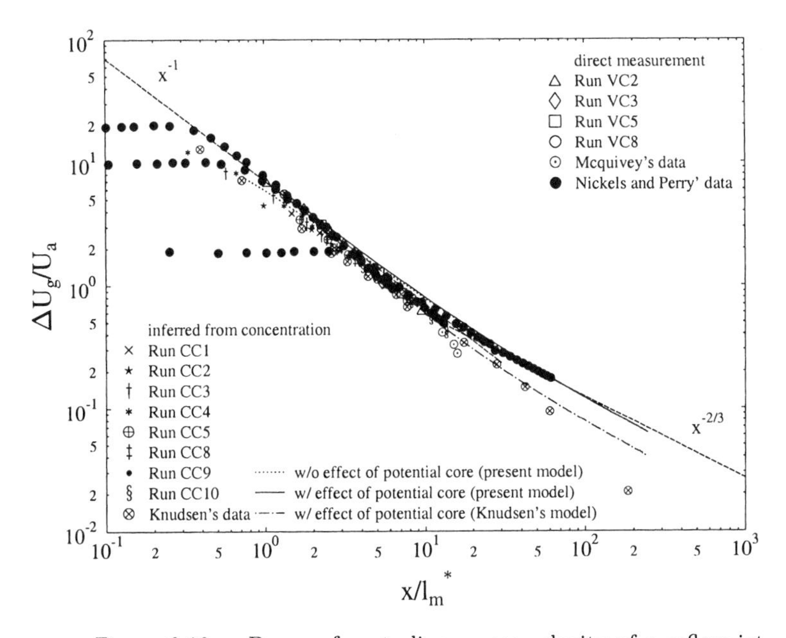

Figure 6.10. Decay of centerline excess velocity of a coflow jet

4.4 THE ENTRAINMENT COEFFICIENT

In many general integral models, the entrainment hypothesis is used. In view of the success of the spreading hypothesis, it is worthwhile to use the result to determine the appropriate entrainment coefficient to be used for the coflowing jet. Using the entrainment hypothesis, the problem can be formulated in terms of top-hat variables as:

$$\frac{d}{dx}(\pi B^2(U - U_a)) = 2\pi B\alpha(U - U_a) \tag{6.42}$$

$$\frac{d}{dx}(\pi B^2 U(U - U_a)) = 0 \tag{6.43}$$

Since $U\frac{dB}{dx} = \beta_s \Delta U = \beta_s(U - U_a)$, it can be shown from the above (Problem 6.4) that:

$$\alpha = \beta_s \frac{\Delta U^2}{U(U + \Delta U)} \tag{6.44}$$

It is seen from Equation 6.44 the entrainment coefficient for a coflowing jet exhibits even greater variability than a buoyant jet in stagnant fluid. For a strong jet, $\Delta U \to U$, $\alpha \to \beta_s/2$ (cf Table 2.3). For a weak-jet, we have $\Delta U \approx 0$, and $\alpha \to 0$. Thus for this basic flow, the entrainment coefficient varies continually along the jet in accordance with the jet velocity; α changes from the stillwater value, 0.057, for a strong jet, and approaches zero in the far field weak jet region. On the other hand, with the Lagrangian spreading hypothesis and the same "universal" spreading coefficient, the simple formulation predicts the jet properties very well in the strong jet region; the agreement is also satisfactory in the far field.

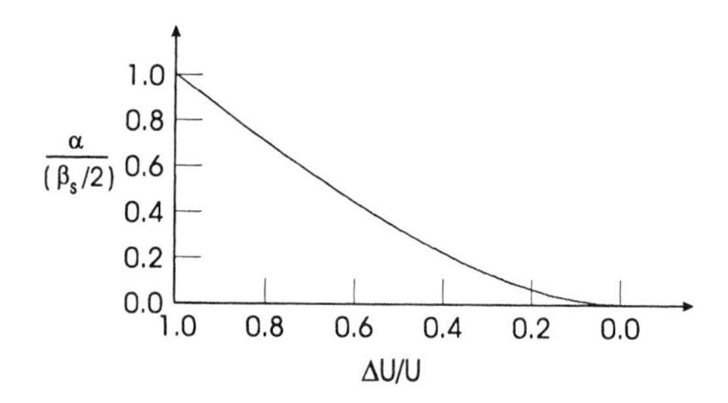

Figure 6.11. Entrainment coefficient of a coflow jet

5. CORRELATION OF MODEL RESULTS WITH EXPERIMENTS

In adopting a top-hat profile, the half-width and average velocity represent the characteristic properties of the dominant eddies. However, in practice, these average properties must be correlated with the experimental observations such as intermittency, visual width, centerline concentration and minimum concentration. Although some of these correlations can be theoretically established based on the observed profile distribution, model validation is necessary.

5.1 VISUAL BOUNDARY

5.1.1 INTERMITTENCY AND TOP-HAT EDGE

The intermittency γ at a point is the proportion of time in which that location is occupied by turbulence; it reflects the fluctuation of the turbulent/non-turbulent boundary. Figure 6.12 shows the intermittency contours at different downstream locations x. The results show good symmetry and similar features as the pure jet case (not shown). The radial variation of γ is shown in Figure 6.12. The intermittency has a peak value of 1 at centerline. It generally stays at about 1 in the central core region, $r/b_{gc} <\simeq 0.6$, and decreases to about zero at $r/b_{gc} >\simeq 1.70$. The size of this core seems to reduce as x/l_m^* increases; this means the ambient fluid is able to penetrate the core more as the turbulence grows in size downstream.

Figure 6.13 shows the comparison between the concentration half-width b_{gc} and the half-width b_i defined at 50% intermittency as a function of downstream distance x/l_m^* for both the coflowing jet and the pure jet (Run CC6 & CC7) cases. The ratio of b_i/b_{gc} generally remains fairly constant for $x/l_m^* < 10$, with an average of about 1.16 and 1.26 respectively; i.e. on average $b_i/b_{gc} \simeq 1.2$. Recall that our top-hat edge $B = \sqrt{2}\, b_g \simeq 1.2 b_{gc}$, (using $\lambda = 1.2$); this suggests the top-hat edge is a close representation of the 50% intermittency contour for pure jet or coflowing jet.

5.1.2 VISUAL BOUNDARY

The LIF image without any further processing can be considered very similar to the normal flow visualization using simple dye injection technique. The visual boundary of the jet region can be identified in the LIF image and extracted by digital image processing techniques (Lee and Chu 1995); it can be defined by a threshold gray level selected based on visual judgement of thresholding an *instantaneous* image against its visual boundary. On the other hand, for a Gaussian profile, $C = C_m e^{-(\frac{r}{\lambda b_g})^2}$, our edge $B = \sqrt{2}\, b_g$ is equivalently defined at concentration $C_m e^{-1.389} \simeq 0.25 C_m$, using $\lambda \simeq 1.2$. Figure 6.14 shows the comparison of the visual boundary and the edge defined at $0.25C_m$ contour for two typical, randomly selected instantaneous LIF images of the coflowing jet, where $C_m = C_g$ is the maximum concentration. The results show excellent agreement between the visually perceived boundary of turbulent region and the $0.25C_m$ contour (i.e. the top-hat edge). Additional examples can be found in Chu (1996).

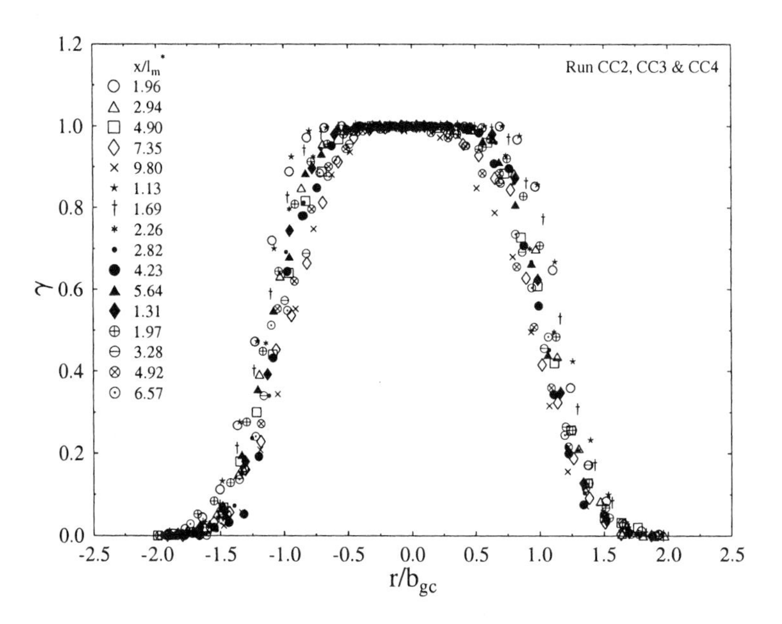

Figure 6.12. Radial profile of concentration intermittency

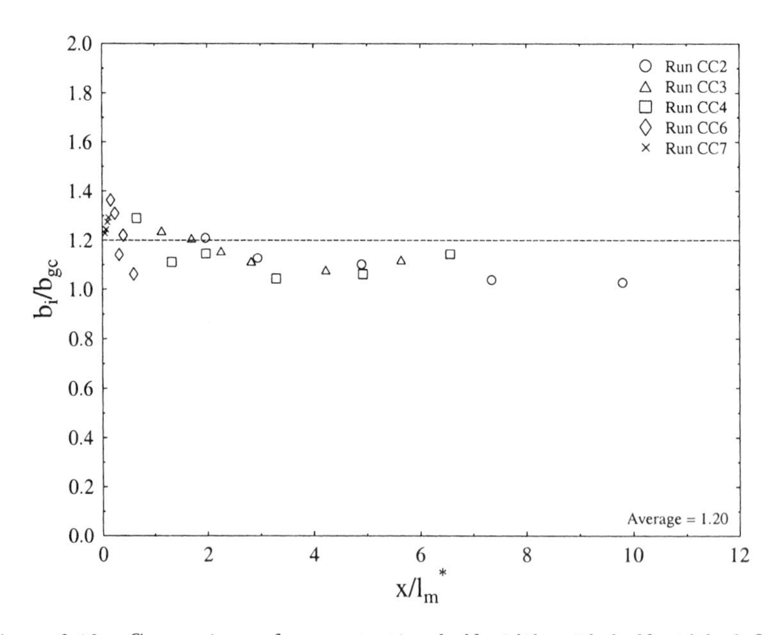

Figure 6.13. Comparison of concentration half-width with half-width defined at 50 percent intermittency

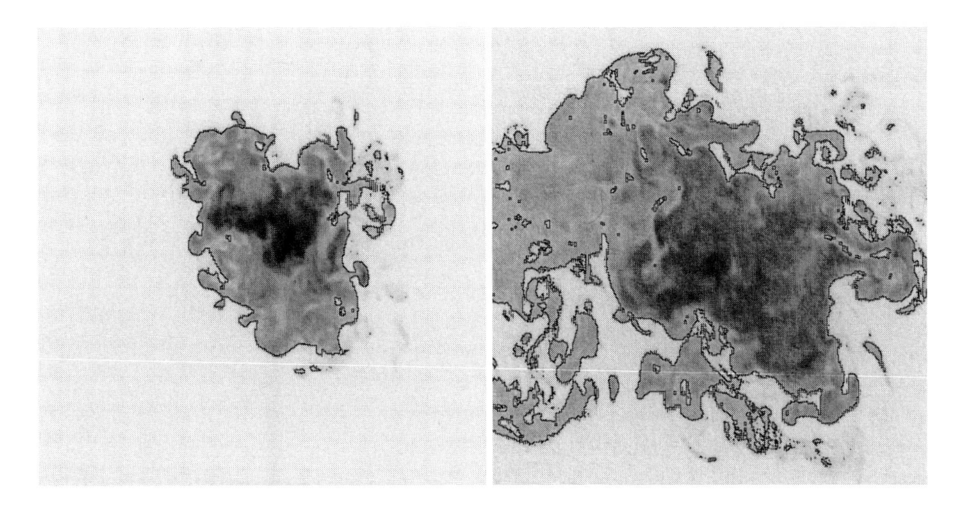

Figure 6.14. Comparison of visual boundary with top-hat boundary in a coflow jet: Run CC2: x/D=49 (left) and x/D=98 (right)

5.2 FLOW-WEIGHTED AVERAGE DILUTION

The flow-weighted average concentration $C_{avg} \simeq Q_o C_o / Q$, where $Q = \int u\, dA$ is the total volume flux, can be defined as $C_{avg} = \int uc\, dA / \int u\, dA$. The flow-weighted average dilution $S_{avg} = C_o/C_{avg}$ can be determined from the integration of the measured scalar concentration and velocity data across the section within the jet boundary. The results are plotted against distance $x/(Q_o/M_e^{1/2})$ (Figure 6.15), and suggest a linear relationship between S_{avg} and $x/(Q_o/M_e^{1/2})$, with a slope of 0.290. The experimentally determined ratio of average to centerline dilution $S_{avg}/S_c = C_m/C_{avg}$ is shown in Figure 6.16. The results show a constant ratio of about 1.66, roughly the same as that of the pure jet in stagnant ambient - for a Gaussian concentration and velocity distribution the theoretical value is $(1+\lambda^2)/\lambda^2 \simeq 1.69$.

EXAMPLE 6.1 *A circular jet discharges an industrial effluent from a pipe of 0.4 m diameter at 3 m/s into a coflowing river velocity of 0.2 m/s. The river is about 10 m deep, and the jet is located at about mid-depth. Compute the variation of jet width, excess jet velocity, and centerline dilution for the first 100 metres downstream of the jet.*

Solution: For this jet, the initial volume flux is $Q_o = 3 \times \pi \times 0.4^2/4 = 0.377\ \mathrm{m}^3/\mathrm{s}$; the excess momentum flux is $M_{eo} = Q_o(U_o - U_a) = 0.377 \times$

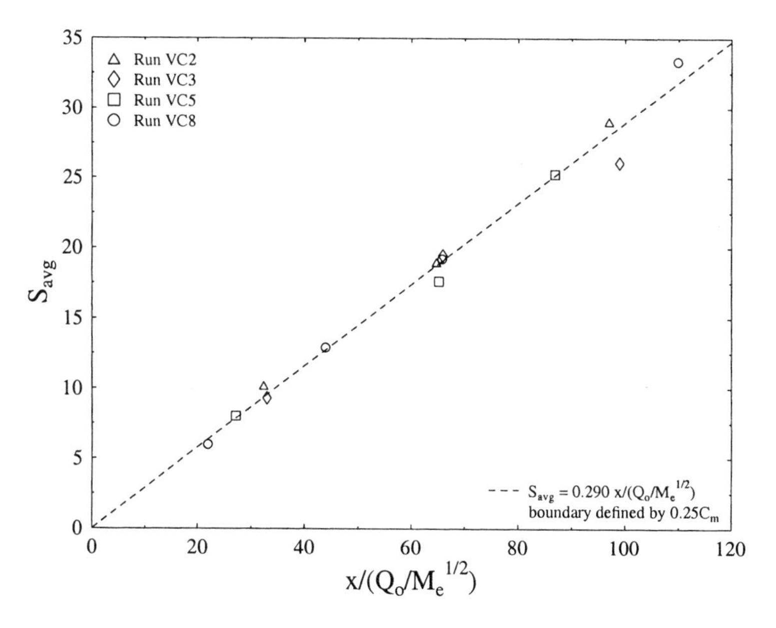

Figure 6.15. Flow-weighted average dilution of a coflow jet

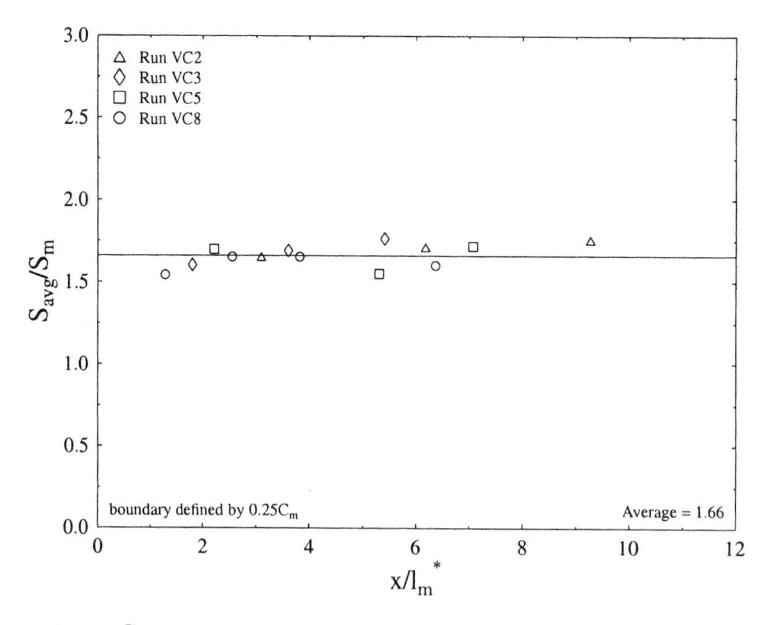

Figure 6.16. Comparison of average and centerline minimum dilution of a coflow jet

$(3 - 0.2) = 1.056$ m^4/s^2. The excess momentum length scale can then be computed as $l_m^* = {M_{eo}}^{0.5}/U_a = \sqrt{1.056}/0.2 = 5.14\ m$. Thus it can be expected that for distances much greater than 25 m ($\approx 5l_m^*$), the jet will be strongly advected by the coflow. Figure 6.17 shows the computed variation of the top-hat half-width, the centerline velocity excess, and the centerline dilution with distance downstream. It is seen that at $x = 100$ m, the half-width is quite narrow ($B = \underline{5.6}$ m) compared with the straight jet value (17m). However, the absolute coflow jet velocity is greater. It can be seen the excess velocity decay is less rapid than the strong jet. At $x = 100$ m, the velocity excess $\Delta U_g = 0.045$ m/s is small compared with the ambient velocity, while the absolute centerline jet velocity is $u_m = 0.245$ m/s. The centerline dilution is $S_c = \underline{43.8}$, only slightly less than the corresponding strong jet value of $S_c = 47.5$. Wastewater effluents are often discharged as nearly non-buoyant coflowing jets into rivers.

5.3 SUMMARY OF EXPERIMENTAL DATA

Table 6.1 shows a summary of time-mean as well as key turbulence properties of the scalar field of a coflow jet. Data on jet in stagnant fluid ($x/l_m^* \to 0$ limit of coflow jet) , or momentum-dominated regime of a vertical buoyant jet (Papanicolaou and List 1987, 1988; Papantonious and List 1989) are also included. It can be seen the instantaneous maximum and minimum concentrations (C_{max}, C_{min}) can differ by more than 50 percent from the time averaged value; the concentration turbulent intensity C_{rms}/C_m is around 0.2, where the root-mean-square concentration $C_{rms} = \sqrt{\overline{c'^2}}$, and $c' = c - \overline{c}$ is the concentration fluctuation. The variabilities in the turbulence properties data reflect the different experimental techniques used and also the limited number of profiles measured. Further details on other statistics such as skewness skewness factor $= \overline{c'^3}/(\overline{c'^2})^{3/2}$ and flatness factor $= \overline{c'^4}/(\overline{c'^2})^2$ can be found in Chu (1996).

6. SUMMARY

The excess momentum of a jet in coflow is conserved. Using a Lagrangian jet spreading hypothesis, an integral model of the jet in coflow has been formulated and validated against experiments. It is demonstrated that the jet spreading hypothesis using the top-hat velocity as the characteristic velocity can be effectively applied to model the time-averaged flow. Theoretical predictions of the jet spreading rate and minimum dilution are in good agreement with experiments.

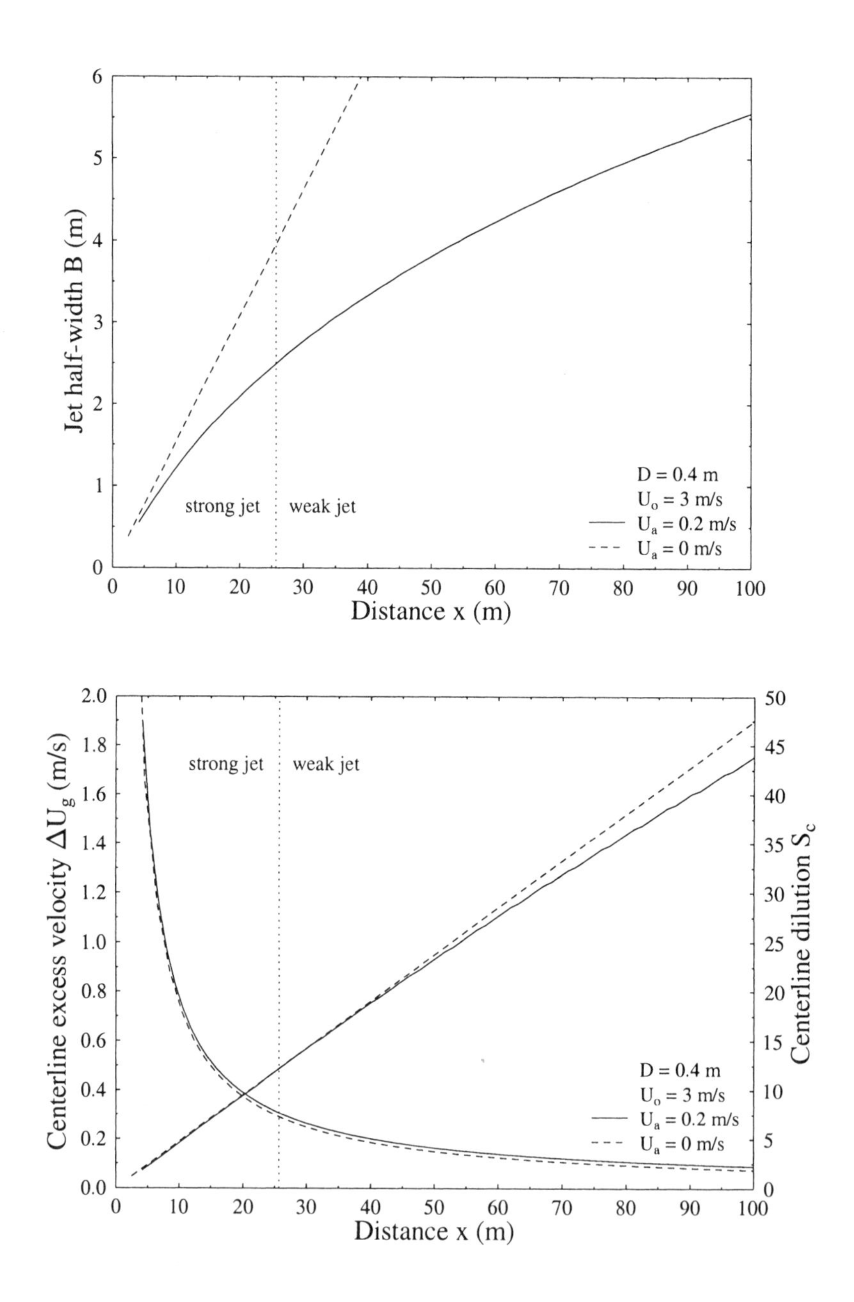

Figure 6.17. Variation of coflowing jet properties with distance from source: effluent discharge in river

The present results support the applicability of the 'top-hat' profile in the integral formulation of buoyant jet in a current. This profile appears to possess a certain physical significance in the description of the active mixing process. It is found that the visual edge, defined by the 25% maximum concentration contour, corresponds to approximately the concentration half-width defined at $r = B = \sqrt{2}\, b_g$. This location also matches the 50% contour of the intermittency factor. A meaningful definition of jet boundary is convenient for many engineering applications (e.g. study of merging of multiple jets). The advantages of the top-hat profile become even clearer for the more complex case of a bent-over jet in crossflow, when the concentration or velocity profile is clearly not Gaussian.

Table 6.1. Summary of experimental results on coflowing jet; previous studies on jet in stagnant ambient are also included

Jet properties	Coflowing jet	Jet in stagnant ambient			
	Cross-sectional LIF measurement[a]	Cross-sectional LIF measurement[b]	Cross-sectional Temperature measurement[c]	Point LIF measurement[d]	Line LIF measurement[e]
db_{gc}/dx	0.125	0.125	0.13	0.139	0.115
$S_m(x/l_Q)^{-1}$	0.174	0.175	0.165	0.147	0.172
C_{max}/C_m	1.4	1.4	1.5	1.65	
C_{min}/C_m	0.45	0.5	0.45	0.2	
$\gamma = 100\%$	$(r/b_{gc} < 0.6)$	$(r/b_{gc} < 0.75)$	$(r/b_{gc} < 0.89)$		$(r/b_{gc} < 0.43)$
$\gamma = 50\%$	$(r/b_{gc} \simeq 1.1)$	$(r/b_{gc} \simeq 1.2)$	$(r/b_{gc} \simeq 1.31)$		$(r/b_{gc} \simeq 1.19)$
C_{rms}/C_m (centreline value)	0.18	0.15	0.165	0.22	0.34
C_{rms}/C_m (peak value)	0.22 $(r/b_{gc} \simeq 0.85)$	0.2 $(r/b_{gc} \simeq 0.75)$	0.25 $(r/b_{gc} \simeq 0.54)$	0.25 $(r/b_{gc} \simeq 0.72)$	0.36 $(r/b_{gc} \simeq 0.43)$

[a] Chu (1996)
[b] Chu (1996)
[c] Papanicolaou and List (1987)
[d] Papanicolaou and List (1988)
[e] Papantoniou and List (1989)

Note that C_{max} and C_{min} are the maximum and minimum concentration; the root-mean-square concentration $C_{rms} = \sqrt{\overline{c'^2}}$, where $c' = c - \bar{c}$ is the concentration fluctuation.

PROBLEMS

6.1 *Jet in Coflow*

The governing equations for an axisymmetric coflowing jet directed in the x-direction can be written in cylindrical co-ordinates (x, r) as:

$$\begin{aligned} \frac{\partial ru}{\partial x} + \frac{\partial rv}{\partial r} &= 0 \\ \frac{\partial ru^2}{\partial x} + \frac{\partial ruv}{\partial r} &= 0 \end{aligned}$$

where the usual boundary layer approximations applicable to a straight jet have been made, and u, v represent the velocity in the x and r-direction respectively. By multiplying the continuity equation by the coflow U_a and subtracting it from the x-momentum equation, show that the jet excess momentum $M_e = \int u u_e dA = \int u(u - U_a)dA$ is conserved along a coflowing jet. $u_e = u - U_a$ = excess velocity.

6.2 The volume and momentum fluxes of a coflow jet are defined respectively as $Q = \int_o^\infty u dA$, and $M = \int_o^\infty u^2 dA$, where the excess velocity profile is given by Eq. 6.3 and Fig. 6.1. Since the integrals Q and M are not bounded, some definition of jet edge is required. If $r = \sqrt{2}b_g$ is defined as the jet edge, show that:

$$\begin{aligned} Q &= \int_0^\infty u dA = \int_0^\infty (u_e + U_a) dA \\ &\approx \int_0^\infty u_{em} \exp(-\frac{r^2}{b_g^2}) dA + \int_0^{\sqrt{2}b_g} U_a dA = \pi b_g^2 (u_{em} + 2U_a) \end{aligned}$$

where $dA = 2\pi r dr$, $u_e = u_{em} \exp(-\frac{r^2}{b_g^2})$ and $u_{em} = \Delta U_g$ is the centerline maximum excess velocity. Similarly show that:

$$M = \frac{\pi}{2} b_g^2 (u_{em} + 2U_a)^2$$

6.3 *Top-hat vs Gaussian profiles*

By referring to Figure 6.1, the excess velocity profile of a coflow jet can be represented in terms of either i) the centerline maximum value and the Gaussian half-width, ΔU_g and b_g respectively (Eq. 6.3); or ii) the average excess velocity $\Delta U = U - U_a$ and half-width B of a 'top-hat' profile.

a) Show by equating the excess volume flux $\int u_e dA$ and the excess momentum flux M_e expressed by the two profiles that:

$$\begin{aligned} \Delta U &= \frac{\Delta U_g}{2} \\ B &= \sqrt{2} b_g \end{aligned}$$

b) Show that with the above equivalence, the jet volume flux, $Q = \int_0^{B=\sqrt{2}b_g} u dA$, is the same regardless of whichever profile is used. Show that the jet momentum flux,

$M = \int_0^B u^2 dA$, can also be equivalently expressed by the two profiles.

c) If the tracer mass concentration for the top-hat profile is $c = C$ for $r \leq B$, establish the following by equating the tracer mass carried by the excess volume flux, $\Gamma_e = \int u_e c dA$:

$$C = \frac{\lambda^2}{1+\lambda^2} C_m$$

where $\lambda = b_{gc}/b_g$. Show that for $\lambda = 1$, the above relation is consistent with conservation of tracer mass flux, $\pi U B^2 C = \int uCdA = Q_o C_o$.

6.4 *Entrainment coefficient of coflow jet*
Consider the governing equations for a jet in coflow by adopting a top-hat profile:

$$\frac{d}{dx}(\pi B^2 \Delta U) = 2\pi B \alpha \Delta U$$

$$\frac{d}{dx}(\pi B^2 U \Delta U) = 0$$

where U, B are the average velocity and half-width of the coflow jet, $\Delta U = U - U_a$ = excess velocity, and α is the entrainment coefficient. β_s = jet spread rate in stagnant fluid. By expanding the excess momentum equation and using the continuity equation and the jet spreading hypothesis, $dB/dx = \beta_s \Delta U/U$ (Eq. 6.11), show that the entrainment coefficient in a coflow jet depends only on the relative excess velocity, $\Delta U/U = \Delta U/(U_a + \Delta U)$:

$$\alpha = \frac{\beta_s}{2} \frac{\Delta U^2}{U(U+\Delta U)}$$

Evaluate the entrainment coefficient for the limiting cases of a strong jet, $\Delta U/U \to 1$ and a weak jet, $\Delta U/U \to 0$.

6.5 *Jet in Counterflow*
Consider a round jet with initial x-velocity u_o and diameter D, discharging into a counterflowing stream of x-velocity $-u_a$. In the presence of the counterflow, the jet is retarded. At some point the jet reaches stagnation at $x = x_p$. The jet effluent issuing into the counterflow occupies a dome shape region. Estimate the penetration distance x_p in terms of the governing parameters.

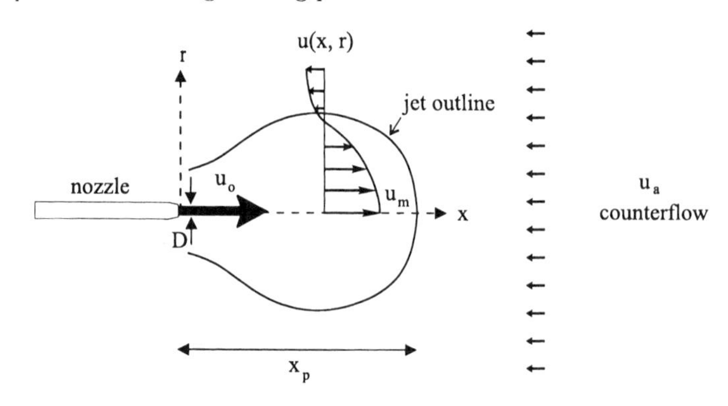

a) Estimate the jet forward velocity u_j in terms of the jet momentum flux $M_o = u_o^2 \pi D^2/4$ and distance from source x. Assume a strong jet, with $u_o \gg u_a$.

b) Assuming heuristically that at $x = x_p$, the jet forward velocity balances the counterflow velocity u_a, show that the penetration distance can be estimated by

$$\frac{x_p}{D} \sim \frac{u_o}{u_a}$$

Both the velocity measurements of Belatos and Rajaratnam (1973) and Lam and Chan (1997) show that $x_p/D = 2.6\ u_o/u_a$.

6.6 Jet Excess Momentum

The excess velocity $u_e(x,r) = u(x,r) - U_a$ of a jet in coflow is given by (Eq. 6.3 and Fig. 6.1):

$$u_e = \Delta U_g \exp(-\frac{r^2}{b_g^2})$$

where $\Delta U_g = (u_c - U_a)$ = centerline velocity excess.

For a jet in a counterflow (of magnitude U_a), the excess velocity is $u_e(x,r) = u(x,r) + U_a$, and the centerline velocity excess is $\Delta U_g = (u_c + U_a)$. Show that the excess momentum flux for either a) a jet in coflow U_a or b) jet in counterflow current U_a is given by:

$$M_e = \int u u_e dA = \frac{\pi b_g^2}{2}(u_c^2 - U_a^2)$$

Chapter 7

JET IN CROSSFLOW: ADVECTED LINE PUFFS

When a jet is directed into a crossflow, the interaction of the jet momentum results in a complicated turbulent shear flow. Environmental mixing calculations are often based on a "stillwater dilution", i.e. the receiving environment is assumed to be stationary. However, both experiments and field studies have shown the presence of even a weak crossflow enhances dilution considerably (Wright 1977a; Lee and Neville-Jones 1987a). The prediction of the trajectory and dilution of a jet in crossflow is a considerably more difficult fluid mechanics problem than the case of the coflowing jet. Vortices of various kinds are being generated close to the source, in particular a pair of twin-vortices in the bent-over phase of the buoyant jet. Fig. 7.1 shows a side view of a round jet perpendicular to the cross flow. Fig. 7.2 shows the cross-sectional view of the jet obtained using the laser-induced fluorescence (LIF) technique (Chu and Lee 1996). Close to the jet source, there are ring vortices characteristic of shear entrainment (similar to jet in stagnant fluid), and wake vortices due to the blockage of the current by the jet itself. In the bent-over phase, however, the jet cross-section is characterised by the vortex pair as seen in the LIF image in Fig. 7.2. The double vortex induces a large scale entrainment that is manifested in the form of tornado vortices drawn into the jet (Fric and Roshko 1994, Chu and Lee 1996). As expected, the mixing is greatly enhanced by the vortex pair — an environmentally friendly gesture of nature as effluents as a rule are discharged into moving fluids. We seek a prediction of the turbulent-mean velocity and scalar field of this flow. The theory must be able to capture the shear entrainment regime close to the source (characterized by Gaussian cross-section tracer concentration distribution), the vortex flow in the bent-over jet, as well as the Gaussian-vortex flow transition.

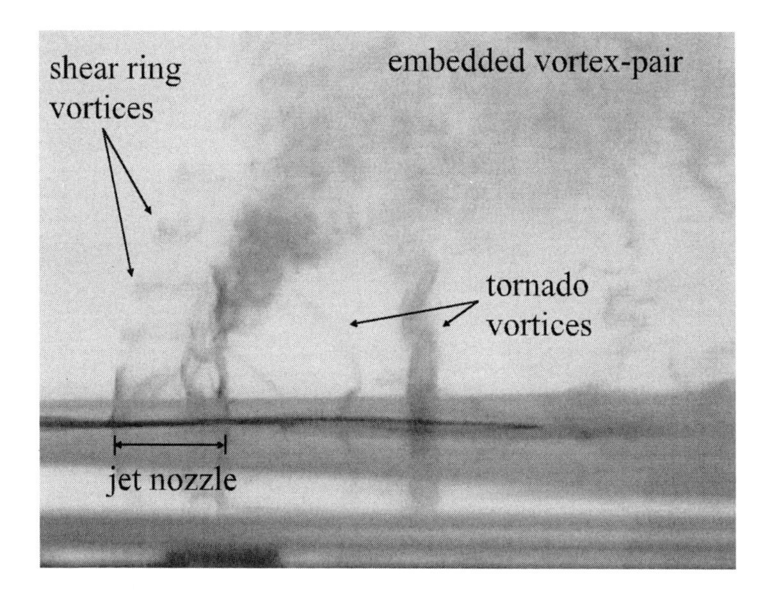

Figure 7.1. Momentum jet in crossflow showing the ring vortices near the nozzle, and the tornado-like vortices in the wake drawn down from the jet to touch the floor of the cross flow. Direction of the cross flow in the figure is from left to the right

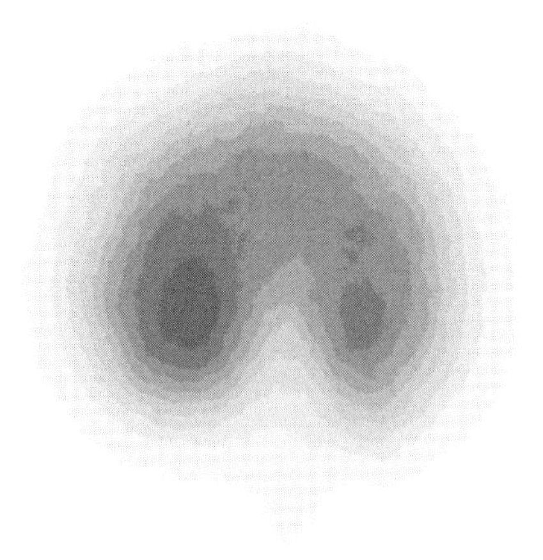

Figure 7.2. Tracer concentration distribution in a cross section of the bent-over jet obtained by Chu and Lee (1996) using LIF technique

There is an extensive literature on the jet in crossflow problem. Many experimental studies (e.g. Keffer and Baines 1963; Kamotani and Gre-

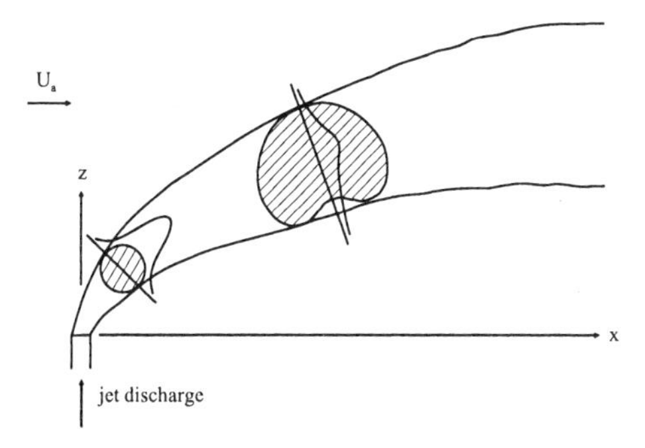

Figure 7.3. Asymptotic flow regimes showing the transition from a Gaussian concentration profile in the near field to the vortex-pair profile in the far field

ber 1972; Rajaratnam 1976; Moussa *et al.*1977; Andreopoulos and Rodi 1984) have revealed some of the essential features of the fluid dynamics, among which the most striking is a transition from an initially vertical jet through a bent-over phase during which the jet becomes almost parallel with the main free stream, to a secondary vortex pair flow in the transverse section of the jet. However, most of these studies were concentrated in a region very close to the source, in the order of 5-10 diameters. On the other hand, for environmental mixing studies, the region of "active mixing" is typically in the order of the water depth, which can extend to beyond 100 jet diameters downstream.

To study the mixing of a passive scalar carried in the jet fluid, visual trajectory and limited tracer concentration measurements in the bent-over phase of a momentum-dominated buoyant jet in crossflow have been made (Ayoub 1971; Chu and Goldberg 1974; Wright 1977). However, most of the concentration measurements were limited to a vertical traverse along the jet centerline. The measured dilution rates also exhibited considerable scatter. On the other hand, numerical studies (Patankar *et al.*1977; Chien and Schetz 1975; Demuren 1983; Sykes *et al.*1986) failed to offer any detailed view of the scalar field, due to the relatively low-resolution and small computational domain used. In the systematic numerical study of Sykes *et al.*(1986), the computational domain extends only 15 jet diameters downstream from the jet source. The computed cross-sectional scalar concentration field fails to reveal any double peak structure. As the only analytical endeavour, a self-similar solution of the bent-over jet in crossflow was obtained by Yih (1981). However, numerical predictions of the scalar field using this free shear layer model are not well-supported by experiments (Lee *et al.*1999). There is scarce data on the scalar field in the bent-over jet. Even the jet boundary could not be clearly defined, and the vortex flow has also not been measured.

In this chapter, we present a comprehensive account of the jet in crossflow. The jet in crossflow problem is examined first using a length scale analysis. In the near field, the effect of the crossflow is shown to be passive. In the far field, the jet behaves essentially as a line puff. In the remainder of the chapter, the emphasis is given to the flow of the advected line puff. Spreading coefficient and added mass coefficient are introduced in a one-dimensional (1D) model as the parameters characterizing the turbulent motion and the irrotational motion surrounding the puff. Two-dimensional (2D) and three-dimensional (3D) turbulence models are developed to compute the turbulent flow in the advected line puff. An extensive series of velocity and tracer concentration measurements obtained in the laboratory are presented. Spreading coefficient and added mass coefficient, the parameters of the 1D model, are evaluated using the 2D and 3D models and the data obtained from the laboratory measurements. The chapter is concluded by a practical example in the final section.

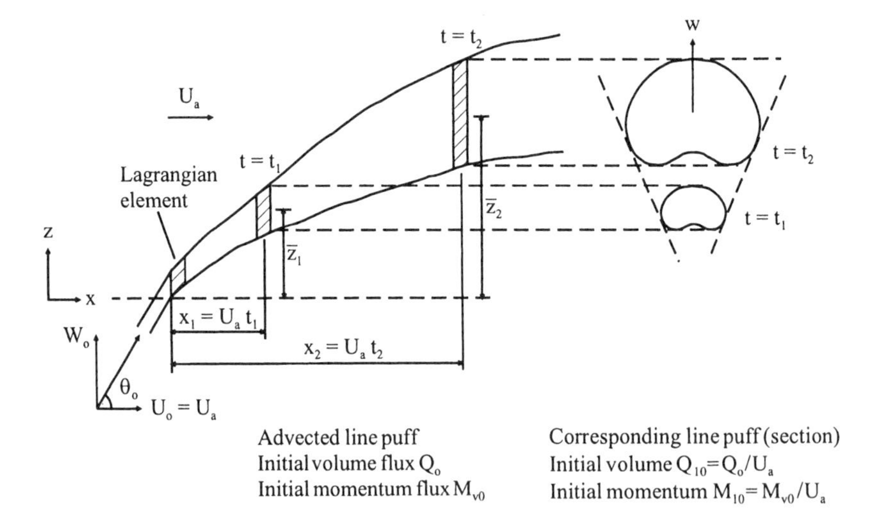

Figure 7.4. Schematic of a jet in crossflow showing the Lagrangian element and its cross section in the analog of the advected line puff

1. LENGTH SCALES AND REGIMES

The problem of the jet in a cross flow is defined schematically in Fig. 7.4. A round turbulent jet, of diameter D, initial velocity U_{jet}, and tracer concentration C_o is discharging from a nozzle into a steady

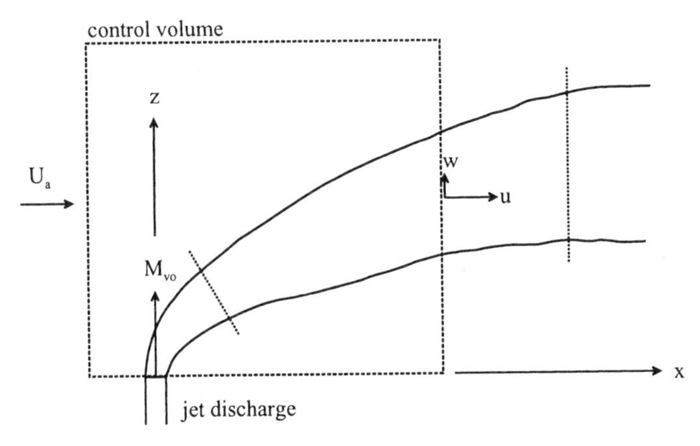

Figure 7.5. Control volume for the momentum flux through a vertical cross section of a turbulent jet in crossflow; vertical momentum flux = $M_v = \int w(udA)$, horizontal momentum flux = $M_h = \int u(udA)$

uniform crossflow U_a. The crossflow is in the horizontal x-direction. The jet is oriented at an angle of θ_o to the x-axis. The initial vertical velocity $W_o = U_{jet}\sin\theta_o$. The initial volume flux from the nozzle is $Q_o = U_{jet}\pi D^2/4$. The vertical jet momentum flux is $M_{vo} = Q_o W_o$. The behaviour of a jet is characterized by its momentum flux. The velocity induced by the vertical momentum varies as $w \sim M_{vo}^{1/2} z^{-1}$. As the jet-induced velocity decreases, the jet becomes bent over due to entrainment of ambient fluid with crossflow momentum. At a vertical distance where $w \approx U_a$, or $z \sim M_{vo}^{1/2}/U_a$, heuristically we would expect the jet to be significantly bent over. Thus a crossflow momentum length scale

$$L_{mv} = \frac{M_{vo}^{1/2}}{U_a} \tag{7.1}$$

can be defined. For $z/L_{mv} = zU_a/M_{vo}^{1/2} \ll 1$, the jet-induced velocity is significantly larger than the ambient current, and the jet is only slightly affected by the current. Thus the mixing is similar to that of a momentum jet in still fluid (Chapter 2). In this region, the jet velocity and concentration are radially-symmetrical, self-similar, and can be well-approximated by Gaussian distributions. By regarding the jet as only advected by the current ($x = U_a t$), the trajectory and dilution relations of the "advected jet" can be obtained from the solutions of a classical pure jet (Wright 1977b; Fischer *et al.*1979; Chu and Lee 1996 - see Section 6.). For $z/L_{mv} \gg 1$, the jet is significantly bent over; it is expected that the horizontal exchange of momentum between the jet and the ambient current is complete - i.e. the x-velocity of the jet is approximately equal to U_a. The jet flow in this bent-over phase is similar to that of a line puff. These two asymptotic regimes (advected jet and line puff) are referred to as the *momentum-dominated near field* (MDNF) and *momentum-dominated far field* (MDFF) respectively (Wright 1977). This chapter concentrates mainly on the MDFF.

1.1 LINE PUFF ANALOGY FOR MDFF

Consider a control volume with outer boundaries far removed from the source, along which the pressure is assumed to be hydrostatic (Fig. 7.5). By momentum conservation, the vertical jet momentum flux across a vertical section, $M_v = \int_{x-section} u\ w\, dA$, is then equal to the initial jet discharge momentum flux M_{vo} - where u, w are the streamwise (x) and vertical (z) velocities. In the bent-over phase, since $u \approx U_a$, we have $M_v \approx U_a \int w\ dA$. The flow in successive vertical x-sections can then be regarded as equivalent to that of a line puff with initial momentum $M_{lo} = M_{vo}/U_a$ imparted in the $+z$ direction, and initial volume $Q_{lo} = Q_o/U_a$, per unit x-length.

Intuitively, the jet in crossflow can be viewed as issuing line momentum puffs at a steady rate (Fig. 7.4). As far as the mixing in the bent-over phase is concerned, the action of the jet is to impart a vertical ($+z$) kinematic momentum flux of M_{vo} to the flow streaming by at U_a. In a continuous manner, the jet distributes a vertical momentum of $M_{vo}\Delta t$ to an ambient fluid element of length $U_a\Delta t$. Each of these Lagrangian elements (advected momentum puffs) receives an impulse of $\frac{M_{vo}\Delta t}{U_a\Delta t}$ per unit x-length as they pass by the source. To an observer moving with the crossflow, the ambient environment is stagnant while the turbulent element rises as a line puff. The mixing of the 3D bent-over jet at successive x-sections (x_1, x_2) then correponds to that of an equivalent line puff at corresponding times (t_1, t_2) via a Galilean transformation. This analog, to be confirmed by the experiments, is expected to hold in the bent-over stage of the jet, $z \gg L_{mv}$.

1.2 SIMILARITY VARIABLES FOR THE LINE PUFF

We consider the mixing created by a cylinder of non-buoyant fluid having initial momentum in an otherwise stagnant fluid. This is illustrated in Fig. 7.6 for an experiment done upside down (Richards 1965). A cylinder of marked fluid having an initial head is released impulsively from a puff device, creating a dyed line patch with significant initial momentum, and mixes with the surrounding fluid in the water tank. Fig. 7.7 shows a cross-section of the line puff (in the $y - z$ plane) with

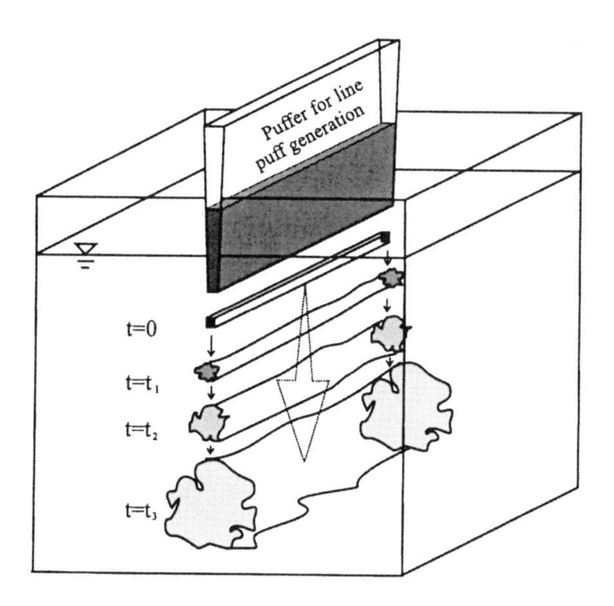

Figure 7.6. Line momentum puff generated by releasing fluid impulsively at excess head from a puffer in a tank

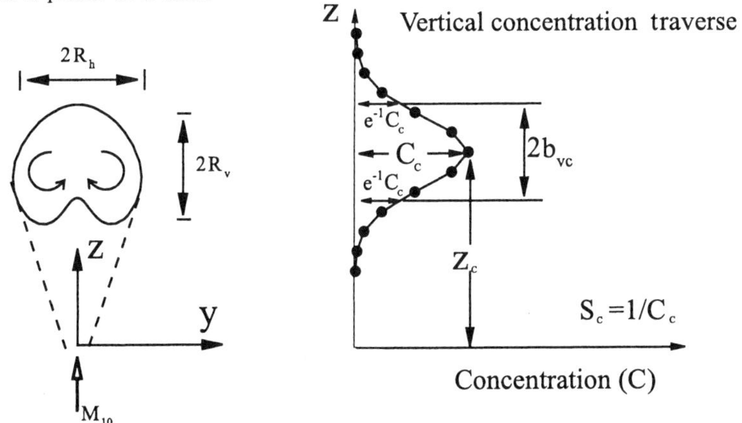

Figure 7.7. Definition sketch showing the crossing section of the line puff and the concentration profile on the center plane of symmetry

an initial volume Q_{lo} $[L^2]$ and an initial momentum M_{lo} $[L^3/T]$ per unit length of the line puff. The tracer concentration in the volume initially is C_o. The unsteady flow of the puff is charaterized by vertical coordinate z, half-width b, vertical velocity W, and concentration C. A time scale of the puff

$$t_c = \frac{Q_{lo}^{3/2}}{M_{lo}} = \frac{Q_o^{3/2}}{U_a^{1/2} M_{vo}} \tag{7.2}$$

is defined for the relative importance of the initial volume Q_{lo} and momentum flux M_{lo}. The related length scale is

$$L_x = U_a t_c = \frac{Q_o^{3/2} U_a^{1/2}}{M_{vo}} \tag{7.3}$$

As the puff is primarily driven by its initial impulse, the effect of the source volume will be negligible for $t/t_c \gg 1$. Therefore, the puff characteristics is assumed to depend only on M_{lo} and the time t. The following similarity relations for the maximum velocity, spreading rate, trajectory, and scalar dilution of the line puff are obtained based on this assumption:

$$W_m \sim M_{lo}^{1/3} t^{-2/3} \quad \text{or} \quad \frac{W_o}{W_m} \sim (t^*)^{2/3} \tag{7.4}$$

$$b \sim z \tag{7.5}$$

$$z \sim M_{lo}^{1/3} t^{1/3} \quad \text{or} \quad z \sim Q_{lo}^{1/2} (t^*)^{1/3} \tag{7.6}$$

where dimensionless time $t^* = t/t_c$; and W_o and W_m stand for the initial and maximum vertical velocity of the jet respectively. Conservation of tracer mass also implies:

$$S_c = \frac{C_o}{C} \sim \frac{b^2}{Q_{lo}} \sim \frac{M_{lo}^{2/3} t^{2/3}}{Q_{lo}} \sim \frac{z^2}{Q_{lo}} \tag{7.7}$$

Given the equivalence between successive vertical x-sections of a bent-over jet and the line puff at corresponding times $t = x/U_a$, the following relations can be obtained for the spreading rate, trajectory, and dilution of the 3D bent-over jet in terms of the crossflow momentum length scale L_{mv}.

$$b_{vc} \sim z_c \tag{7.8}$$

$$\frac{z_c}{L_{mv}} \sim (\frac{x}{L_{mv}})^{1/3} \tag{7.9}$$

$$\frac{S_c Q}{U_a L_{mv}^2} \sim (\frac{z_c}{L_{mv}})^2 \tag{7.10}$$

$$\frac{W_o}{W_m} \sim (x^*)^{2/3} \tag{7.11}$$

where b_{vc} is the vertical centerline half-width defined by the $e^{-1}C_c$ points, z_c is the vertical location defined by the maximum centerline concentration C_c, $S_c = C_o/C_c$ is referred to as centerline minimum dilution, as illustrated in Fig. 7.7; and the dimensionless downstream distance

$x^* = x/L_x$. For the sake of reference, Eqs. 7.9 and 7.10 can be expressed alternatively as:

$$z_c \sim (Q_o/U_a)^{1/2}(x^*)^{1/3} \quad (7.12)$$

$$S_c \sim z_c^2/(Q_o/U_a) \quad (7.13)$$

The above similarity relations furnish a useful means of correlating experimental data; the proportionality constants can be derived by best-fitting experimental data or numerical model results.

2. 1D MODEL OF LINE PUFF

We begin with the analysis of line puff by a one-dimensional (1D) model. In the absence of body forces, the momentum of the puff, $M_p(t) = \pi R^2 W$ is conserved, where $R = \sqrt{R_h R_v}$ is a characteristic radius of the puff, and W is the average vertical velocity. Application of the general Lagrangian spreading hypothesis (following the puff element or dominant eddy as in Chapter 4 to 6) and momentum conservation to the puff in Fig. 7.7 then gives:

$$\frac{dR}{dt} = \beta_n W \quad (7.14)$$

$$\frac{d}{dt}((1 + k_n)\pi R^2 W) = 0 \quad (7.15)$$

where β_n is the puff spread rate and k_n is an added mass coefficient to account for the motion of the irrotational flow outside of the puff. As the puff (the colored patch) moves through the surrounding fluid, energy is expended to form the external flow. The presence of the ambient fluid may be represented by an additional "virtual mass" to the inertia of the puff element (see Prob. 8.3).

Noting that $W = dz/dt$, the average flow or 'global' solution is then given by:

$$R = (\frac{3\beta_n M_{lo}}{(1 + k_n)\pi})^{1/3}\ t^{1/3} \quad (7.16)$$

$$W = (\frac{M_{lo}}{(1 + k_n)9\pi\beta_n^2})^{1/3}\ t^{-2/3} \quad (7.17)$$

$$z = (\frac{3M_{lo}}{(1 + k_n)\pi\beta_n^2})^{1/3}\ t^{1/3} \quad (7.18)$$

$$\frac{C_o}{C} = \frac{\pi B^2}{Q_{lo}} = \frac{\pi B^2 U_a}{Q_o} \quad (7.19)$$

The solution for the puff location translates via a Galilean transformation into the path of the bent-over jet in crossflow:

$$R = \beta_n \ z \tag{7.20}$$

$$z = (\frac{3}{(1+k_n)\pi\beta_n^2})^{1/3}(\frac{M_{vo}x}{U_a^2})^{1/3} \tag{7.21}$$

or

$$\frac{z}{L_{mv}} = C_{2p}(\frac{x}{L_{mv}})^{1/3} \tag{7.22}$$

with $C_{2p} = (\frac{3}{(1+k_n)\pi\beta_n^2})^{1/3}$. Several characteristic parameters of the turbulent-mean puff flow can hence be determined to within a constant without knowledge of the detailed velocity or concentration distributions. There are two model parameters, β_n and k_n, which must be determined independently for a consistent prediction of the puff spread rate, trajectory, and dilution. To obtain these parameters, we turn to numerical computations and measurements of the flow and tracer concentration field of a jet in crossflow.

3. 2D MODEL OF LINE PUFF

Numerical computations have been conducted for the line puff using a two-dimensional (2D) model and for the general problem of jet in crossflow using a three-dimensional (3D) model. The results obtained for the line puff using a two-dimensional k-ϵ turbulence model is presented in this section. The 2D model equations for the line puff is the Reynolds-averaged equations, which for constant density incompressible flow, are:

$$\frac{\partial U_i}{\partial x_i} = 0 \tag{7.23}$$

$$\frac{\partial U_i}{\partial t} + U_j\frac{\partial U_i}{\partial x_j} = -\frac{1}{\rho}\frac{\partial p}{\partial x_i} + \frac{\partial \tau_{ij}}{\partial x_j} \tag{7.24}$$

where U_i = fluid velocity in (x, y, z) direction, ρ = density, p = dynamic pressure, τ_{ij} = stress tensor. τ_{ij} is equal to the sum of the viscous and Reynolds-stresses. The eddy-viscosity model (Boussinesq hypothesis) provides the following expression for the stresses:

$$\tau_{ij} = \nu_{eff}(\frac{\partial U_j}{\partial x_i} + \frac{\partial U_i}{\partial x_j}) - \frac{2}{3}k\delta_{ij} \tag{7.25}$$

where the effective viscosity $\nu_{eff} = \nu + \nu_t$, ν is the molecular viscosity, $\nu_t = C_\mu k^2/\epsilon$ is the turbulent viscosity, k is the turbulence kinetic energy (TKE), ϵ is referred to as dissipation rate (DR) of k. We adopt the standard two-equation $k - \epsilon$ model of Launder and Spalding (1974) for turbulence closure:

$$\frac{\partial k}{\partial t} + U_j \frac{\partial k}{\partial x_j} = \frac{\partial}{\partial x_i}(\frac{\nu_t}{\sigma_k}\frac{\partial k}{\partial x_i}) + \tau_{ij}\frac{\partial U_i}{\partial x_j} - \epsilon \quad (7.26)$$

$$\frac{\partial \epsilon}{\partial t} + U_j \frac{\partial \epsilon}{\partial x_j} = \frac{\partial}{\partial x_i}(\frac{\nu_t}{\sigma_\epsilon}\frac{\partial \epsilon}{\partial x_i}) + C_{1\epsilon}\frac{\epsilon}{k}\tau_{ij}\frac{\partial U_i}{\partial x_j} - C_{2\epsilon}\frac{\epsilon^2}{k} \quad (7.27)$$

with the following standard empirical model coefficients: $C_\mu = 0.09$, $C_{1\epsilon} = 1.44$, $C_{2\epsilon} = 1.92$, $\sigma_k = 1.0$, $\sigma_\epsilon = 1.3$ (Rodi 1980). In the above, the turbulent closure is achieved by computing the turbulence kinetic energy locally; the generation, transport, and decay of turbulence can hence be accounted for. TKE is computed by a transport equation for k, while the equation for ϵ is by and large empirical. The eddy viscosity $\nu_t \sim k^2/\epsilon$ is then obtained by dimensional analysis.

In addition to the flow equations, the study of puff characteristics necessitates the calculation of a passive scalar field from the tracer mass conservation equation:

$$\frac{\partial C}{\partial t} + U_j \frac{\partial C}{\partial x_j} = \frac{\partial}{\partial x_j}(\frac{\nu_{eff}}{Sc_t}\frac{\partial C}{\partial x_j}) \quad (7.28)$$

where Sc_t stands for the turbulent Schmidt number, taken to be 0.75 as found appropriate for a related problem (Sykes *et al.* 1986). Note the above governing equations are applicable for both the 2D line puff and the 3D advected line puff.

3.1 NUMERICAL SIMULATION OF LINE PUFFS

The nature of the line puff motion can be illustrated by a numerical 'thought experiment'. A finite momentum is imparted to a two-dimensional marked patch of incompressible fluid, and the time evolution of the mean flow and passive scalar field of such a puff is investigated. Fig. 7.8 shows a volume of water initially at rest in a tank of width W_T and depth H. Over a time interval, $t = 0$ to $t = t_s$, an impulse $M_o = M_{lo}$ in the vertical (z) direction is applied to the marked square patch of fluid (of length L_o), which has a uniform tracer concentration C_o and is located sufficiently far away from the solid boundaries. The momentum source gives rise to vorticity. Turbulent flow is produced and is advected in the direction of the imparted momentum while mixing with the ambient fluid. The motion of this turbulent puff (Fig. 7.9) has been computed using the 2D $k - \epsilon$ model (Lee, Rodi and Wong 1996). The following parameters are adopted: $W_T = 120$ cm, $H = 90$ cm,

$t_s = 0.01\ s$, $M_o = 1712\ cm^3/s$, and $L_o = 4\ cm$. The marked patch is initially located at $Z_o = 24\ cm$ from the bottom boundary.

At $t = 0$, zero velocities and pressure are prescribed. The scalar concentration is given the value of $C_o = 1$ inside the marked patch, and zero elsewhere. The puff is defined as the region inside the 0.01 C_m contour, where $C_m(t)$ is the maximum computed concentration, and L defined such that L^2 is the area within that contour. The governing equations are solved using the finite volume method (Patankar 1980). Details of the numerical solution procedure can be found in Lee *et al.*(1996). The main features of the flow and scalar field are summarized below.

The momentum source gives rise to sharp velocity gradients, leading to the formation of two vortices at the lateral edges of the puff (Fig. 7.10). The force input also generates large positive pressures in front of, and large negative pressures behind the puff. As a result of this pressure interaction, the puff vertical momentum $M_p = \int W dA$ drops to half of the nominal impulse after source introduction. Fig. 7.11 shows that the puff momentum initially decreases, and attains an asymptotic value of 0.5 M_o beyond t* $\approx$ 20. The injection of the momentum puffs into the crossflow requires extra work done to form the ambient irrotational flow; some of the initial impulse is contained in the puff flow, while the rest is carried by the ambient fluid outside of the turbulent flow (the added mass effect). The result implies that the line puff motion has an added mass coefficient of $k_n \approx 1$.

The numerical solution shows that the shape of the flow and pressure field, as well as the vorticity, is well-preserved for $t^* \geq 20$. Fig. 7.12 shows clearly the puff flow is self-similar and characterized by a vortex-pair. The computed concentration field and turbulence kinetic energy also display approximate self-similarity as the flow field (Lee *et al.*1996). From the computed flow and passive scalar field, characteristic puff variables such as the maximum vertical velocity $W_m(t)$ and concentration C_m, elevation z, width R_h, R_v, and aspect ratio R_h/R_v can be determined along with the turbulent kinetic energy k_m. For example, the computed time variation of maximum vertical velocity W_m and front location of the puff z_f are in excellent agreement with the -2/3 and 1/3 power law respectively. The puff front is also linearly proportional to the maximum radius R. It is then possible to determine the dimensionless puff mixing paramters in the self-similar relations from the numerical solution in the asymptotic stage $t^* \gg 1$. These are summarized in Table 7.1 (SKE-standard $k - \epsilon$ model). Also shown are the ratio of the maximum to mass weighted average vertical velocity, $W_m/\overline{W}$, and the circulation around one half of the puff, defined as $\Gamma = \int_o^H W(0, z)dz$. The properties of the self-similar puff are rather insensitive to the exact

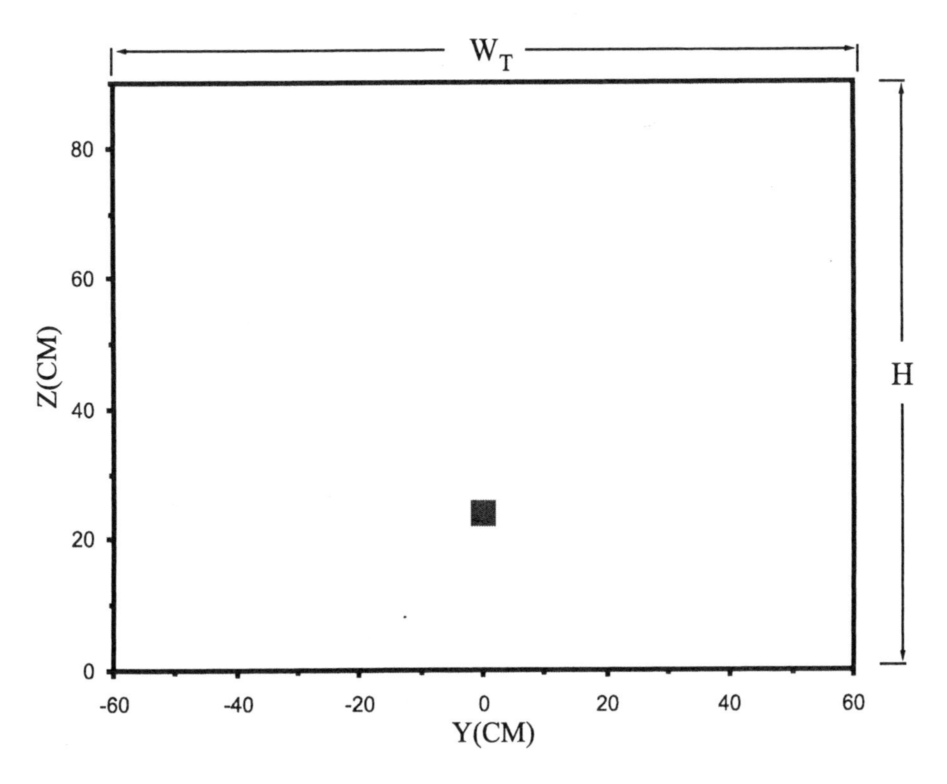

Figure 7.8. Numerical experiment of turbulent line momentum puff; an initial impulse is applied in +z direction to marked patch of fluid

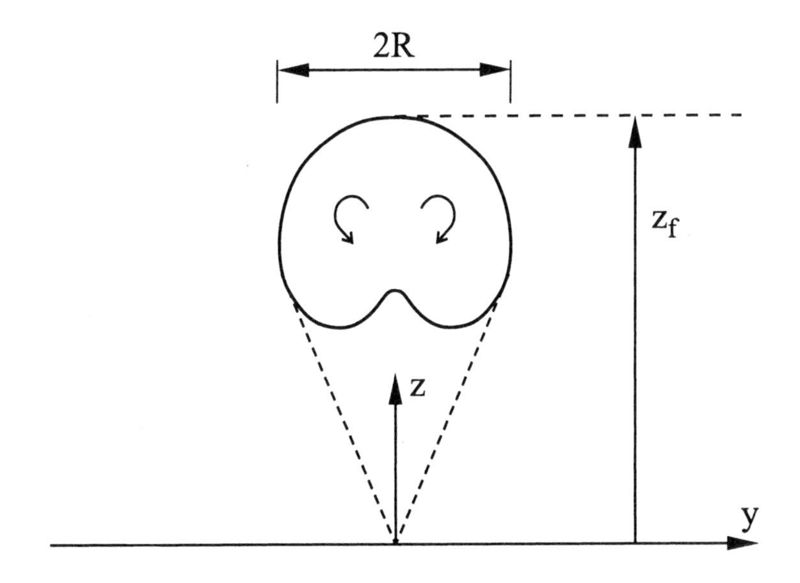

Figure 7.9. Definition of puff front and radius

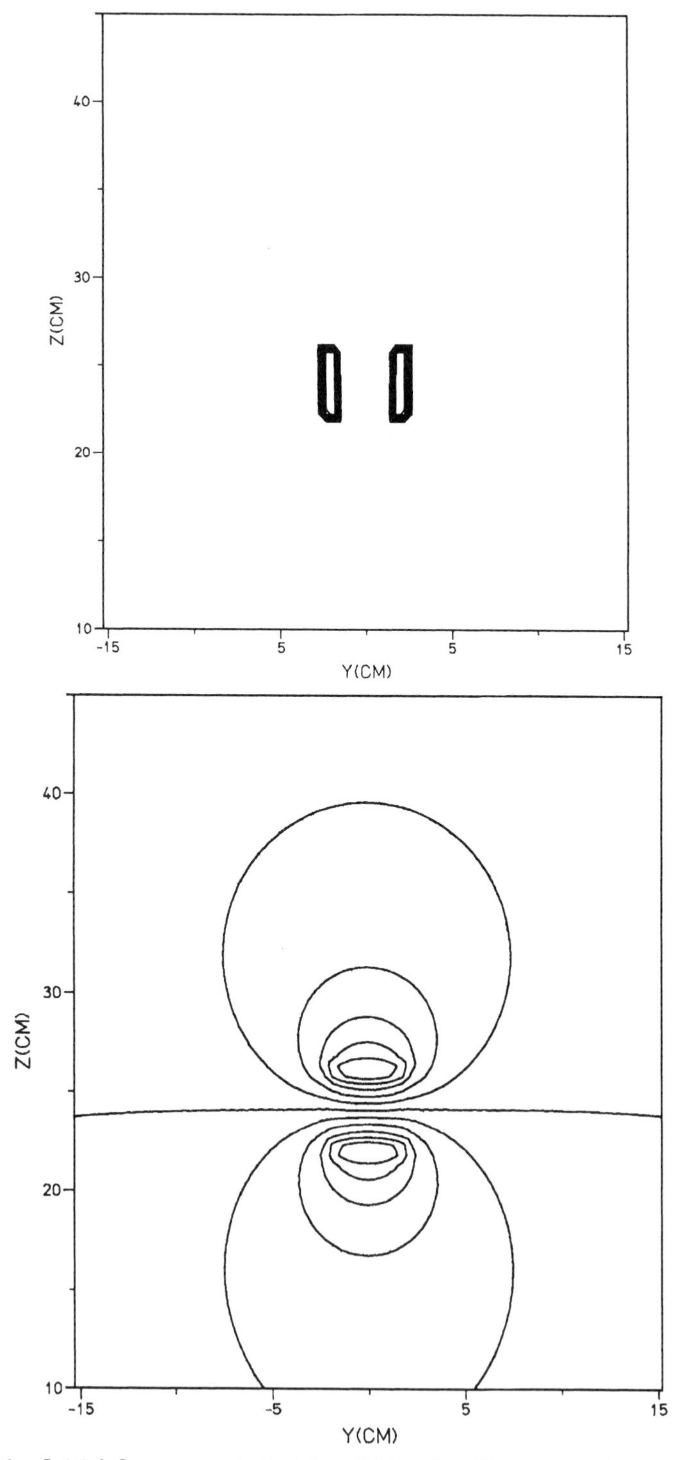

Figure 7.10. Initial flow generated at $t = 0.5t_s$. The line momentum source gives rise to vorticity concentrated into two rings (top); b) pressure field (from Lee, Rodi and Wong 1996)

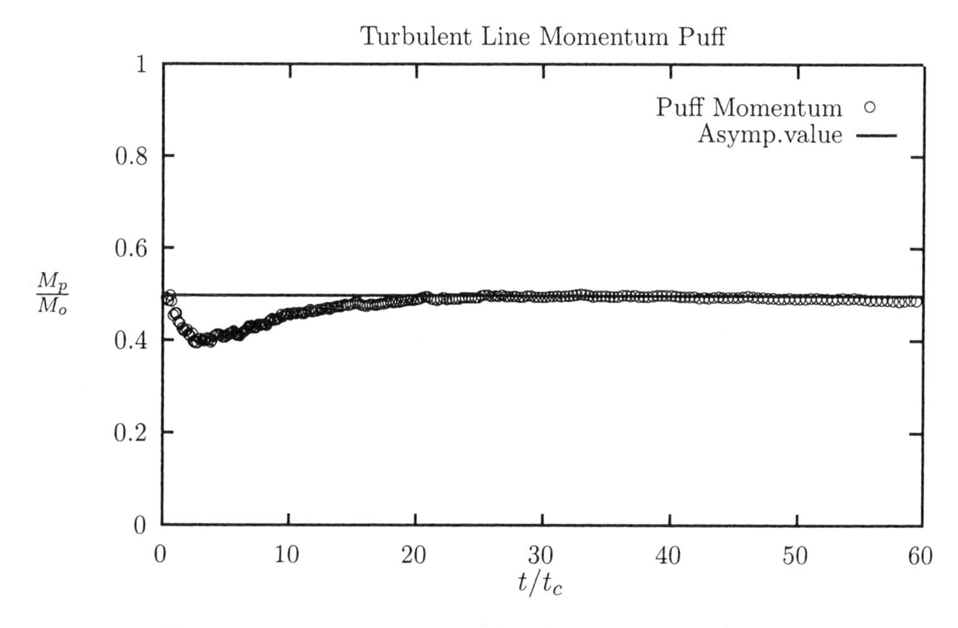

Figure 7.11. Time variation of puff momentum M_p/M_o

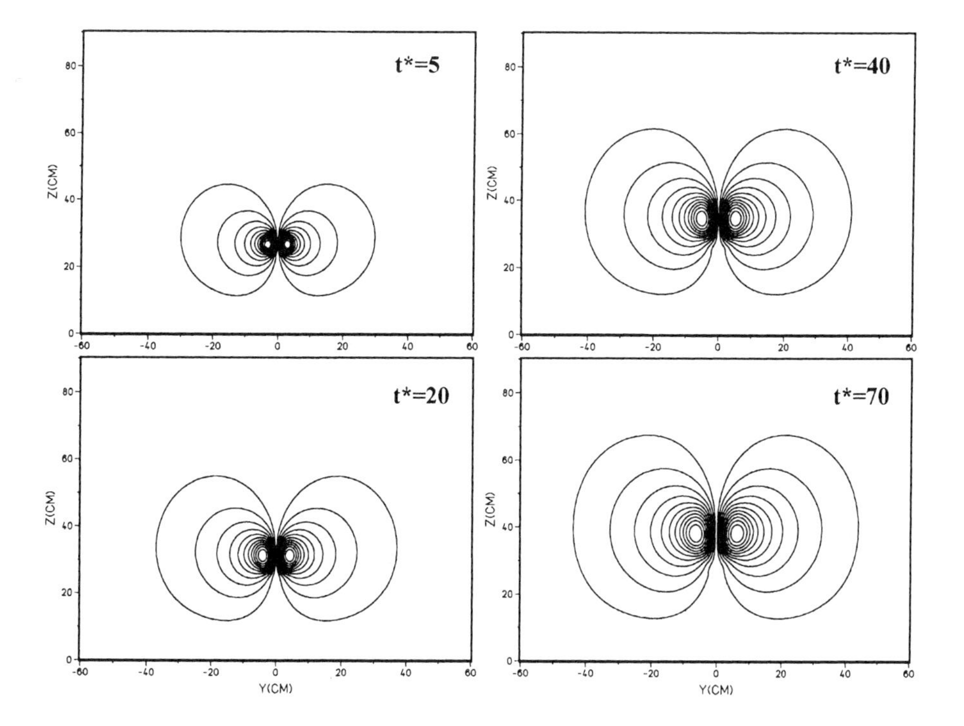

Figure 7.12. Computed streamlines of line puff at $t^* = 5, 20, 40, 70$; maximum normalized stream function is $\Psi_m^* \approx 0.18$ (contour interval $\Psi_m^*/12$)

initial conditions of the k and ϵ equations assumed. Defining the puff by the 0.05 C_m contour also produced negligible changes in the results. However, one distinct feature of the 2D line puff is the absence of bifurcation of the tracer concentration field in the asymptotic phase; i.e. the aspect ratio of the pear-shaped line puff is approximately one - in contrast to observations of the bent over jet in crossflow. The result is obtained by the standard $k-\epsilon$ model (Lee *et al.*1996) and also by a modified $k-\epsilon$ model based on ReNormalization Group theory (RNG model in Table 7.1, Lee and Chen 1998). This result also shows that a vortex-pair flow does not necessarily imply the bifurcation of the time-averaged concentration field.

4. 3D MODEL OF JET IN CROSSFLOW

4.1 THE ADVECTED LINE PUFF

The advected line puff (ALP) has been introduced by Lee and Wong (1993) as an experimental analog for the study of the bent-over jet in crossflow. To best approximate the line puff condition, the round jet is discharged at an angle θ_o so that the initial excess momentum is zero relative to the crossflow. This is done by setting the angle of discharge θ_o such that $U_o = U_{jet}\cos\theta_o = U_a$ as shown in Fig. 7.4. The jet to ambient velocity ratio is then $W_o/U_a = tan\theta_o$. Since the component of the initial excess momentum in the direction of the cross flow is zero, the vertical motion of the ALP can be viewed in a coordinate system moving with the crossflow. In this moving coordinate system, the ambient is stationary and the tubulent fluid in the ALP rises in the vertical direction essentially as pure line puff. Source conditions are much better controlled in the ALP than the source conditions of the pure line puff produced by a line source of momentum in still fluid. Experimental data obtained in the ALP are much less scattered compared with those obtained in the pure line puff experiment or previous MDFF data of vertical jet in crossflow.

4.2 3D MODEL OF ADVECTED LINE PUFF

A more accurate picture of the scalar field can be obtained from a three-dimensional $k-\epsilon$ model calculation of the scalar field in a jet in crossflow (Lee *et al.*2002). A 3D numerical experiment of the advected line puff in Fig. 7.4 can be performed. A near vertical non-buoyant jet is discharged at no excess horizontal momentum into a horizontal crossflow. The scalar field is illustrated for the case of a water jet (diameter 2 cm) discharging in a numerical channel of length 1.2 m, width 0.8 m and depth 0.6 m. The parameters are $U_{jet} = 1.086$ m/s, $U_a = 0.1887$ m/s, and $\theta_o = 80^o$. The computational domain extends to 55 jet diameters downstream. The jet to ambient current ratio is $W_o/U_a = 5.67$.

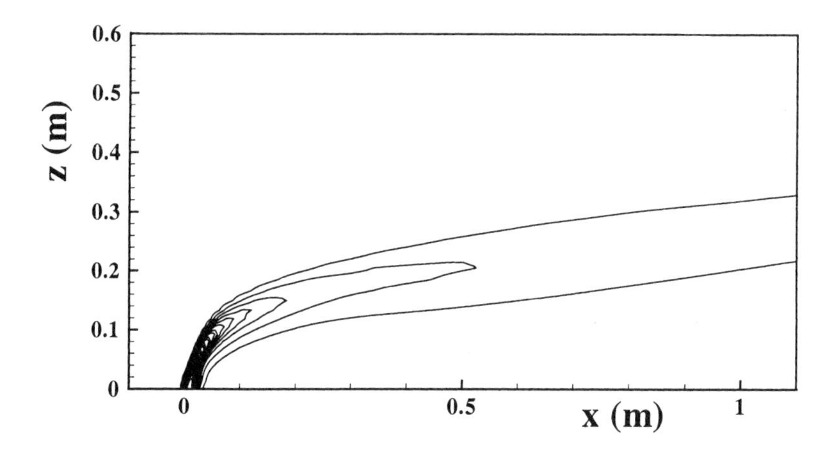

Figure 7.13. Computed scalar field in the center plane of symmetry of a jet in crossflow

Fig. 7.13 shows contour plots of the computed scalar field in the plane of symmetry at $y = 0$. The trajectory of the maximum concentration shows a rapid rise to about $z \approx 0.13$ m at $x = 0.1$ m corresponding to $x^* \approx 14$, whereafter the jet rises very slowly, reaching $z \approx 0.27$ m at $x = 1.0$ m corresponding to $x^* \approx 142$. The bent-over phase, during which the scalar contours become almost parallel with the mainstream, takes place essentially in the interval of $10 < x^* < 40$.

Fig. 7.14 shows transverse cross sections of the scalar field at different downstream locations. As the salient feature of bent-over jet observed in experiments, a double peak concentration distribution can generally be observed in the kidney-shaped sectional puff, which becomes approximately self-similar for $x^* > 20$.

4.2.1 CHARACTERISTICS OF ADVECTED LINE PUFF

Analysis of the 3D model results reveals the following characteristics:

i) *Puff aspect ratio* - As a basic parameter to characterize the sectional puff shape, the puff aspect ratio is defined as the ratio of horizontal half-width R_h over vertical half-width R_v of the contour $C = e^{-1}C_m$, where C_m stands for the maximum concentration over the puff section

(Fig. 7.7). The results show that the puff aspect ratio is roughly equal to 1.20 for the main stage of $20 < x^* < 70$, but reduces, with a minimum value of 1.05, for the earlier and later stages of $x^* \leq 10$ and $x^* \geq 100$. This is consistent with the experimental results of Chu (1996) who found a variation of puff aspect ratio from about 1.05 for the early and later stages to about 1.25 for the bent-over stage, and of Wong (1991) who found a value of 1.23 with probe-based measurements; and comparable to the value of 1.3 suggested by Pratte & Baines (1967) for the developed phase from an analysis of flow visualization.

ii) *Maximum concentration ratio* - As observed from the concentration contours, the maximum concentration is not located in the center line. It is desirable to correlate the centerline maximum concentration C_c to the sectional maximum concentration C_m, since often only centerline concentration was measured. It is found that the ratio of C_m/C_c varies, from a smaller value of 1.05 for the initial and later stages of $x^* \leq 10$ and $x^* \geq 130$, to a higher value of 1.35 for the central phase of $30 \leq x^* \leq 120$, with an averaged value of 1.2. This is consistent with the range of 1.0 - 1.5 with an average of about 1.2 found by Chu (1996), and compares well with the range of 1.1 to 1.5 found by Wong (1991).

iii) *Centerline trajectory* - Fig. 7.15 shows the puff trajectory z_c/L_{mv} plotted against downstream distance x/L_{mv}. The results follow the power similarity law given in Eq. 7.9. Corresponding coefficient is found as $C_{2p} = 1.56$. This value is very close to the experimental values of 1.56 and 1.63 in Wong (1991) and Chu (1996), respectively, and comparable to the value of 1.77 from the buoyant jet experiment of Ayoub (Lee 1989).

iv) *Centerline half-width* - Fig. 7.16 shows the centerline half-width b_{vc} plotted against centerline trajectory z_c. For $z_c/L_{mv} \geq 1.2$, corresponding to $x^* \geq 10$, the results follow the linear similarity law. The corresponding coefficient, referred to as the centerline half-width spreading rate, is found as $C_{1p} = 0.297$. This spreading rate is close to the experimental value of 0.276 by Wong (1991) and 0.299 by Chu (1996).

v) *Centerline dilution* - Fig. 7.17 shows the dimensionless centerline dilution $S_cQ/(U_aL_{mv}^2)$ plotted against centerline trajectory z_c/L_{mv}. The results follow the power similarity law of Eq. 7.10. The corresponding minimum centerline dilution constant is found as $C_{3p} = 0.488$. No notable effect of virtual origin is observed. This value can be compared to experimental values of 0.409-0.484 (Wong 1991; Chu 1996).

A summary of the self-similar properties of the 3D advected line puff is given in Table 7.1. The main conclusions of the 3D modelling are as follows:

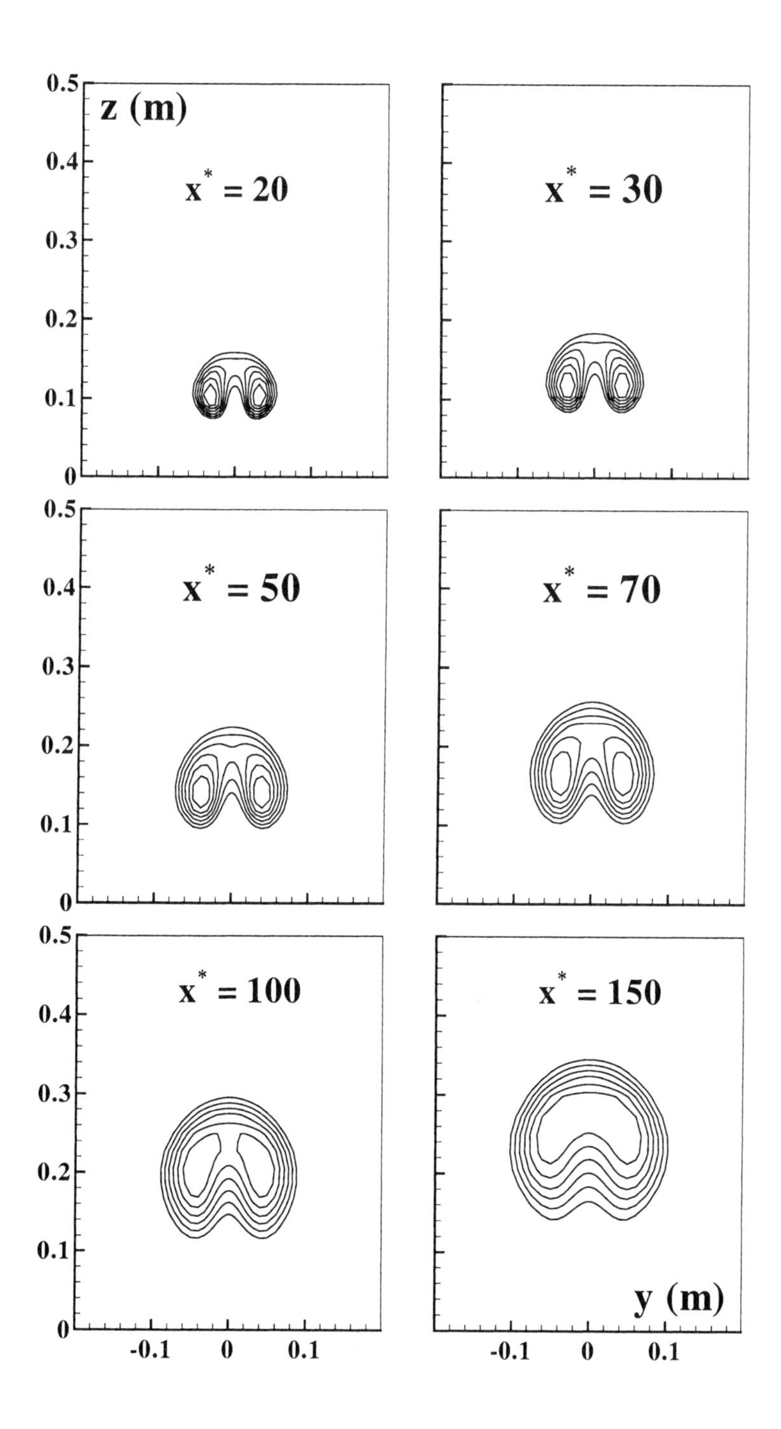

Figure 7.14. Computed scalar field at different downstream locations of a bent-over jet in crossflow (contour interval = $(1 - e^{-1})C_m/6$, outmost concentration contour = $e^{-1}C_m$)

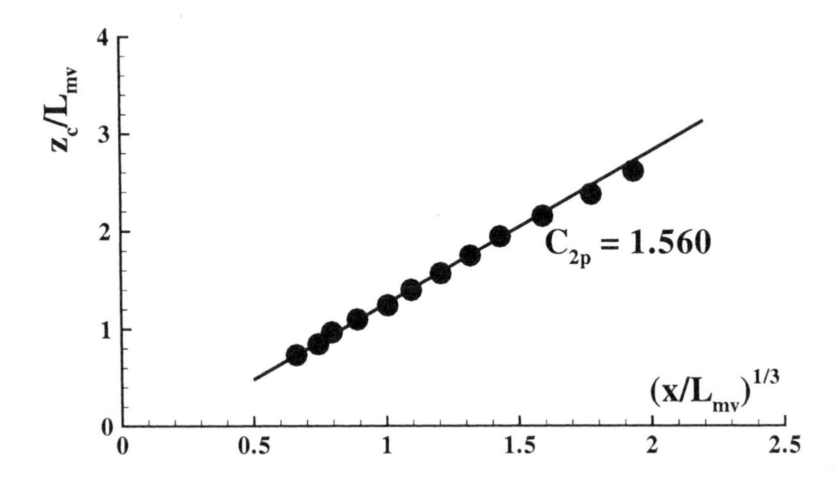

Figure 7.15. Puff trajectory z_c/L_{mv} vs downstream distance x/L_{mv}

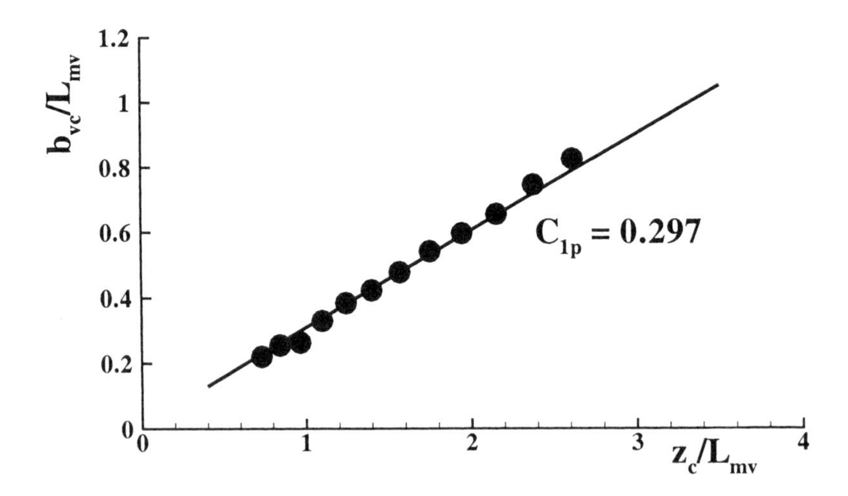

Figure 7.16. Centerline half-width b_{vc} vs centerline trajectory z_c

- The 3D advected line puff (bent-over jet) is longitudinally characterized by a bent-over phase taking place essentially in the interval of $10 < x^* < 40$, during which the fluid flow and scalar contours become almost parallel with the free stream. The flow is approximately self-similar beyond a dimensionless length of $x^* = 20 - 30$.

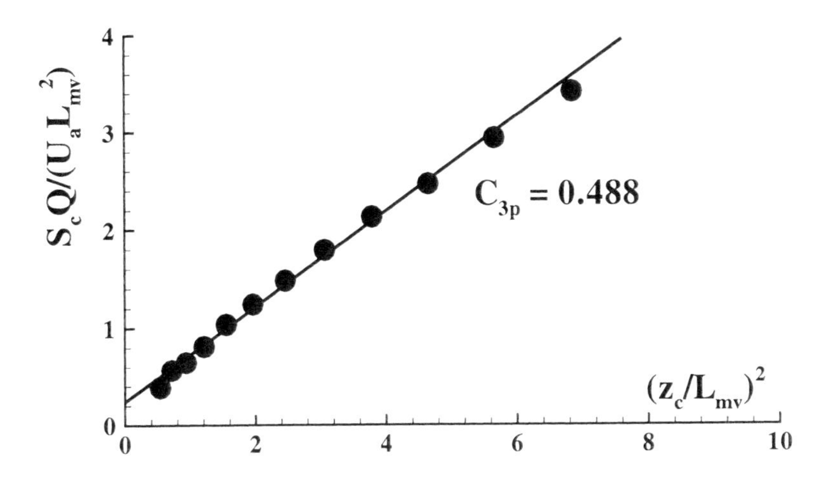

Figure 7.17. Centerline dilution $S_cQ/(U_aL_{mv}^2)$ vs centerline trajectory z_c/L_{mv}

- Similar to the line puff, the flow in the jet cross-section is characterized by a vortex pair structure, with an added mass coefficient of approximately 1.

- The scalar field is sectionally characterized by a kidney-shaped outline containing a double peak of concentration maxima. The puff aspect ratio (R_h/R_v) and the maximum concentration ratio (C_m/C_c) are both around 1.2.

One practical advantage of a field solution is the determination of the added mass coefficient k_n and the puff spreading rate β_n. The numerical work clearly demonstrates the existence of added mass. The substantial loss of nominal impulse is also supported by the ALP experiments (see next section). The asymptotic value of 0.5 for the puff momentum, a fundamental property, can be interpreted as an added mass coefficient of $k_n = 1$ for the puff motion (Chu 1985; Turner 1986). Half of the nominal impulse accounts for the irrotational flow of the surrounding fluid set up by the moving puff. It is interesting to note that (Table 7.1) approximately the same value of k_n is obtained in all the 2D and 3D numerical calculations, regardless of the turbulence model. From Table 7.1, the parameters for the simple Lagrangian model can be inferred - $k_n \approx 1$ and $\beta_n \approx \sqrt{2}\frac{db_{vc}}{dz} \approx 0.4$. The predicted maximum turbulence intensity $u_m/W_m = 0.37$ is also supported by the observed value of 0.42 based on concentration measurements (Chu 1996).

Table 7.1. Similarity relations of 2D line puff and 3D Advected Line Puff (ALP) derived from numerical solution and laboratory experiment; numbers refer to dimensionless coefficients of equations or ratios.

	Parameter	SKE	RNG	3D JET	Experiment
Puff momentum	M_p/M_{lo}	0.497	0.502	0.48	0.47 (Chu 1996) 0.3 (vortex ring; Glezer & Coles 1990)
Vertical velocity	$W_m = C_v M_{lo}^{1/3} t^{-2/3}$	1.61	1.90	1.18	1.21 (Chu 1996)
Puff front	$z_f = nR_h$	3.03	3.09		2.6-5.7 (Richards 1965) 3.1 (Chu 1996)
Trajectory	$z_c = C_{2p} M_{lo}^{1/3} t^{1/3}$	1.70	1.54	1.565	1.44 (Chu 1974) 1.56 (Wong 1991) 1.63 (Chu 1996)
Length scale	$L = C_{4p} M_{lo}^{1/3} t^{1/3}$	1.17	1.19		
Centre of mass	$L \sim \overline{z}$	0.75	0.74		0.71 (Chu 1996)
Scalar concentration	$C_o/C_c = C_{3p} z_c^2/Q_{lo}$	0.28	0.28	0.489	0.484 (Wong 1991) 0.42 (Chu 1996)
	C_m/C_c	1.02	1.02	1.35	1.0-1.56 (Wong 1991) 1.1-1.5 (Chu 1996)
Centerline half-width	$b_{vc} = C_{1p} z_c$	0.269	0.355	0.274	0.276 (Wong 1991) 0.294 (Chu 1996)
Aspect ratio	R_h/R_v	1.03	1.00	1.35	1.23 (Wong 1991) 1.25 (Chu 1996) 1.3 (Pratte & Baines 1967)
Velocity ratio	$W_m/\overline{W}$	3.18	3.73		3.17 (Chu 1996)
Circulation	$\Gamma/(\overline{W}L)$	2.67	2.71		2.98 (Chu 1996)
Momentum factor	$M_p/(\overline{W}L^2)$	0.69	0.64		0.83 (Chu 1996)
Turbulent viscosity	$\nu_{tm}/(W_m L)$	0.018	0.017		
Turbulent intensity	u_m/W_m	0.39	0.37	0.37	0.42 (Chu 1996)

5. MEASUREMENTS IN ADVECTED LINE PUFFS

Wong (1991) measured the cross-section tracer concentration (using salt as tracer) in the ALP using a probe-based techniques. Chu (1996) measured the ALP scalar field using non-intrusive LIF techniques, and the velocity field by LDA. The LIF measurements enable instantaneous and time-average measurements of the cross-section scalar field; from this many parameters of engineering interest (e.g. visual boundary, vertical and horizontal half-width, center of mass, vertical centerline minimum dilution and its location, aspect ratio) can be extracted. In particular, the added mass can be directly determined, and some idea of the turbulence structure gained. As shown in Fig. 7.18, the puff region is bounded by the puff edge defined by the concentration contour of $C = 0.25C_m$ where C_m is the cross-sectional maximum scalar concentration. Similar to the coflow jet (Chapter 6), it is found that this edge definition gives good agreement with the visual boundary. $2R_v$,$2R_h$ are the maximum vertical and horizontal width. C_c is the centerline maximum concentration; z_c is the location of C_c. The measured vertical z-velocity across the puff region is integrated to give the actual vertical jet momentum flux M_v, or the momentum of the equivalent line puff $M_p = M_v/U_a$.

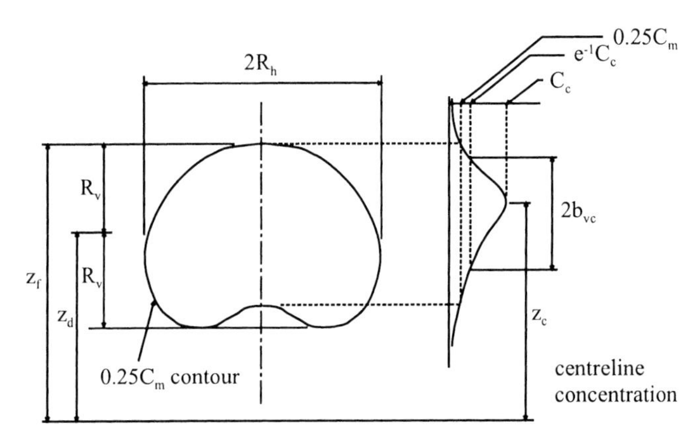

Figure 7.18. Definition diagram of characteristic properties of an advected line puff

Fig. 7.19 shows instantaneous images of the cross-sectional scalar field of ALP. The scalar field is highly intermittent; rotating patches with highly irregular rough boundaries are observed to move around the puff central region. However, unlike a pure jet or coflow jet, a distinct core turbulence structure concentrated near the centre does not exist; the intermittency factor is generally less than unity throughout the entire puff region. The entrainment of the ambient fluid into the core can be clearly observed in the surroundings of the rotating turbulent patches consisting of eddies of different scales. In addition to mixing near the puff front, significant entrainment takes place at the rear bottom of the puff. The engulfing of ambient fluid and traces of fluorescent dye (transported by wake vortices shed near the source) towards the puff centre from beneath can be clearly observed. This results in the bifurcation of patches and the counter-rotating vortex-pair flow structure. The time-averaged image of the scalar field is smooth and kidney-shaped. A double peak concentration distribution can generally be observed. The mean scalar field is marked by a lobe-like appendix in the rear; this is related to wake vortices shed near the jet exit, which are re-entrained into the rear of the main vortex-pair downstream (Fig. 7.1). The self-similar puff flow grows slowly with elevation.

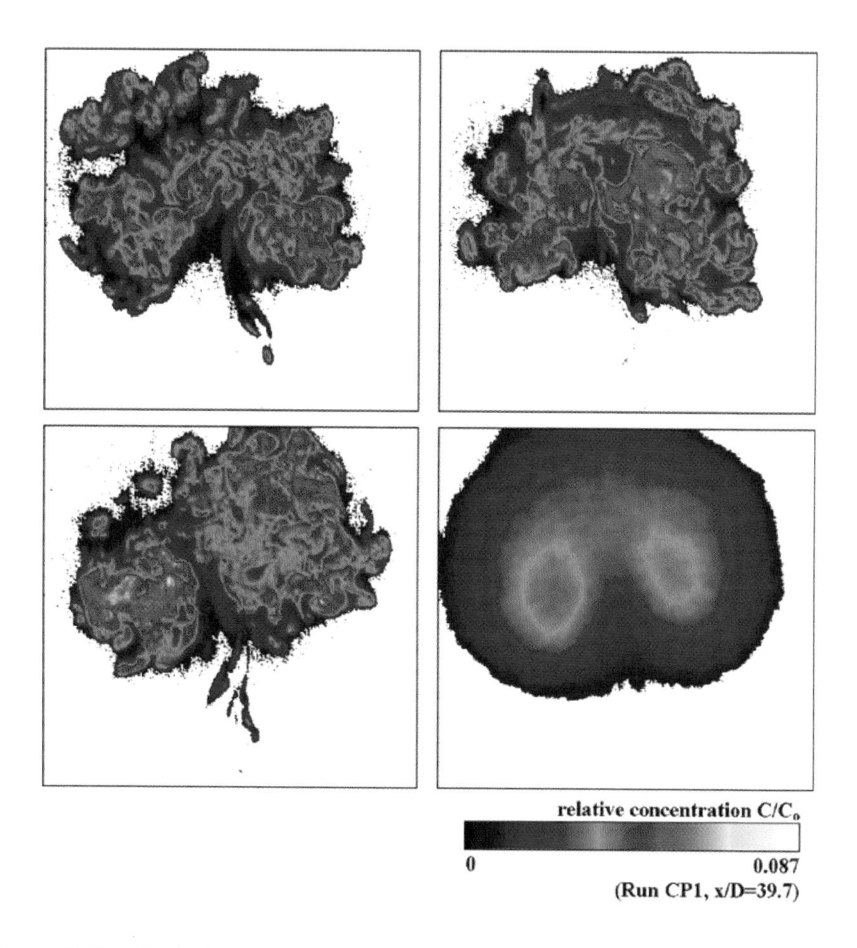

Figure 7.19. Typical instantaneous and time-averaged LIF images in the cross-section of the advected line puff ($U_o = 57.5$ cm/s, $U_a = 5.0$ cm/s, $\theta_o = 85^o$, $D = 0.75$ cm). Concentration lower than the visual threshold not shown.

5.1 TIME-AVERAGED PROPERTIES

Fig. 7.20 shows measured centerline half-width b_{vc}, jet trajectory z_c, and centerline minimum dilution S_c (Wong 1991). It can be seen the linear spread rate, the 1/3 trajectory law and the z^2 power dilution law are well supported by the data. Despite the narrow range of z/L_{mv} within which the puff can be measured, the data clearly supports the similarity relations, with $C_{1p} = 0.276$, the trajectory constant $C_{2p} = 1.56$, and the centerline dilution constant $C_{3p} = 0.46$. Similar values have been obtained in the LIF experiments. It is evident from Fig. 7.14, Fig. 7.18 and Fig. 7.19 that the trajectory z_c defined by the centerline maximum concentration is located generally higher than the visual centerline - a typical feature of the pear-shaped concentration structure. The visual trajectory z_d (defined as mid-way between top and bottom boundaries) is found to correspond exactly to the location of the centre of mass (Fig. 7.21); z_c however is on average about 7 % higher than z_d. This result can also be deduced from the numerical model results of the line puff (Problem 8.2). The jet spreads linearly with vertical distance, with $R_v = 0.36\ z_d$, and the aspect ratio is $R_h/R_v \approx 1.2$; the puff area is also found to agree perfectly with $A = \pi R_h R_v$. The puff front varies as $z_f = z_d + R_v = 3.52\ R_v$. Using an average aspect ratio of $R_h/R_v = 1.2$, we have $z_f = 3.06\ R_h$; this is contrary to a value of around 2.0 reported earlier by Richards (1965), whose data show a great deal of scatter. In terms of the jet radius in the Lagrangian model, $R = \sqrt{R_h R_v}$, this would translate to a puff spread rate of $dR/dz = \sqrt{2} \times 0.36 = 0.39$. The centerline dilution varies as $S_c \sim z_d^2$, with a dilution constant of 0.46. The ratio of flux-averaged to centerline minimum dilution, $\overline{S}/S_c$, is on average around 1.45. Unlike previous experiments of line puffs (Richards 1965), the constants for the similarity laws can be successfully correlated from the *collective* experimental data. The empirical trajectory and mixing rates obtained by different investigators are shown for comparison in Table 7.1 and Table 7.2.

Having defined the location of the ALP, it is possible to map the velocity field within the visually observed puff cross-section. Fig. 7.24 shows a typical three minute-averaged $v - w$ velocity distribution in the jet cross-section. Tracer concentration contours at $C/C_m = 0.05$, 0.135 and 0.25 are overlapped on the figure to show the extent of the scalar field. Fig. 7.25 shows the ratio of the measured vertical momentum flux M_p to initial vertical momentum M_{lo} of the equivalent line puff (= ratio of momentum fluxes M_v/M_{vo}) plotted against dimensionless distance $x^* = x/L_x$ ($= t^* = t/t_c$). In the developed stage, $x^* \geq 75$, the ratio shows a constant value of about 0.42 (s.d. = 0.058). This demonstrates the vertical momentum of the advected line puff is conserved once the flow is developed. The vertical momentum contained in the puff region bounded by the puff edge can be estimated to account for only about 91% of the total vertical momentum of the puff. The ratio of total ver-

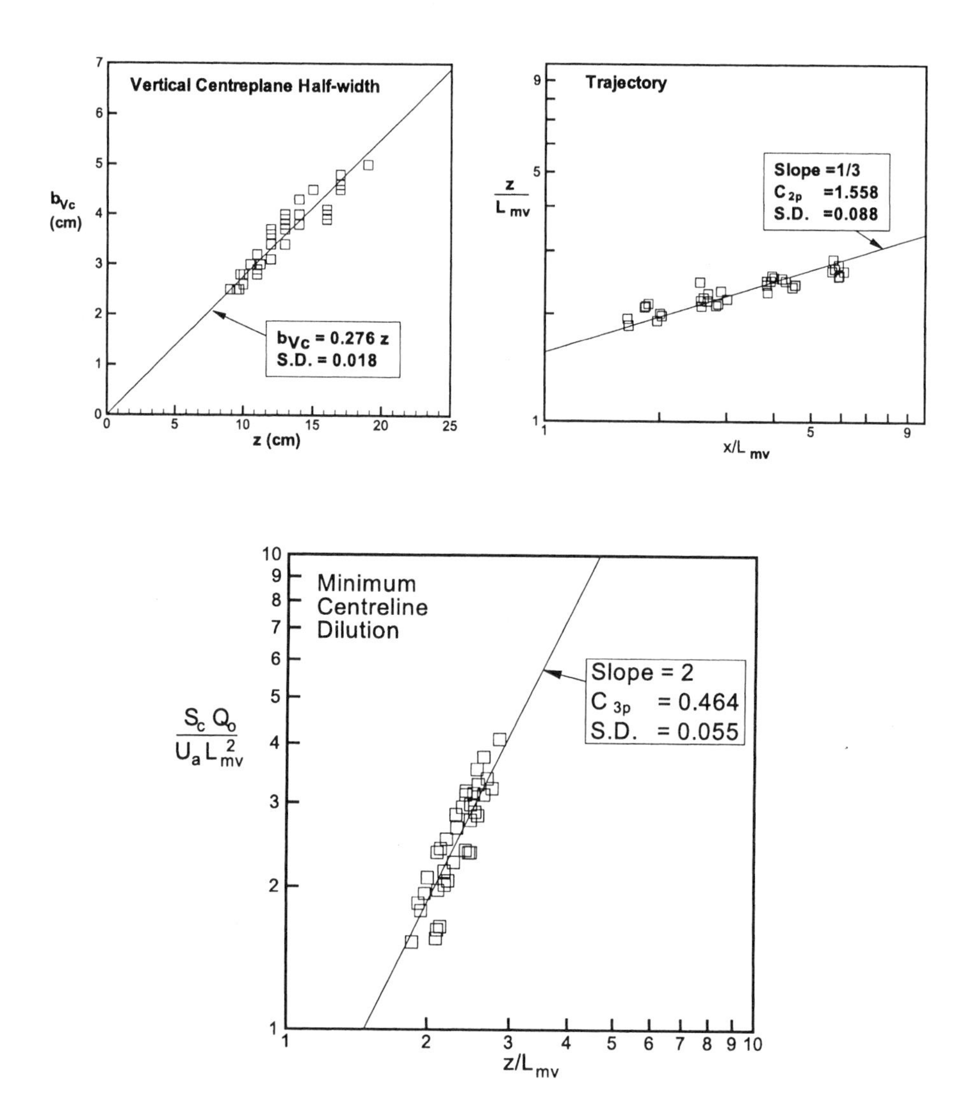

Figure 7.20. Comparison of self-similar ALP relations with measured width, trajectory, and dilution in ALP experiments by Wong(1991)

tical momentum of the advected line puff in the developed phase to the initial vertical momentum is equal to $0.424/0.91 \simeq 0.47$. In other words, the vertical jet momentum flux within the visible jet boundaries has been directly determined to be approximately half of the jet discharge momentum flux; the existence of added mass has been demonstrated. This experimentally determined puff momentum ratio of about 0.5 can

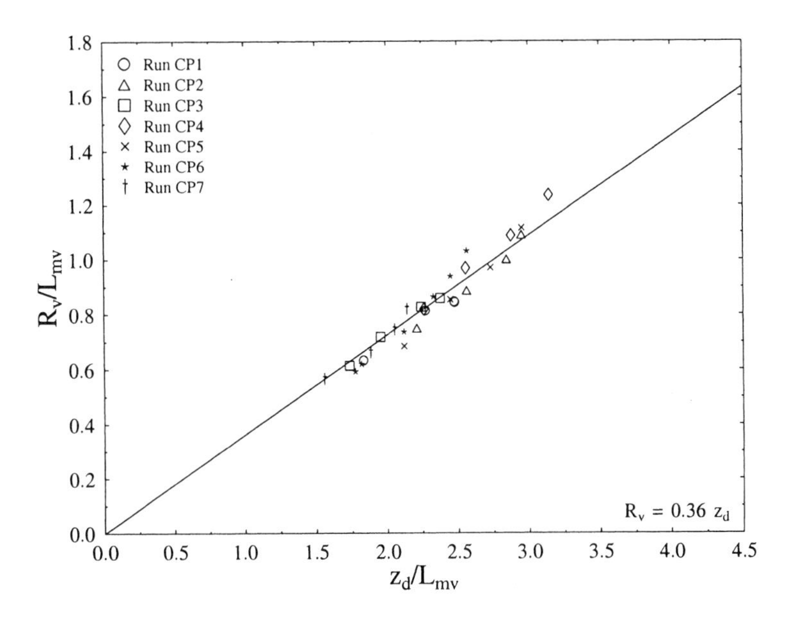

Figure 7.21. Puff half-width vs puff trajectory

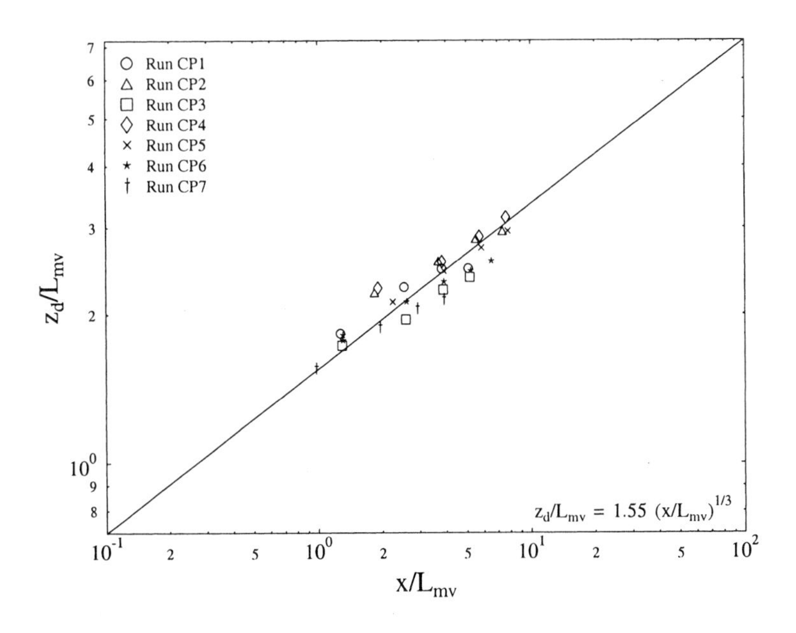

Figure 7.22. Puff trajectory vs horizontal distance

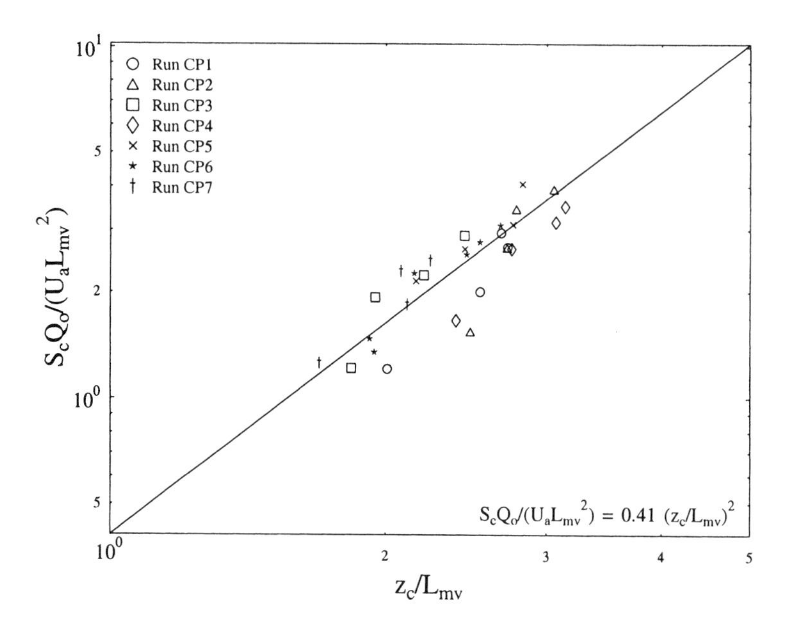

Figure 7.23. Minimum centerline dilution vs centerline trajectory

Table 7.2. Summary of experimental constants (visual spreading rate, visual and centreline trajectory, centreline and minimum dilution) of ALP

Previous studies	Advected line puffs (MDFF)				
	Spreading rate	Trajectory		Dilution	
	β	z_d	(z_c)	S_c	(S_m)
	C_{1p}	C_{2p}	(C_{2pc})	C_{3p}	$(C_{3pm}$
Richards (1965)	0.27[a]	2.07[a]			
Hoult and Weil (1972)	0.42[b]	1.38			
Pratte and Baines (1967)	0.39				
Chan, Lin and Kennedy (1976)		1.44			
Chu and Goldberg (1974)	0.35[b]	1.56			
Wright (1977b)		1.6	(2.1)	0.38	
Chu (1979)	0.34	1.64			
Lee (1989)			(1.77)	0.32	
Wong (1991)	0.43[c]	1.32	(1.56)	0.46	
JETLAG (Lee and Cheung 1990)	0.45[b]	1.3		0.49[d]	
Chu (1996) - shadowgraphy	0.45	1.54			
Chu (1996) - LIF	0.39	1.55	(1.63)	0.41	(0.34
1D Lagrangian model ($\beta_n = 0.4$ and $k_n = 1$)	0.4[e]	1.44	(1.51)	0.39	(0.33
				0.43[d]	(0.36

[a] using $R_h/R_v \simeq 1.2$
[b] deduced from entrainment coefficient using $k = 1$
[c] combined the results of visual and concentration half-width spreading rate
[d] based on puff trajectory
[e] selected value

be compared to values of $0.48 - 0.50$ predicted by the turbulence model calculations.

Velocity measurements have also been performed for a vertical jet in crossflow ($\theta_o = 90^o$). The results show good agreement with those of the advected line puff; a momentum loss coefficient of about 0.5 has also been observed (Chu 1996).

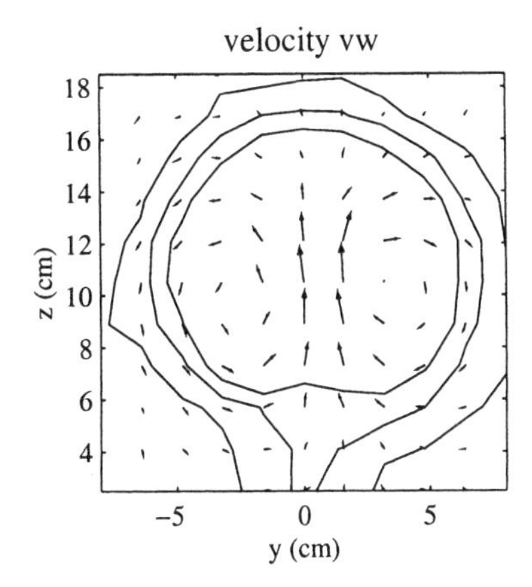

Figure 7.24. Typical time-averaged velocity field in vertical section of advected line puff ($U_o = 57.5$ cm/s; $U_a = 5.0$ cm/s; $\theta_o = 85^o$; $D = 0.52$ cm, $t/t_c = 147$)

5.2 TURBULENCE PROPERTIES

The measured horizontal and vertical profiles of the rms concentration fluctuation intensity, C_{rms}, are shown in Fig. 7.26. It is seen the turbulence intensity is much greater than that of the straight jet, with a value of around 0.4 in the core. Fig. 7.26 a) shows the centerline vertical profile of C_{rms} normalized by the centerline maximum concentration C_c. An asymmetric distribution with double peaks can be noted. The peak in the upper part of the puff is located at $(z - z_c)/b_{vc} \approx 0.75$, with turbulence intensity of 0.35-0.5. The lower peak is located at $(z - z_c)/b_{vc} \approx -0.6$, with intensity in the range of 0.26-0.6. The peak in turbulent intensity in the puff front (upper part) is consistent with turbulence model predictions (Lee *et al.*1996). The peak in the bottom part appears to reflect the significant entrainment of ambient fluid in the puff rear (cf Fig. 7.1 and 7.19). Fig. 7.26 b) shows the horizontal profile of C_{rms}, normalized by the cross-section maximum concentration C_{cm}, at the elevation of the center of mass of the puff. The turbulent intensity is characterized by two peaks located at $y/b_{vc} \approx \pm 0.8$, with values in

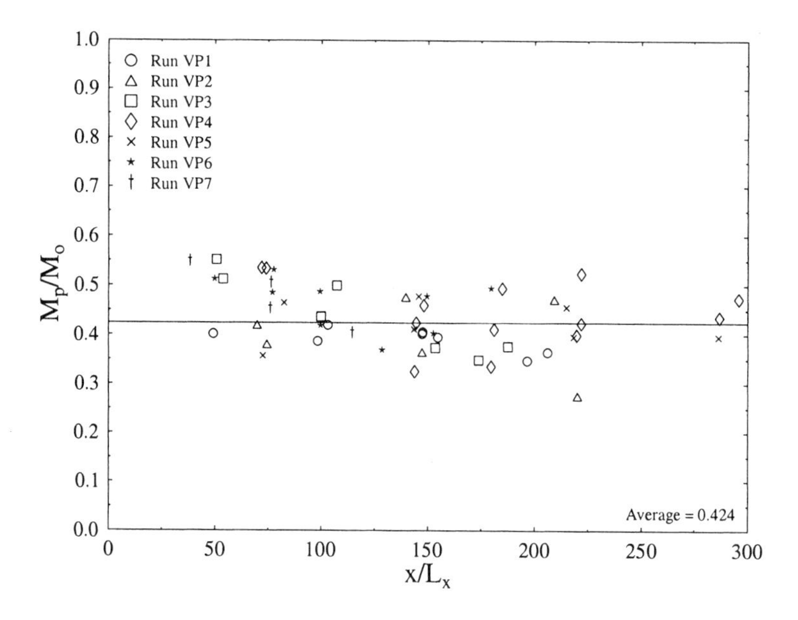

Figure 7.25. Ratio of measured puff vertical momentum to initial jet momentum

the range of 0.3-0.52. In view of such fluctuations, it is not surprising then that the instantaneous maximum concentration can be two times higher than that of the centerline maximum concentration, while the minimum concentration stays close to zero (Fig. 7.27). Fig. 7.28 shows the horizontal profile of the concentration intermittency at the level of the center of mass of the advected line puff. The intermittency has double peaks located at $y/b_{vc} \approx \pm 0.5$, and has a value of 0.7-1.0 near the centerline. In contrast to a coherent jet core (with an intermittency of one), the measured intermittency shows that parts of the puff core has an intermittency of less than 1; at these locations there is always a probability of ambient fluid present.

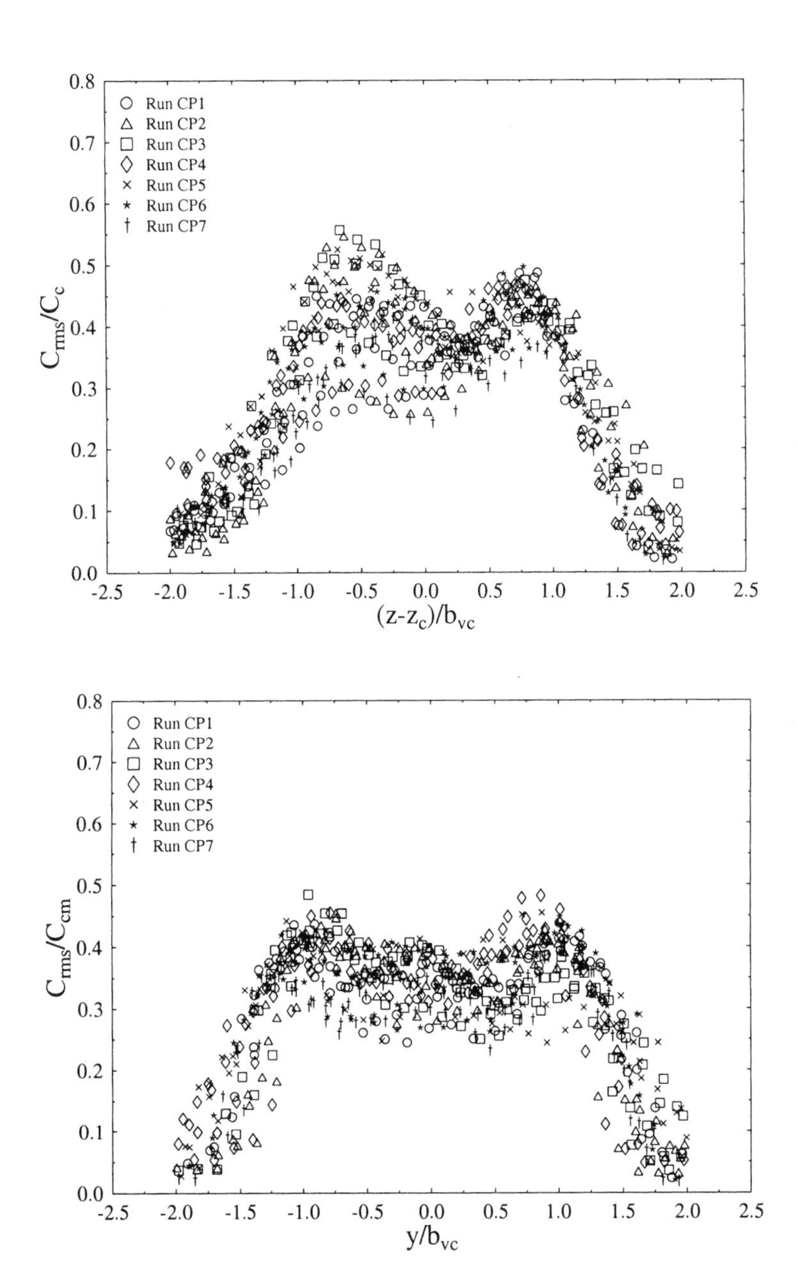

Figure 7.26. RMS concentration fluctuation of advected line puff: a) vertical centerline profile (top); b) horizontal profile at level of center of mass (bottom)

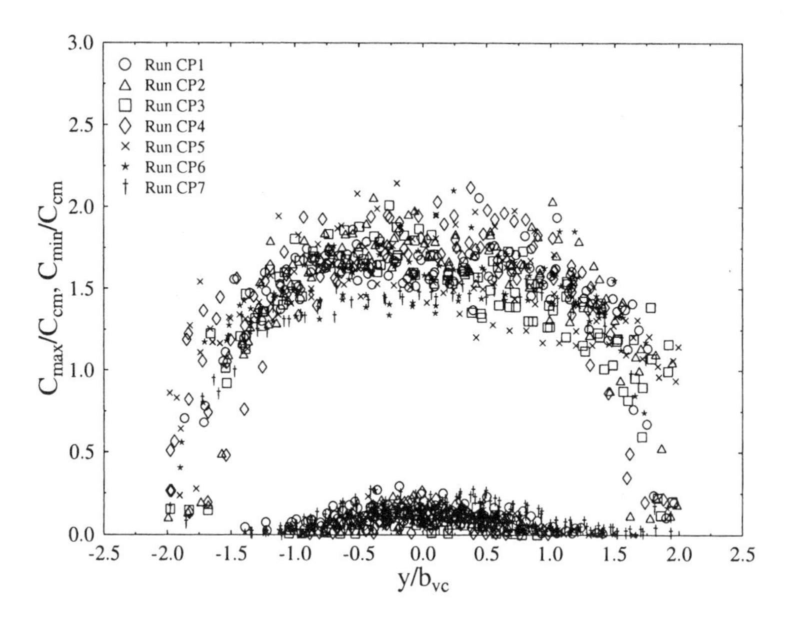

Figure 7.27. Horizontal profile of instantaneous maximum and minimum concentration of ALP

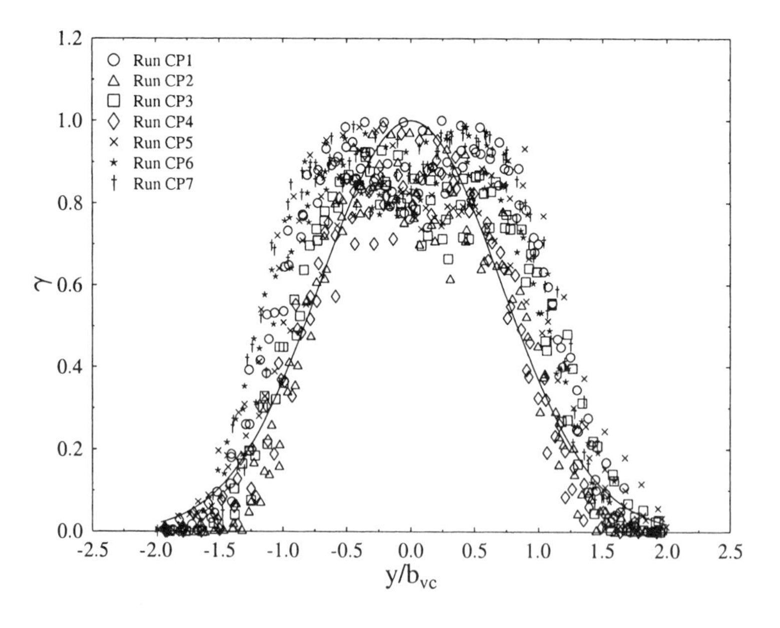

Figure 7.28. Horizontal profile of intermittency of ALP

6. PRACTICAL APPLICATION

For prediction of jet mixing in a crossflow, the numerical model results can be cast in a more convenient form. For completeness the well-known solution for the MDNF is also included.

Momentum-dominated Near Field (MDNF)

The classical relations for the characteristic properties of a round jet in stagnant fluid are (Chapter 2):

$$W_m = 7.0 \, M_{vo}^{1/2} \, z_c^{-1} \tag{7.29}$$

$$b_{vc} = 0.114 \, z_c \tag{7.30}$$

$$S_c = 0.16 \, \frac{M_{vo}^{1/2} \, z_c}{Q_o} \tag{7.31}$$

In the MDNF the jet is only weakly advected; mixing is similar to that of a jet in stagnant fluid. Note that $\frac{dz}{dx} = \overline{W}/U_a$, and for a Gaussian profile, the ratio of centerline maximum to average velocity can be shown to be $W_m/\overline{W} = 2$ by equating mass and momentum fluxes. The following equations can then be obtained for the advected jet:

$z/L_{mv} \leq 1$ **(MDNF):**

$$\frac{z_c}{L_{mv}} = 2.65 \, (\frac{x}{L_{mv}})^{1/2} \tag{7.32}$$

$$b_{vc} = 0.114 \, z_c \tag{7.33}$$

$$\frac{S_c Q_o}{U_a L_{mv}^2} = 0.16 \, (\frac{z_c}{L_{mv}}) \tag{7.34}$$

For example, the trajectory equation, Eq. 7.32, follows directly from invoking a kinematic relation and using Eq. 7.29 (with $\overline{W} = W_m/2$). The above simple advected jet model for the MDNF is verified by the detailed LIF experiments of a vertical jet in weak crossflow (Davidson and Pun 1999) which give a trajectory coefficient of 2.68 - very close to the value of 2.65 given in Eq. 7.32. This supports the use of the average velocity in defining the jet trajectory (see also related discussion on coflow jet formulation in Chapter 6 and BDNF in Chapter 8). Jet trajectory predictions by two Lagrangian models based on the use of average velocity are also in excellent agreement with data for a wide range of situations (see Chapters 9 and 10).

If one thinks of jet mixing as a diffusion process purely effected by small scale eddies (say of the order of 0.1 b), then it may appear that the jet trajectory (defined by the locus of centerline maximum concentration) is determined by the centerline maximum velocity. The experimental results however suggest that the jet mixing is governed by the large scale motion (the 'dominant eddy') which is responsible for the mass transfer between the center region and the edge of the jet, while the many small eddies (carried along by the dominant eddy) smooth the sharp concentration gradients created by the large scale motion. The fluid and scalar material are transported at a range of velocities with the mean jet path defined by the average of these velocities.

According to theory, the MDNF asymptotic flow regime is characterized by $z/L_{mv} \ll 1$; in practical application, however, experiments have shown that $z/L_{mv} \leq 1$ can often be used to delimit the region. It should be noted that none of the asymptotic relations are strictly applicable in the transition region.

Momentum-dominated Far Field (MDFF)

Based on the ALP numerical model and experimental results, the following equations for the advected line puff are recommended:

$z/L_{mv} > 1$ **(MDFF):**

$$\frac{z_c}{L_{mv}} = 1.56 \, (\frac{x}{L_{mv}})^{1/3} \tag{7.35}$$

$$b_{vc} = 0.28 \; z_c \tag{7.36}$$

$$\frac{S_c Q_o}{U_a L_{mv}^2} = 0.46 \, (\frac{z_c}{L_{mv}})^2 \qquad \text{or}$$

$$S_c = 0.46 \frac{U_a \; z_c^2}{Q_o} \tag{7.37}$$

The jet in crossflow has also been studied using a semi-empirical approach by laboratory and field experiments of a Canadian river (Hodgson and Rajaratnam 1991). It can be shown that the dilution prediction for the asymptotic stage can be expressed as a function of distance downstream (x) from a jet discharge in the following form (Lee 1993):

$$S_c = \frac{C_o}{C_m} = 0.64 \, (\frac{W_o \; x}{U_a D})^{2/3} \tag{7.38}$$

Fig. 7.29 shows the comparison of Eq. 7.38 with the experiments of Hodgson and Rajaratnam and their field dilution data.

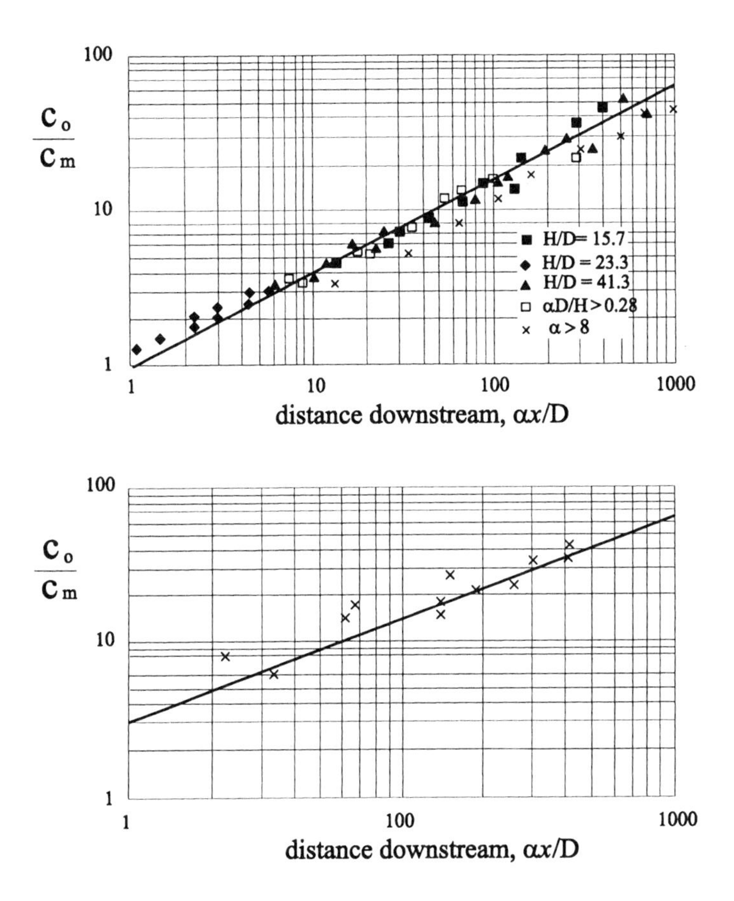

Figure 7.29. Comparison of Eq.28 with the laboratory and field data (upper/lower plot respectively) of Hodgson and Rajaratnam (1992); $\alpha = W_o/U_a$

EXAMPLE 7.1 *Consider a single jet discharge into a river with* $Q = 0.0138\ m^3/s$. *The jet diameter is* $D = 0.1\ m$. *The river depth is* $H = 20\ m$, *and the river velocity is* $U_a = 0.2\ m/s$. *The initial density difference (buoyancy) of the effluent is negligible. Estimate the initial dilution, vertical jet velocity, and width at an elevation of* $z = 12\ m$. *Also estimate the corresponding downstream distance of the jet at this location.*

Solution: The discharge is modelled as a turbulent jet in crossflow. Compute jet velocity and momentum fluxes and obtain the crossflow momentum length scale L_{mv} as:

$$W_o = \frac{0.0138}{\pi \times 0.1^2/4} = \underline{1.76}\ m/s$$

$$M_{vo} = Q_o W_o = 0.0138 \times 1.76 = \underline{0.0243}\ m^4/s^2$$

$$L_{mv} = \frac{M_{vo}^{1/2}}{U_a} = \frac{\sqrt{0.0243}}{0.2} = \underline{0.78}\ m$$

At an elevation of $z = 12$ m, $z/L_{mv} = 12/0.78 = 15.4 \gg 1$, apply the MDFF equations to obtain the dilution, trajectory, and width.

$$S_c = 0.46 \frac{0.2 \times 12^2}{0.0138} = \underline{960}$$

This occurs at a distance given by the trajectory equation:

$$x = 0.78 \times (15.4/1.560)^3 = \underline{750}\ m$$

Note the relatively long distance that it takes for the jet to travel to this elevation. Using $B = \sqrt{R_h R_v} = 0.4z = \underline{4.8}\ m$ and $R_h/R_v = 1.2$, the vertical half-width is $R_v = 4.8/\sqrt{1.2} = \underline{4.4}\ m$, and $R_h = \underline{5.3}\ m$. Typically a good estimate of the visible width of a jet is approximately $\sqrt{2}b$; this suggests that the bent-over jet is quite well-mixed over the cross-section at this location. The vertical velocity at this location can also be shown to have decayed to negligible values:

$$W_m = 1.33 \times (0.0243/0.2)^{1/3}/(750/0.2)^{2/3} = \underline{2.73 \times 10^{-3}}\ m/s$$

The efficiency of mixing in a current ($S_c = 960$) can best be appreciated by computing the stillwater dilution at this elevation (Eq. 7.31):

$$S_c(U_a = 0) = 0.16 \times \frac{0.0243^{1/2} \times 12}{0.0138} = \underline{21.7}\ !$$

Alternatively, the moving water dilution for this problem can be obtained from the semi-empirical equation based on the field data of Hodgson and Rajaratnam (Eq. 7.38):

$$S_c = 0.64 \times (\frac{1.76 \times 750}{0.2 \times 0.1})^{2/3} = \underline{1046}$$

EXAMPLE 7.2 *A town discharges a highly toxic flow of* $Q_o = 0.442$ m^3/s *through a 0.92 m square culvert at a river bank. The river depth is*

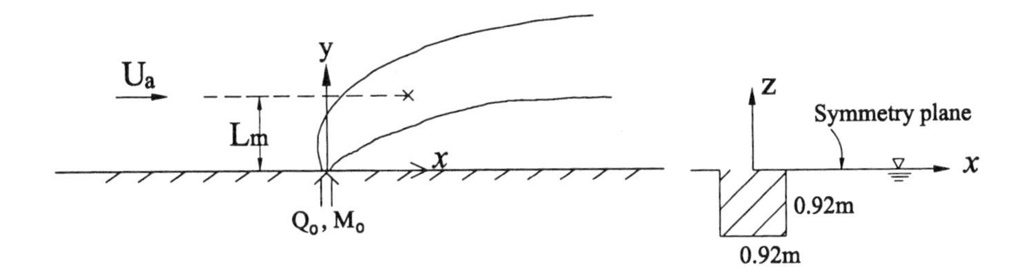

H = 7.6m, and the design river velocity is $U_a = 0.0137$ m/s. *The effluent is just submerged below the surface. Estimate the initial dilution 20 m downstream from the discharge point.*

Solution: The toxic effluent is discharged as a surface jet. Assuming zero momentum and mass fluxes across the free surface, the velocity and concentration gradient normal to the free surface is zero. The surface jet can be modelled as one-half of a momentum jet in crossflow, with the free surface approximating a plane of symmetry. The jet velocity, momentum and volume fluxes can be computed to be: $U_o = 0.52$ m/s, $Q_o = 2 \times 0.442$ m^3/s, $M_o = 0.884 \times 0.52 = 0.46$ m^4/s^2 The crossflow momentum length scale is then $L_m = M_o^{1/2}/U_a = 49.5$ m. The equivalent hydraulic diameter is $D = \sqrt{4/\pi \times 2 \times 0.92^2} = 1.47$ m. At $x = 20$ m, use Eq. 7.35 to obtain $z/L_{mv} = 1.2$. Hence use the MDFF Eq. 7.38 to obtain: $S_c = 0.64(\frac{0.52}{0.0137}\frac{20}{1.47})^{2/3} = \underline{41}$.

7. SUMMARY

The bent-over phase of a turbulent jet in crossflow has been analysed numerically and experimentally using the advected line puff (ALP) analogy. Numerical computation results obtained using 2D and 3D turbulence models are compared with probe-based as well as non-intrusive LIF and LDA measurements of the advected line puff in the laboratory. Simple equations for mixing analysis are developed. The spread rate and added mass coefficient for the advected line puff are:

$$\text{Spreading rate}: \quad \frac{dR}{dz} = \beta_n = 0.4$$

$$\text{Added mass coefficient}: \quad k_n = 1.0$$

Chapter 8

PLUME IN CROSSFLOW: ADVECTED LINE THERMALS

A variety of turbulent flows observed in nature are driven by sources of buoyancy. The interaction of such sources with the surrounding flow is generally characterized as plumes in crossflow. Fig. 8.1 is an example. The plume in crossflow is produced in the laboratory by discharging buoyant water from a pipe of diameter $D = 1$ cm at a velocity $U_o = 0.25$ m/s into a crossflow of velocity $U_a = 0.063$ m/s. The LIF (Laser Induced Fluorescence) images in the figure show the structure of the turbulence and the concentration profile on the plane of symmerty. The initial relative density difference of 2.57% between the source fluid and the crossflow is created by salinity differences ($\Delta\rho_o/\rho_a = 0.0257$). The source densimetric Froude number $Fr = U_o/\sqrt{g\Delta\rho_o/\rho_a D} = 5.0$ is quite small in this case. Buoyancy becomes the dominant effect in a short distance from the source. The buoyant effluent is seen in the figure to contract slightly and then mixed with the surrounding fluid to form the plume in cross flow. Close to the source, the flow behaves essentially as a plume. In the far field, the plume is bent-over to form the advected line thermal. In a co-ordinate system moving with the crossflow, the surounding fluid is stationary and the plume rises essentially as a line thermal in stagnant fluid. A cross-section of the thermal is a vortex pair as shown in Fig. 8.2. The average tracer concentration profile in the cross-section of the thermal is similar to that of the advected line puff examined in the previous chapter. While the average flow pattern is similar, the eddying motion in the thermal is more vigorous — the turbulent intensity is much greater when compared with the puff. The study of the thermal is most significant as most of the flows observed in lakes, oceans and the atmosphere are driven by buoyancy. Given sufficient depth for the buoyancy force to act, any buoyant jet in crossflow will tend to behave like plumes and thermals in the ultimate stage of mixing with the surounding fluid.

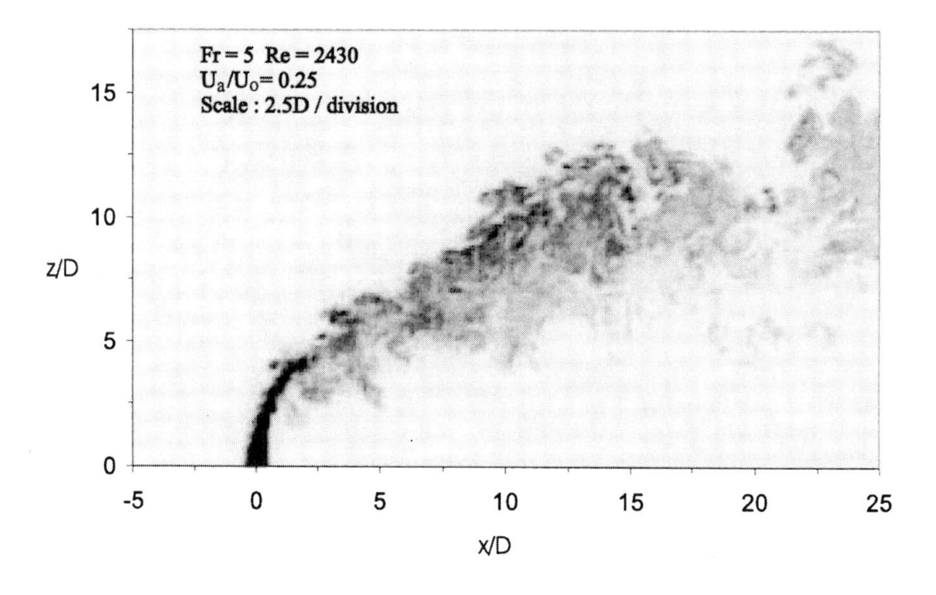

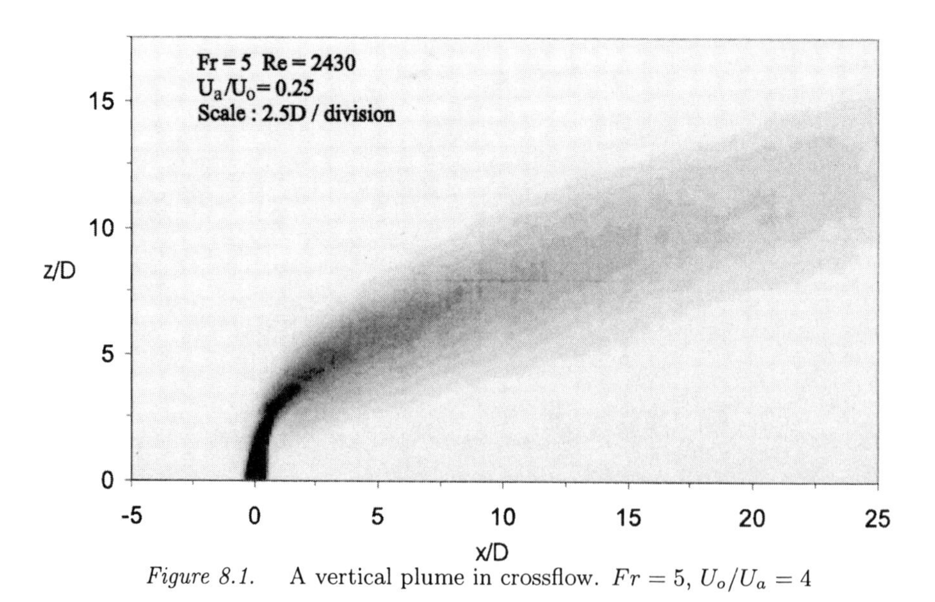

Figure 8.1. A vertical plume in crossflow. $Fr = 5$, $U_o/U_a = 4$

Buoyant plumes in crossflow have been extensively studied. Most of the previous studies are based on probe-based measurements of the

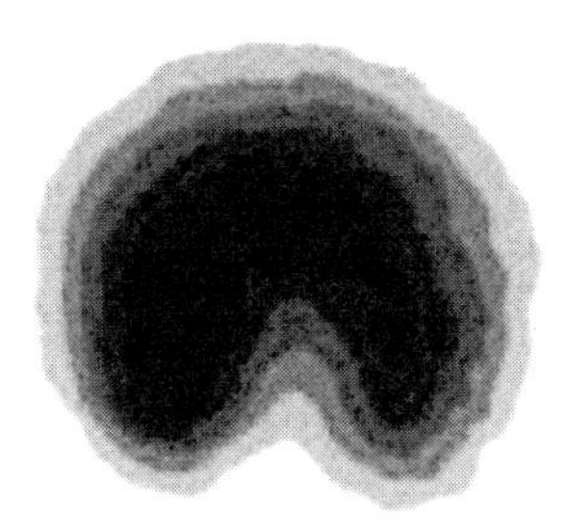

Figure 8.2. Tracer concentration distribution in cross section of bent-over plume in crossflow ($x/D = 12$, $C_m/C_c = 1.2$)

trajectory and the dilution on the center plane of symmetry (e.g. Scorer 1959, 1978; Csanady 1965; Chu and Goldberg 1974; Fischer *et al.* 1979). Asymptotic relations of the plume trajectory in the near and far fields have been established from these measurements. Detailed calculations and measurements in the cross-section of a bent-over plume have also been made (e.g. Lilly 1964; Yih 1981; Knudsen 1988; Wong 1991; Wood 1993; Gaskin and Wood 2001; Chen and Lee 2002).

In this chapter, the laboratory data are presented and compared with the results of the computational models. The plume-in-crossflow problem is examined first using a length scale analysis. In the near field, the effect of the discharge buoyancy dominates. In the far field, the plume behaves essentially as a line thermal. In the remainder of the chapter, the emphasis is given to the flow of the line thermal. Spreading coefficient and added mass coefficient are introduced in a one-dimensional (1D) model as the parameters characterizing the turbulent motion and the irrotational motion surrounding the line thermal. Two-dimensional (2D) and three-dimensional (3D) mathematical models are developed to compute the turbulent flow. Measurements and numerical computations on the flow and tracer concentration field of advected line thermals are presented. Spreading coefficient and added mass coefficient, the parameters of the 1D model, are evaluated using the 2D and 3D models and the data obtained from the laboratory measurements.

1. LENGTH SCALES AND REGIMES

The problem of the plume in cross flow is delineated in Fig. 8.3. The plume is shown to be produced by a vertical round source of diameter D, initial velocity U_o, and tracer concentration C_o discharging into a steady

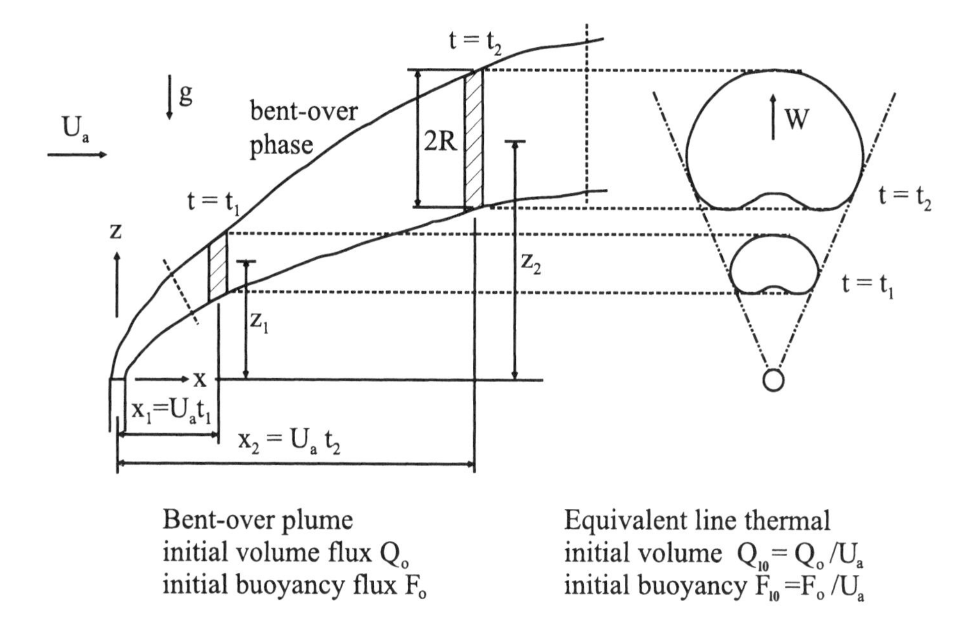

Figure 8.3. Schematic of a plume in crossflow showing the Lagrangian element and its cross section in the analog of the advected line thermal

uniform crossflow U_a in the horizontal $+x$ direction. The initial volume flux of the discharge is $Q_o = U_o \pi D^2/4$. The specific buoyancy flux is $F_o = Q_o g_o'$. Neglecting the effect of the initial momentum flux, the behaviour of the plume is determined by its buoyancy flux F_o and the crossflow velocity U_a. Asymptotic regimes of the plume can be defined in terms of the length scale

$$L_b = \frac{F_o}{U_a^3}. \tag{8.1}$$

The vertical velocity in the plume, W, varies with the height z above the source as a minus one-third power law: $W \sim F_o^{1/3} z^{-1/3}$. As the vertical velocity diminishes with height, the plume is bent over and moves primarily in the horizontal direction of the crossflow. The length scale L_b, defined by the buoyancy flux F_o and the crossflow velocity U_a, is the height of the plume when the buoyancy-induced velocity W has decreased to a value comparable to the velocity of the crossflow ($W \sim U_a$ and $z \sim F_o/U_a^3$). Asymptotic flow regimes are defined by the dimensionless height z/L_b. For dimensionless height $z/L_b = zU_a^3/F_o \ll 1$, the buoyancy-induced velocity W is much greater than the crossflow velocity. In this near-field region, the plume is essentially vertical and the turbulent motion in the region is similar in structure to that of the plume in stagnant fluid, but slightly advected. For dimensionless height $z/L_b \gg 1$, the vertical velocity W is small compared with the horizontally velocity of the crossflow, U_a. The behaviour in this far field is similar to that of an advected line thermal (Scorer 1979; Wood 1993). Wright (1977a) used the terms BDNF (buoyancy-dominated near field) and BDFF (buoyancy-dominated far field) to characterize the "advected plume" and the line thermal regimes respectively. Plume in stagnant fluid was considered in Chapter 3. The transition from BDNF to BDFF will be an example of the general formulation to be described in Chapter 9 and 10. This chapter concentrates mainly on the BDFF.

1.1 LINE-THERMAL ANALOGY FOR BDFF

The plume in the BDFF can be modeled using a line-thermal analogy. Consider the control volume in Fig. 8.4, balancing the tracer mass flux of the inflow and the mass flux of the outflow through a cross-section of the plume, we have $Q_o c_o = \int_{x-section} u \, c \, dA$, where u and c are the velocity and the concentration at the cross-section respectively. Assuming a linear equation of state, the buoyancy flux $F(x)$ passing a vertical section is then equal to the initial buoyancy flux, with $F(x) = \int_{x-section} u \, g' \, dA = F_o = Q_o g'_o$, where g' = reduced gravity. In the BDFF, $u \simeq U_a$. Hence, $F_o \simeq U_a \int_A g' \, dA$. The flow in successive vertical x-sections can then be approximated by a line thermal with buoyancy $F_{lo} = \int_{x-section} g' \, dA = F_o/U_a$ (per unit x-length) released in the $+z$ direction, and initial volume $Q_{lo} = Q_o/U_a$.

For an observer in a coordinate system moving with the crossflow, the line thermal rises in a stagnant ambient fluid and the plume is issuing the buoyancy to the line element of the thermal at a steady rate. As far as the mixing in the BDFF is concerned, the action of the plume over a period of time Δt is to apportion the source of buoyancy $F_o \Delta t$ to the line thermal over a length equal to $U_a \Delta t$. Therefore, the buoyancy per unit length of the line thermal is F_o/U_a. The plume distributes the buoyancy in a continuous manner to the thermal. The Lagrangian element in the line thermal has a length equal to $U_a \Delta t$ and a buoyancy per unit length of the element equal to $F_o \Delta t / U_a \Delta t$. The bent-over plume at successive cross-sections, located at $x = x_1$ and x_2, corresponds to the line thermal observed at successive times (t_1, t_2), respectively. The space and time relations are $x_1 = U_a t_1$ and $x_2 = U_a t_2$ via the Galilean transformation. This thermal analog of the plume in BDFF is to be confirmed by the experiments and is expected to hold in the bent-over stage of the plume, where $z \geq L_b$. The flow according to the analogy is referred to in the present context as ALT (advected line thermal).

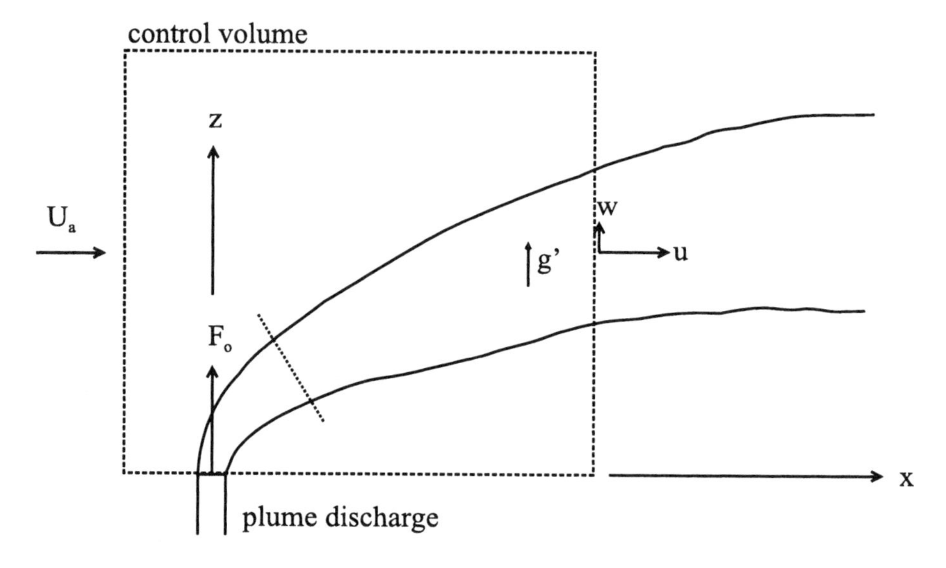

Figure 8.4. Control volume for plume in crossflow. The buoyancy flux entering the control volume is F_o. The buoyancy flux through a cross-section perpendicular to the crossflow is $F(x) = \int_A g'(udA)$

1.2 SIMILARITY VARIABLES FOR THE LINE THERMAL

A line thermal is the two-dimensional motion formed when a line of buoyant fluid is released from rest instantaneously. This is illustrated in Fig. 8.5 for an experiment done upside down (Richards 1963). An isolated mass of dyed negatively buoyant fluid is released from a cylindrical trough extending perpendicularly between parallel vertical walls in a water tank. The release is located sufficiently far away from boundaries. Consider a line thermal with initial volume Q_{lo} (per unit x-length) and buoyancy F_{lo} at $t = 0$ (Fig. 8.6). The initial tracer concentration is C_o. Fig. 8.6 shows the two-dimensional unsteady flow in a cross section on the y-z plane of the thermal. The thermal motion is characterized by a vortex pair with its concentration maxima located at the turbulent cores. The geomerty of the thermal is characterized by the radii of the cross section R_v and R_h, the leading edge at the front z_f, the mass center z_d. The motion of the thermal is characterized by its vertical velociy, W, and the concentration profile on the center plane of symmetry by the half-width b_{vc}, the maximum concentration C_c and its location Z_c. The flow in the thermal is driven by the line source buoyancy. The thermal

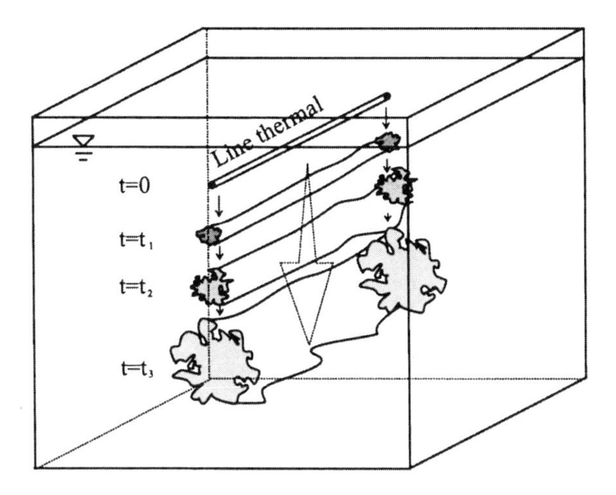

Figure 8.5. Line thermal generated by releasing a cylinder of negatively buoyant fluid instantaneously in a tank

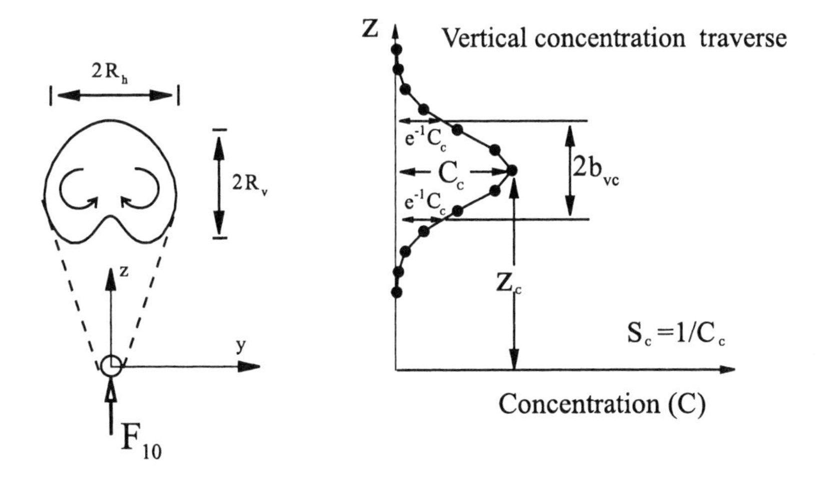

Figure 8.6. Definition sketch showing the cross-section of the line thermal and the concentration profile on the center plane of symmetry

characteristics are function of the buoyancy flux F_{lo} and time t. The initial volume, Q_{lo}, could also be important. The time scale

$$t_c = \frac{Q_{lo}^{3/4}}{F_{lo}^{1/2}} = \frac{Q_o{}^{3/4}}{U_a^{1/4} F_o^{1/2}} \tag{8.2}$$

and the corresponding length scale

$$L_x = U_a t_c = \frac{(Q_o U_a)^{3/4}}{F_o^{1/2}} \tag{8.3}$$

are measures of the relative importance of the initial volume Q_{lo} and buoyancy flux F_{lo}. If the flow is self-similar, the following relations for velocity and width are expected from dimensional analysis:

$$\frac{W_m}{W_r} \sim f(\frac{t}{t_c}) \tag{8.4}$$

$$\frac{b}{Q_{lo}^{1/2}} \sim f(\frac{t}{t_c}) \tag{8.5}$$

where W_m and W_r stand for the maximum and reference vertical velocity of the thermal respectively. The dimensionless time is $t^* = t/t_c$. The vertical velocity of the thermal is proportional to $W_r = F_{lo}^{1/2}/Q_{lo}^{1/4}$. The effect of the initial volume Q_{lo} is negligible for fully developed thermal at large time. The asymptotic relations for velocity, width and trajectory of the thermal for $t/t_c \gg 1$ are:

$$W_m \sim F_{lo}^{1/3} t^{-1/3} \quad \text{or} \quad \frac{W_m}{W_r} \sim (t^*)^{-1/3} \tag{8.6}$$

$$b \sim F_{lo}^{1/3} t^{2/3} \tag{8.7}$$

$$b \sim z \quad \text{or} \quad z \sim Q_{lo}^{1/2} (t^*)^{2/3} \tag{8.8}$$

Conservation of tracer mass also implies:

$$S_c = \frac{C_o}{C} \sim \frac{b^2}{Q_{lo}} \sim \frac{F_{lo}^{2/3} t^{4/3}}{Q_{lo}} \sim \frac{z^2}{Q_{lo}} \tag{8.9}$$

Given the equivalence between successive vertical x-sections of a bent-over plume and the line thermal at corresponding times $t = x/U_a$, the following relations can be obtained for the spreading rate, trajectory, and dilution of the 3D bent-over plume in terms of the crossflow buoyancy length scale L_b.

$$b_{vc} \sim z_c \tag{8.10}$$

$$\frac{z_c}{L_b} \sim (\frac{x}{L_b})^{2/3} \tag{8.11}$$

$$\frac{S_c Q}{U_a L_b^2} \sim (\frac{z_c}{L_b})^2 \tag{8.12}$$

$$\frac{W_r}{W_m} \sim (x^*)^{1/3} \tag{8.13}$$

where b_{vc} is the vertical centerline half-width defined by the $e^{-1}C_c$ points, and z_c is the vertical location defined by the maximum centerline concentration C_c. $S_c = C_o/C_c$ is the centerline minimum dilution, as illustrated in Fig. 8.6; and the dimensionless downstream distance $x^* = x/L_x$. For the sake of reference, Eqs. 8.11 and 8.12 can be expressed alternatively as:

$$z_c \sim (Q_o/U_a)^{1/2}(x^*)^{2/3} \tag{8.14}$$

$$S_c \sim z_c^2/(Q_o/U_a) \tag{8.15}$$

The proportionality constants in the above similarity relations can be derived by best-fitting experimental data or numerical model results.

2. 1D MODEL OF LINE THERMAL

Similar to the line puff, it is instructive to first analyse the line thermal motion by a one-dimensional model. The formulation is based on the Lagrangian spreading hypothesis and the concept of the dominant eddies described in Chapter 4. The dominant eddies in the thermal rises in a stagnant environment. The elevation of the dominant eddies is z. The rising velocity is

$$W = \frac{dz}{dt}. \tag{8.16}$$

The size of the dominant eddies is defined by R_v and R_h as shown in Fig. 8.6. The thermal radius is assumed to increase at a rate proportional to the rising velocity:

$$\frac{dR}{dt} = \beta_n W, \tag{8.17}$$

where $R = \sqrt{R_h R_v}$ is the radius. The proportionality constant is the spreading coefficient, β_n. Since $W = dz/dt$, it follows the thermal radius increases linearly with height (neglecting initial radius):

$$R = \beta_n z \tag{8.18}$$

The volume per unit length of the line thermal is $Q_l = \pi R^2 = \pi(\beta z)^2$. The momentum associated with this thermal volume has two parts. The part associated with the turbulent motion is $\pi R^2 W$. The part associated with the irrotational motion is $k_n \pi R^2 W$. The total momentum is

$$M_t(t) = (1 + k_n)\rho\pi R^2 W; \tag{8.19}$$

the added mass coefficient k_n is the fraction of the momentum in the irrotational surrounding. The thermal momentum increases at a rate equal to the buoyancy force, i.e.,

$$\frac{dM_t}{dt} = \frac{d}{dt}[(1 + k_n)\pi R^2 W] = F_{lo} = \frac{F_o}{U_a} \tag{8.20}$$

where F_o/U_a is the buoyancy force on unit length of the line thermal. Integrating Eq. 8.20 with time gives

$$(1 + k_n)M_t = F_{lo}t \tag{8.21}$$

Further integration with time leads to:

$$R = (\frac{3\beta_n F_{lo}}{2(1 + k_n)\pi})^{1/3} \ t^{2/3} \tag{8.22}$$

$$W = (\frac{4F_{lo}}{(1 + k_n)9\pi\beta_n^2})^{1/3} \ t^{-1/3} \tag{8.23}$$

$$z = (\frac{3F_{lo}}{2(1 + k_n)\pi\beta_n^2})^{1/3} \ t^{2/3} \tag{8.24}$$

Tracer mass conservation results in:

$$\frac{C_o}{C} = \frac{\pi R^2}{Q_{lo}} = \frac{\pi R^2 U_a}{Q_o} \tag{8.25}$$

The line thermal solution translates via a Galilean transformation into the well-known two-third power trajectory law for a bent-over plume in crossflow:

$$z = (\frac{3}{2(1 + k_n)\pi\beta_n^2})^{1/3}(\frac{F_o x^2}{U_a^3})^{1/3} \tag{8.26}$$

or in dimensionless form:

$$\frac{z}{L_b} = C_{2t}(\frac{x}{L_b})^{2/3} \tag{8.27}$$

where

$$C_{2t} = [\frac{3}{2(1 + k_n)\pi\beta_n^2}]^{1/3} \quad \text{and} \quad L_b = \frac{F_o}{U_a^3}. \tag{8.28}$$

Since $R = \beta_n z$, and the average dilution $S = C_o/C$,

$$S = \frac{\pi\beta_n^2 z^2 U_a}{Q_o}. \tag{8.29}$$

where z is given by Eq. 8.27.

There are two model parameters, β_n and k_n. They can be obtained from numerical computations and measurements of the flow and tracer concentration field of the plume in crossflow. The remainder of this chapter explain how these 1D-model parameters are independently determined for consistent prediction of the spread rate, path, and dilution in the BDFF.

3. 2D MODEL OF LINE THERMAL

3.1 NUMERICAL SIMULATION OF LINE THERMALS

Numerical calculations are performed for the line thermal using a similar procedure as the numerical experiment for the line puff as described in the previous chapter (see also Lee and Chen 2002). Fully three-dimensional calculations for the plume in crossflow also are conducted (Chen and Lee 2002). The results of the two-dimensional calculations using the k-ϵ model are presented in this section.

Computations are based on the following Reynolds-averaged equations that are derived from the assumptions of (1) small density differences, $(\rho_a - \rho_o)/\rho_a \ll 1$, and (2) linear equation of state, $\rho = \rho_a(1 + \beta C)$:

$$\frac{\partial U_i}{\partial x_i} = 0 \tag{8.30}$$

$$\frac{\partial U_i}{\partial t} + U_j \frac{\partial U_i}{\partial x_j} = -\frac{1}{\rho_a}\frac{\partial p}{\partial x_i} + \frac{\partial \tau_{ij}}{\partial x_j} + \beta C g_i \tag{8.31}$$

$$\frac{\partial C}{\partial t} + U_j \frac{\partial C}{\partial x_j} = \frac{\partial}{\partial x_j}\left(\frac{\nu_{eff}}{Sc_t}\frac{\partial C}{\partial x_j}\right) \tag{8.32}$$

where $U_i = (V, W)$ = fluid velocity in y and z direction respectively, p = dynamic pressure, $\beta = \frac{1}{\rho_a}\frac{d\rho}{dC}$ is a constant fluid property, C = concentration, τ_{ij} = stress tensor, is equal to the sum of the viscous and Reynolds-stresses (Eq. 7.25), and $g_i = (0, -g)$ is the gravitational acceleration vector. The gradient transport analogy (Boussinesq hypothesis) is assumed for the stresses and tracer mass (buoyancy) fluxes. The effective viscosity $\nu_{eff} = \nu + \nu_t$, where $\nu_t = C_\mu k^2/\epsilon$ is the turbulent viscosity. k is the turbulence kinetic energy (TKE), ϵ is the dissipation rate (DR) of k; the effective diffusivity is given by ν_{eff}/Sc_t, where Sc_t is the turbulent Schmidt number taken as 0.7 as commonly recommended. The buoyancy-extended two-equation $k - \epsilon$ model (Rodi 1980) is adopted for turbulent closure:

$$\frac{\partial k}{\partial t} + U_i \frac{\partial k}{\partial x_i} = \frac{\partial}{\partial x_i}\left(\frac{\nu_t}{\sigma_k}\frac{\partial k}{\partial x_i}\right) + \tau_{ij}\frac{\partial U_i}{\partial x_j} + G_b - \epsilon \tag{8.33}$$

$$\frac{\partial \epsilon}{\partial t} + U_i \frac{\partial \epsilon}{\partial x_i} = \frac{\partial}{\partial x_i}\left(\frac{\nu_{eff}}{\sigma_\epsilon}\frac{\partial \epsilon}{\partial x_i}\right) + \frac{\epsilon}{k}\left(C_{1\epsilon}\tau_{ij}\frac{\partial u_i}{\partial x_j} + C_{3\epsilon}G_b\right) - C_{2\epsilon}\frac{\epsilon^2}{k} \tag{8.34}$$

where the production of k due to buoyancy is given by:

$$G_b = -g_i \frac{\nu_t}{Sc_t}\frac{1}{\rho}\frac{\partial \rho}{\partial x_i} \tag{8.35}$$

with the following empirical model coefficients: $C_\mu = 0.09$, $C_{1\epsilon} = 1.44$, $C_{2\epsilon} = 1.92$, $\sigma_k = 1.0$, $\sigma_\epsilon = 1.3$. and $C_{3\epsilon} = C_{1\epsilon}$ when $G_b \leq 0$ (and $C_{3\epsilon} = 0$ when $G_b \geq 0$).

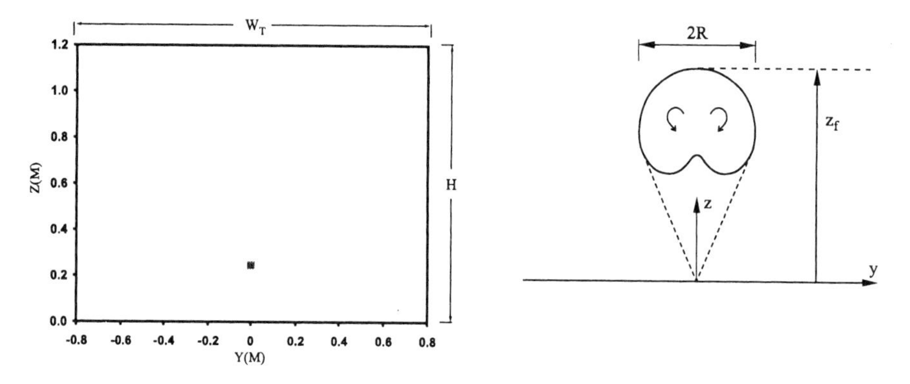

Figure 8.7. Numerical experiment of line thermal; an initial buoyancy is applied in $+z$ direction to marked patch of fluid

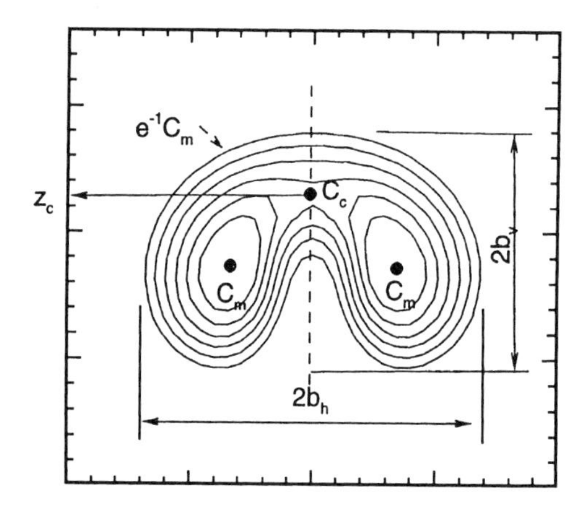

Figure 8.8. Definition of line thermal parameters

The line thermal flow can be revealed via a numerical experiment. Fig. 8.7 shows a volume of water initially at rest in a tank of width W_T and depth H. At time $t = 0$, a significant buoyancy impulse in the vertical (z) direction is applied to a square patch of fluid (the 'colored' patch) of length L_t. The patch is initially labelled at uniform passive scalar concentration C_o corresponding to density ρ_0 and located sufficiently far away from the solid boundaries. The buoyancy gives rise to fluid acceleration and vorticity; turbulent motion is produced and the thermal moves in the direction of the buoyancy. The following parameters are adopted: $W_T = 1.6$ m, $H = 1.2$ m, $L_t = 0.04$ m. $Q_{lo} = 1.6 \times 10^{-3}$ m^2 and

$F_{lo} = 6.43 \times 10^{-5}$ m^3/s^2 - corresponding to an initial density difference of around 0.4 percent. Accordingly, we have $t_c = 1.0\ s$ and $W_r = 0.04\ m/s$. The lower end of the square patch is located at $z = 0.245$ m from the bottom boundary. Both the tank dimensions and initial thermal size are comparable with previous experiments of line thermals and advected line thermals. At $t = 0$ zero velocities and pressure are prescribed; $C_o = 1$ inside the marked patch, and zero elsewhere. The governing equations are solved using the finite volume method; details of the numerical solution can be found in Lee and Chen (2002). The main features of the flow and scalar field are summarized herein.

Fig. 8.9 shows the computed stream function for the line thermal. A vortex pair flow similar to the line puff is clearly revealed. The shape of the flow field is well-preserved for $t^* \geq 10$. Fig. 8.10 shows the computed tracer concentration field. A symmetrical double maxima kidney-shaped structure is formed very early at about $t^* = 3$, and becomes approximately self-similar for $t^* > 10$. This simple kidney-shaped structure is quite different from that for a line puff, for which there is no double peak maxima for the asymptotic phase (Lee and Chen 2002). The concentration maximum C_m is typically as much as 2 times the maximum concentration C_c along the centerline. This value of C_m/C_c is in accord with previous observations in the line thermal (Richards 1963), though higher than the value of 1.7 (Fan 1967) and 1.4-1.6 (Lee and Cheung 1991) obtained for bent-over plumes in crossflow. To characterize the sectional thermal shape, the thermal aspect ratio is defined as the ratio of horizontal half-width b_h to vertical half-width b_v of the contour $C = e^{-1}C_m$ (Fig. 8.8). The results show that the thermal aspect ratio increases very quickly from its initial value of 1 at source to an asymptotic value of around 1.45 when $t^* > 3$. From the shape of the thermal, it is clear that the location of the centerline maximum concentration (z_c) lies above the visual centerline (z_d, mid-point of upper and lower boundary), $z_c > z_d$ (see also Prob. 8.2).

In Fig. 8.11, the vertical momentum inside the thermal (as defined by the 0.01 C_m contour), $M_t = \int\int_{x-section} W dydz$, is plotted against time. The best fit of the numerical results gives $M_t = 0.5 F_{lo} t$. This implies that the added mass coefficient of the line thermal is $k_n = 1$ (Eq. 8.21); half of the momentum is applied to the ambient irrotational flow around the thermal. To investigate self-similarity more fully, a large number of thermal characteristics are computed at each time step. The maximum vertical velocity W_m, trajectory z_c and maximum scalar concentration C_m well satisfies the -1/3, 2/3 and -4/3 similarity laws (Eqs. 8.6, 8.8, 8.9) respectively, for $t^* \geq 10$. It is interesting to note that the vertical velocity is roughly constant in the initial phase, with $W_m/W_r \approx 0.5 - 0.55$

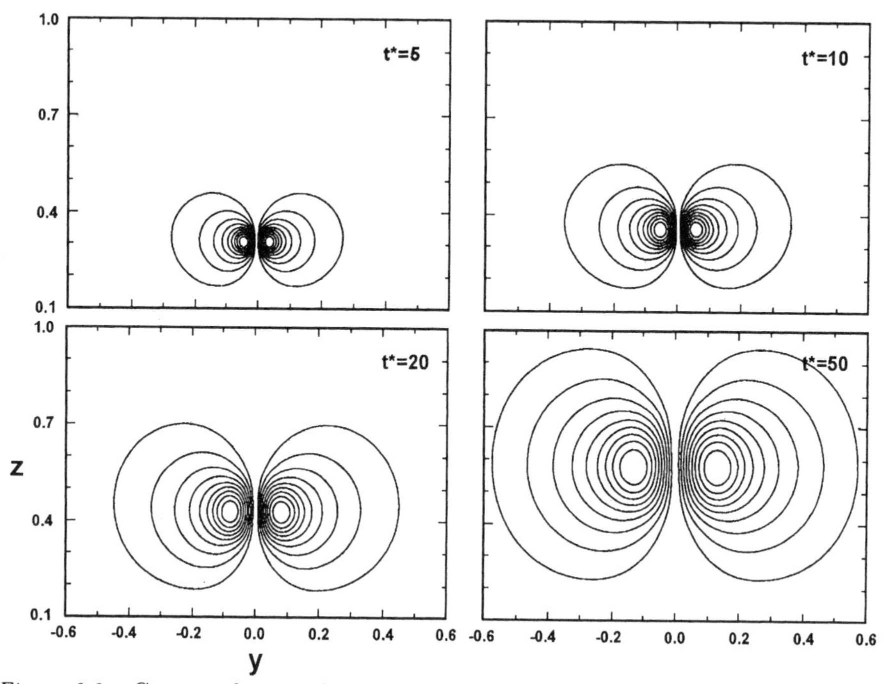

Figure 8.9. Computed streamlines of line thermal at $t^* = 5, 10, 20, 50$ (contour interval $\Psi_m^*/10$)

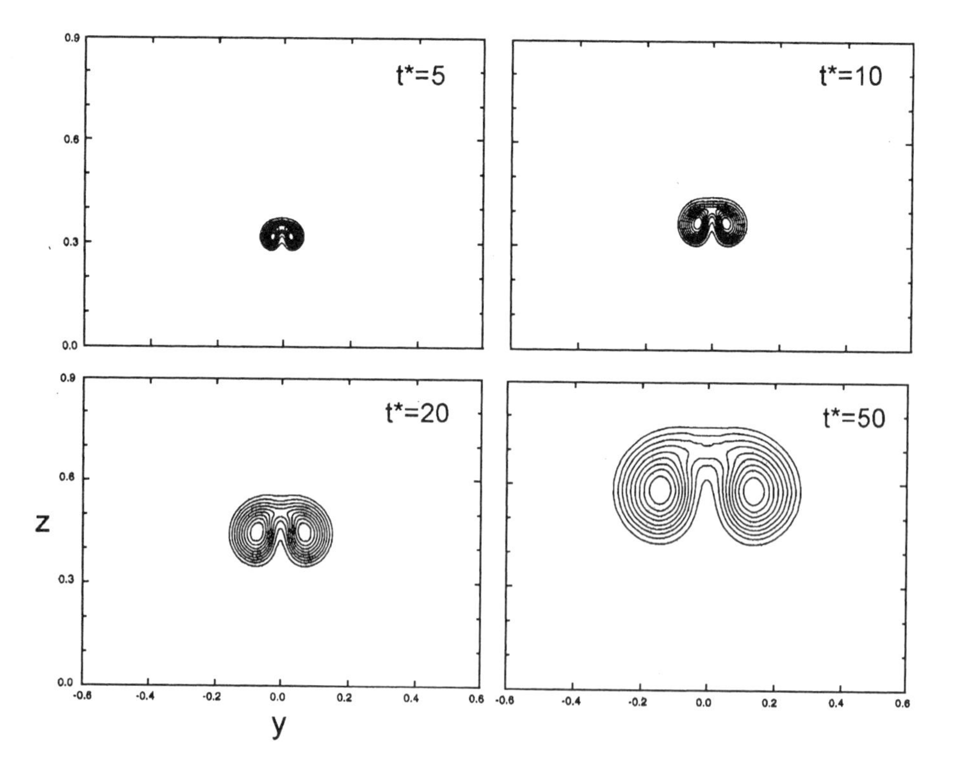

Figure 8.10. Tracer concentration field of line thermal

in the range of $t^* = 2 - 8$. The linear variation of the vertical location of the thermal front Z_f with the maximum horizontal radius R can also be seen. The thermal is gradually developed from the buoyancy; it can grow to about 20 times its original size by $t^* \approx 80$. The similarity relations in the asymptotic stage $t^* > 10$ are summarized in Table 8.1. Also shown are a length scale L, defined such that L^2 is the area within the 0.01 C_m contour, the ratio of the maximum to tracer mass-weighted average vertical velocity $(W_m/\overline{W})$, and the circulation around one half of the thermal, defined as $\Gamma = \int_0^H W(0,z)dz$. The vortex pair of the computed line thermal remains connected with time and is never completely divided; the circulation ($\Gamma \sim WL \sim t^{1/3}$) increases with time, in accordance with the measurements of Richards (1963). Similar to the line puff, the derived coefficients in the self-similar relations are not sensitive to the initial conditions assumed over a reasonable range of values.

Fig. 8.13 shows the computed vorticity and turbulent kinetic energy which also display self-similarity for $t^* \geq 10$; a simple pair of vortex eyes can be seen. The computed turbulent kinetic energy shows that the region of high TKE is located towards the front of the patch, where large spatial gradients can be found. In the self-similar stage, the maximum turbulent intensity is $u_m/W_m \approx 0.39$, where $k_m = 3/2u_m^2$. In the main body of the thermal, the turbulent viscosity, very similar in shape to k, varies spatially by about a factor of 4, with the maximum value $\nu_t/(W_m L) = 0.02$, and ν_t/ν typically in the order of 100. In Fig. 8.14, the computed streamlines are shown against concentration contours for the self-similar line thermal. It can be seen the double concentration maxima are located close to the vortex eyes but with a smaller separation distance than that between the vortex eyes. This computed feature is supported by experiments (Fig. 11 of Tsang 1971).

Analysis of the computed velocity field reveals the flow outside of the thermal can be modelled by a doublet; the stream function may be approximated by that produced by a circular cylinder having the same cross-section area as the thermal and moving with the velocity of the thermal centre, with strength of 0.53. This is close to the experimental value of 0.522 given by Tsang (1971). The 2D model results and Richards' experiment both suggest that almost all of the fluid presented to the thermal cross-section is entrained into the thermal - this is related to a Projected Area Entrainment hypothesis employed in the Lagrangian model JETLAG (see Chapter 10). The entrainment coefficient (α) as used in models employing the entrainment assumption (Turner 1986) can be deduced to be $\alpha = 0.46$ (Prob. 8.4). This value of α is very close to the typical value of 0.5 obtained for buoyant forced plumes (Chu and

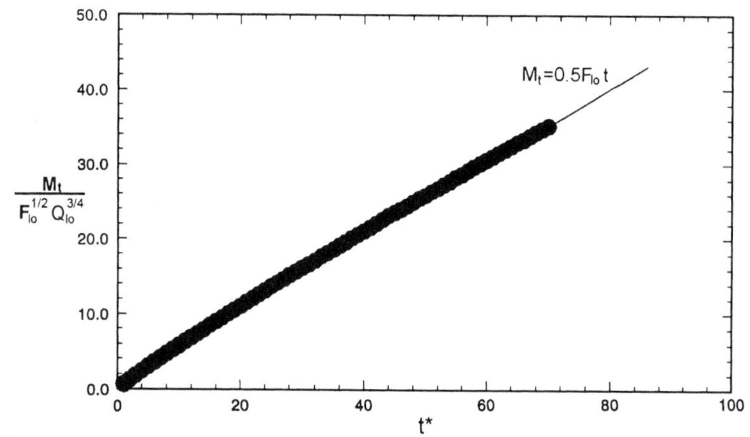

Figure 8.11. Time variation of vertical momentum M_t of line thermal

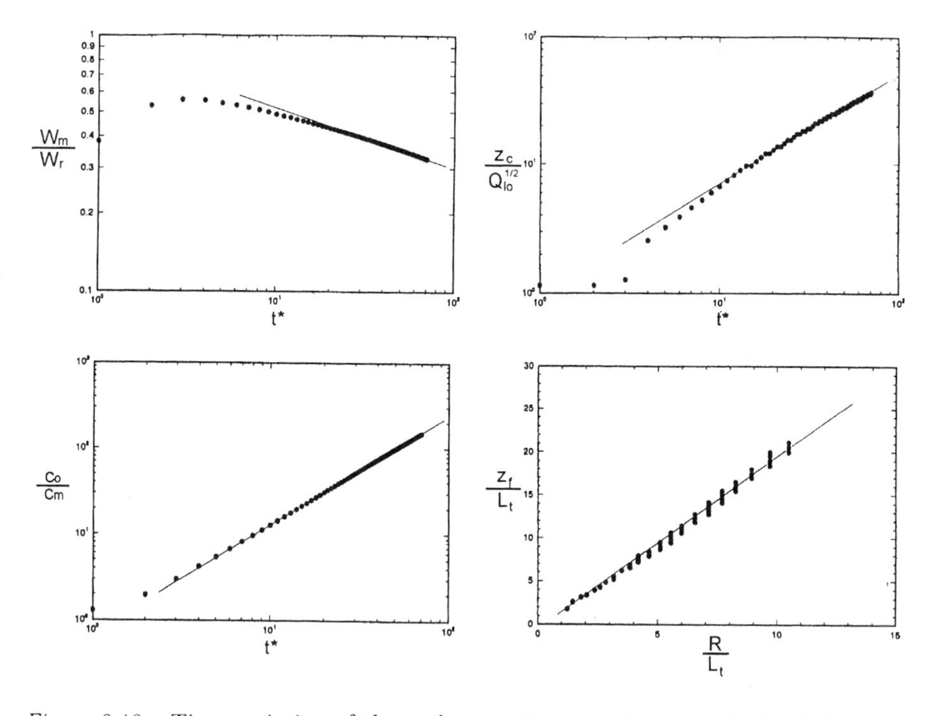

Figure 8.12. Time variation of thermal properties: maximum vertical velocity, trajectory, centerline maximum concentration, thermal front location (solid line is best fit self-similar relation)

Goldberg 1974), and is similar to the value for a line puff (Lee *et al.*1996) and a range of $0.25 - 0.67$ obtained by Chu (1985) in an analysis of data for jets in crossflow.

In summary, the computed flow is very similar to the observed mean fluid motion of line thermals, and offers a realistic picture and many insights of the thermal. Csanady (1965) presented an analytical model of the linearized weak thermal by assuming constant diffusivity; the computed flow and scalar field differ greatly from observations. Lilly (1964) solved the full nonlinear equations of motion for a line thermal within confined boundaries, and Yih (1981) presented a semi-analytical model for the bent-over plume in crossflow by assuming $u = U_a$. Both Lilly and Yih essentially adopted an unrealistic eddy viscosity distribution based on a free shear layer hypothesis; the predicted scalar field is not supported by experiments.

4. 3D MODEL OF PLUME IN CROSSFLOW

4.1 THE ADVECTED LINE THERMAL

The flow in the bent-over phase of a buoyancy-dominated jet in crossflow is analogous to that of a line (2D) thermal. An experimental analog of the line thermal is the advected line thermal (ALT) proposed by Knudsen and Wood (1985). Fig. 8.15 shows an advected line thermal produced by discharging a horizontal round buoyant jet with velocity U_o, density ρ_o, and diameter D into a steady coflow (density ρ_a) with the same velocity, $U_a = U_o$. Since there is no initial relative velocity, the flow simulates a vertically rising line thermal (advected at U_a) with initial volume per unit x-length $Q_{lo} = Q_o/U_a$, and buoyancy per unit length $F_{lo} = F_o/U_a$. The flow at two successive vertical sections (x_1, x_2) then correspond to those of the equivalent line thermal at two successive times by a Galilean tranformation. With this method, repeatability of source conditions can be achieved with much less variation than in experiments in which a cylinder of buoyant fluid is released from rest (Richards 1963, Tsang 1971). The data obtained in the latter experiments typically show great scatter. With the advected line thermal, the ensemble-averaged properties at a given time (or x) can be conveniently studied.

The advected line thermal in Fig. 8.16 is produced by a horizontal heated water jet in a coflow of the same velocity. Flow visualization of the dyed plume clearly shows that close to the source, within $\approx 5D$, a train of buoyancy-generated vortex rings is formed, which become stretched and diffused to form turbulent eddies further downstream. Fig. 8.17 shows an instantaneous and a time-averaged image of the side view of an advected line thermal. The upper boundary of the thermal is highly irregular and fluctuating, contributed by occcasionally large buoyant elements protruding from the jet proper, while the lower edge

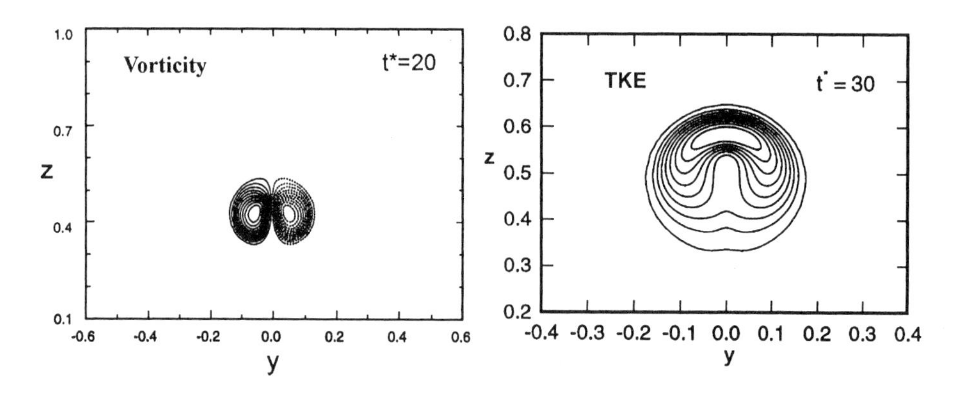

Figure 8.13. Vorticity and turbulence kinetic energy of line thermal

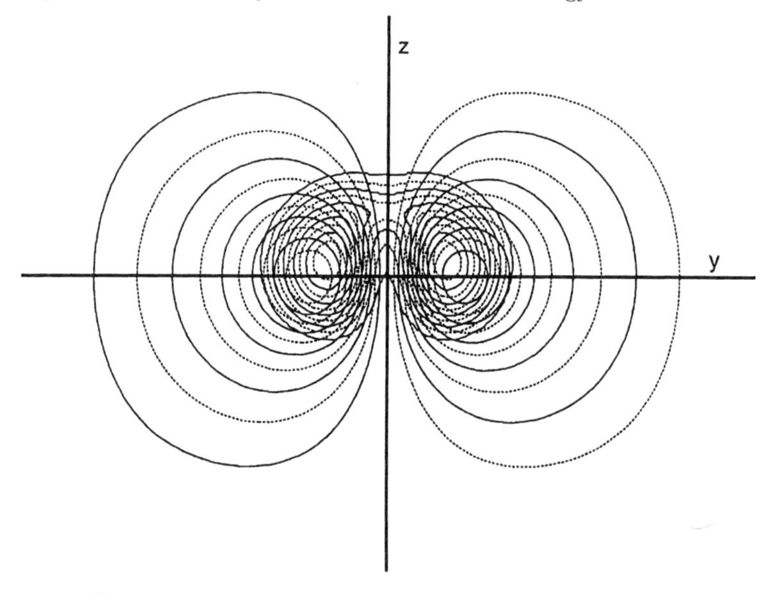

Figure 8.14. Computed streamlines shown against concentration contours for $t^* = 30$

is relatively smooth. The path of the advected thermal is initially rather flat, before being bent-over by the crossflow.

4.2 3D MODEL OF ADVECTED LINE THERMAL

The spatial evolution of an advected line thermal has been studied using the standard two-equation $k - \epsilon$ model with a buoyancy expansion (Chen and Lee 2002). The governing equations outlined in the previous section are solved for the three dimensional flow. A numerical

Table 8.1. Similarity relations of 2D line thermal and 3D advected line thermal (ALT) derived from numerical solution and laboratory experiment; numbers refer to dimensionless coefficients of equations or ratios.

	Parameter	2D thermal	3D ALT	Experiment	2D puff
Added mass coefficient	k_n	1.0	1.0		1.01
Velocity	$W_m = C_v F_{lo}^{1/3} t^{-1/3}$	1.34	1.37		1.61
Thermal front	$Z_f = nR$	2.0		1.3-3.2 (Richards 1963) 2.2-2.5 (Chu 1996)	3.03
Trajectory	$Z_c = C_{2t} F_{lo}^{1/3} t^{2/3}$	0.80	1.30	$0.82t^{*0.767}$ (Wong 1991) 1.19 (Wright 1977) 1.26 (Knudsen 1988) 1.32 (Gaskin 1995)	1.51
Length scale	$L = C_{4t} F_{lo}^{1/3} t^{2/3}$	1.12			1.17
Scalar concentration	$C_o/C_c = C_{3t} Z_c^2 / Q_{lo}$	1.14	0.68	0.56 (Wong 1991) 0.41 (Knudsen 1988) 0.44 (Wright 1977) 0.56 (Chu 1979) 0.51 (Cheung 1991)	0.35
	C_m/C_c	2.0	1.5	1.4-1.6 (Cheung 1991) 1.7 (Fan 1967)	1.0
Characteristic radius	$b = C_{1t} Z_c$	0.44	0.43	0.42 (Wong 1991) 0.39 (Chu 1996) 0.39 (Knudsen 1988) 0.42 (Gaskin 1995)	
Centerline halfwidth	$b_{vc} = C_{1t'} Z_c$	0.29			0.269
Aspect ratio	b_h/b_v	1.45	1.43	1-1.38 (Wong 1991) 1.23 (Richards 1963) 1.19 (Gaskin 1995)	1.0
Velocity ratio	$W_m/\overline{W}$	2.63			3.18
Circulation	$\Gamma/(\overline{W}L)$	2.35			2.67
Turbulent viscosity	$\nu_t/(W_m L)$	0.020			0.018
Turbulent intensity	u_m/W_m	0.38			0.39

experiment of an advected thermal is performed adopting the following parameters: $U_o = U_a = 0.1\,\text{m/s}$, $D = 1$ cm, $F_o = 1.155 \times 10^{-6} \text{m}^4/\text{s}^3$, and $Q_o = 7.85 \times 10^{-6}\, \text{m}^3/\text{s}$; $(\rho_a - \rho_0)/\rho_a = 1.5 \times 10^{-2}$. The jet densimetric Froude number $Fr = 2.6$. The characteristic length and time scales have values of $L_b = 1.15 \times 10^{-3}$ m, $L_x = 2.45 \times 10^{-2}$ m and $t_c = 0.245$ s, respectively. The computational domain has a length of $L = 1.25$ m, depth $H = 0.4$ m and width $W = 0.3$ m, and extends to $x/D \geq 120$ and

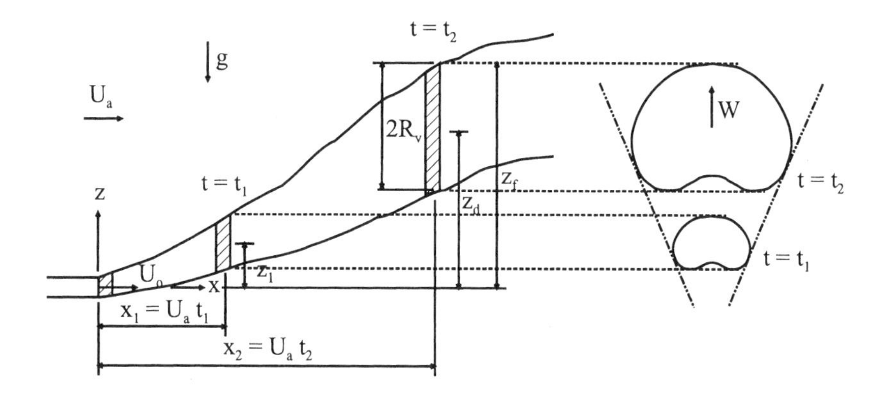

Advected line thermal
$Q_o = U_o \pi D^2/4$
$F_o = Q_o(\rho_a - \rho_o)g/\rho_a$

Equivalent line thermal (section)
Initial volume/length $= Q_o/U_a$
Initial buoyancy/length $= F_o/U_a$

Figure 8.15. Schematic diagram of an advected line thermal ($U_o = U_a$)

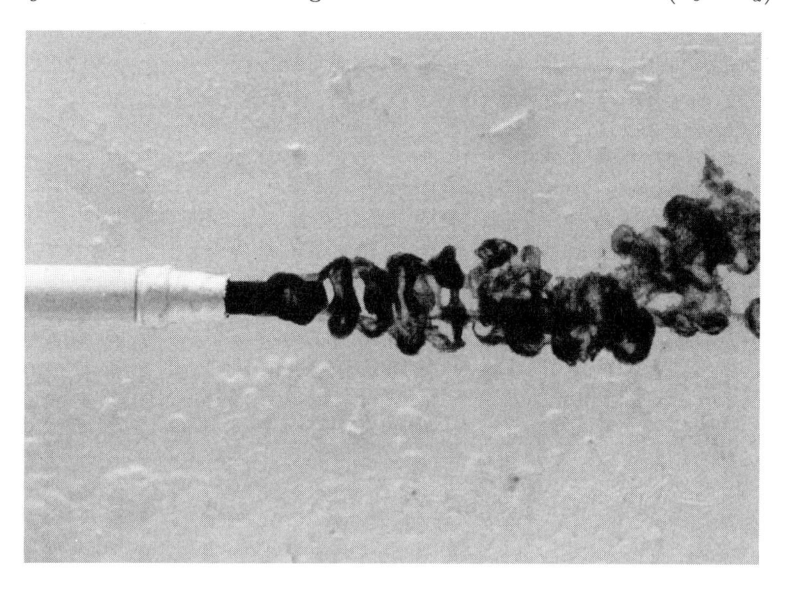

Figure 8.16. An advected line thermal produced by a horizontal buoyant jet discharging with no initial excess momentum in a coflow (top view)

$x^* > 50$ downstream of the source. Details of the computation can be found in Chen and Lee (2002).

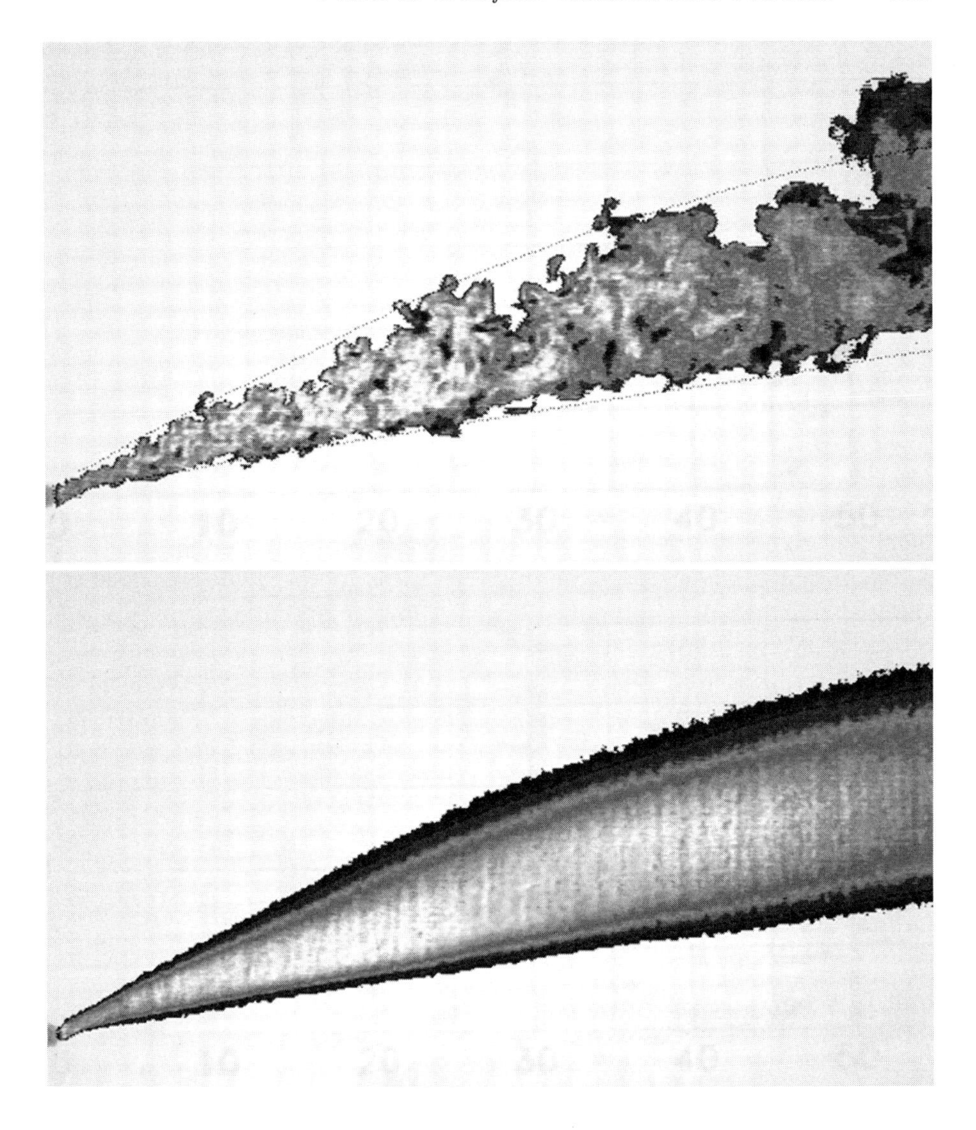

Figure 8.17. Instantaneous and time-averaged shadowgraph pseudo-color image (side view) of an advected line thermal ($D = 1.02$ cm; $U_o = 4.4$ cm/s; $U_a = 4.6$ cm/s; $g'_o = 0.1177$ m/s^2)

Fig. 8.18 shows contour plots of the computed scalar field in the plane of symmetry at $z = 0$. The trajectory of the maximum concentration looks very flat on first glance, like a straight line. The process of scalar dilution takes place essentially within 40 diameters downstream, corresponding to $x^* < 15$. The computed thermal trajectory resembles the observed path (Fig. 8.17).

Fig. 8.19 shows transverse cross sections of the scalar field at different downstream locations. A double peak concentration distribution can be observed in the kidney-shaped thermal, which becomes approximately self-similar for $x^* > 10$. Qualitatively the computed scalar field is very similar to that of the line thermal (Fig. 8.10), but with slightly more rounded front edges. The double-peak structure is also similar to the numerical results for advected line puffs (Fig. 7.14). The self-similar relations for the 3D advected line thermal are given in Table 8.1.

4.2.1 CHARACTERISTICS OF ADVECTED LINE THERMAL

Analysis of the three-dimensional numerical results reveal the following characteristics:

i) *Thermal aspect ratio* - In Fig. 8.19, the thermal aspect ratio b_h/b_v increases very quickly from its initial value of 1 at the exit to around 1.43 when $x^* > 3$. This is approximately the same as the value obtained for the line thermal, and comparable with the corresponding experimental result of Wong (1991) who found a variation of thermal aspect ratio from 1.1 - 1.3, with an average of 1.2. Analysis of thermal boundaries in the classical study of line thermals (Richards 1963) suggests an average aspect ratio of 1.23; analysis of $e^{-1}C_m$ concentration contours measured by Gaskin (1995) in the developed phase of advected line thermals ($Fr \geq 1$) in developed phase suggests an average value of 1.19. This value is also comparable to the value of 1.3 suggested by Pratte & Baines (1967) from an analysis of flow visualization and that found for advected line puffs (Table 7.1).

ii) *Maximum concentration ratio* - As observed from the concentration contours, the maximum concentration C_m is not located in the centerline. The maximum of the ratio of the cross-section to centerline maximum concentration, C_m/C_c, is found to be 1.5, comparable to the value of 1.7 (Fan 1967) and 1.4-1.6 (Lee and Cheung 1991) observed in bent-over plume in crossflow. It is also interesting to note the case of advected line puffs: $C_m/C_c = 1.35$ in the developed phase and values of $1.1 - 1.5$ from experiments.

iii) *Characteristic half-width and trajectory* - Fig. 8.20 shows the characteristic radius $b = \sqrt{b_h b_v}$ plotted against centerline trajectory z_c; the linear spread is predicted. The width spread rate, $C_{1t} = 0.43$, is close to the value of 0.42 experimentally found by Wong (1991). It is however notably higher than the theoretical and experimental values of 0.27-0.29 for advected line puffs (defined somewhat differently). Fig. 8.20 also shows the puff trajectory z_c/L_b plotted against downward distance x/L_b. The 2/3 power law is confirmed, with a coefficient of $C_{2t} = 1.30$. This can be

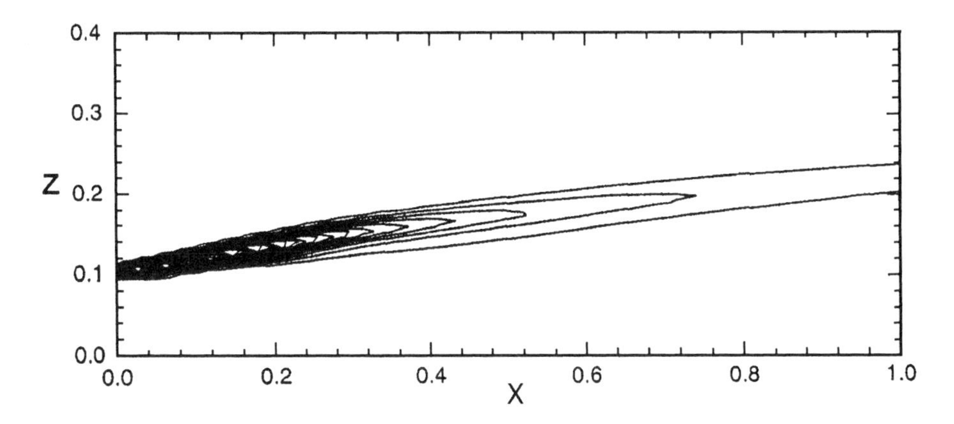

Figure 8.18. Computed scalar concentration contours in the plane of symmetry at $y = 0$. Contour interval is $1/100$.

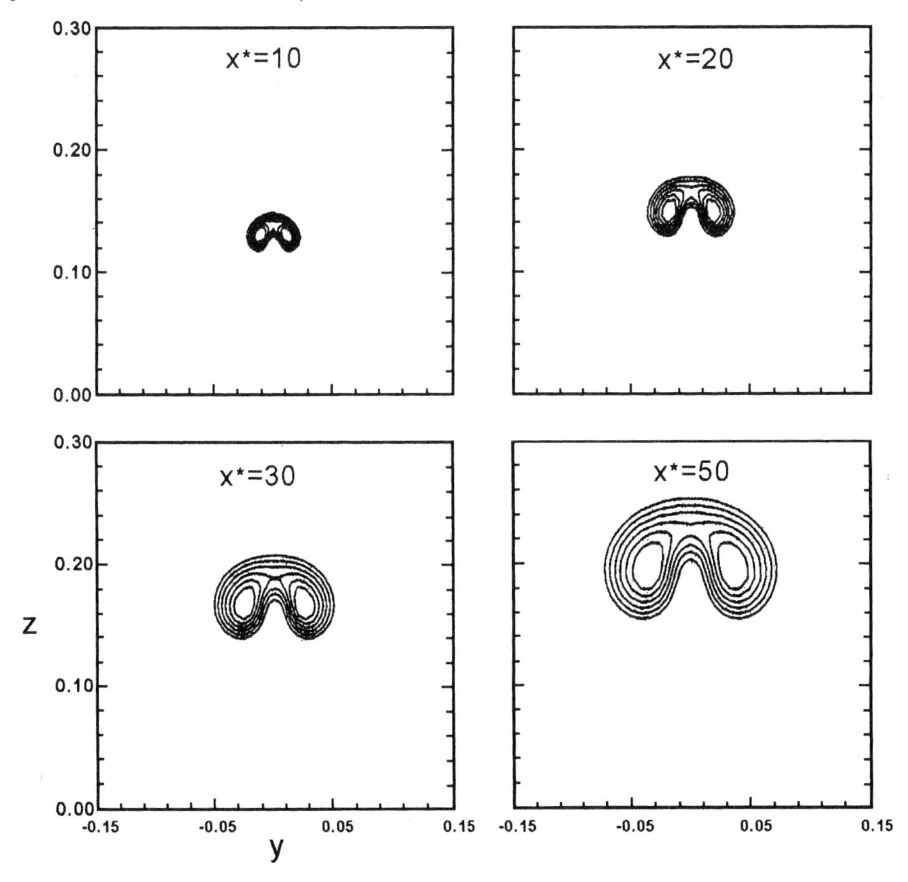

Figure 8.19. Computed tracer concentration in different vertical cross-sections of advected line thermal. Contour interval is $(1 - e^{-1})C_m/6$. Outermost concentration contour is $e^{-1}C_m$.

compared with experimental values of 1.19 (Wright 1977b), 1.26 (Knudsen 1988), and 1.40 in Wong (1991). The effect of virtual origin is found to be small.

iv) *Centerline dilution* - Fig. 8.21 shows the centerline dilution $S_c Q/(U_a L_b^2)$ plotted against trajectory z_c/L_b. The results follow the square power similarity law, with a dilution coefficient of $C_{3t} = 0.68$, This value has been obtained by Chu (1979) in his analysis of data of Fan (1967) for the bent-over phase of a buoyant jet, and is comparable to the experimental value of 0.56 in Wong (1991) for advected line thermals. In Chu's analysis, the visual trajectory z_d instead of z_c was used in the correlation. Since $z_c \approx 1.1 z_d$, Fan's data suggests a dilution coefficient of $0.68/1.1^2 = 0.56$.

v) *Vertical momentum and added mass* - Conservation of buoyancy flux (Eq. 8.21) leads to the following equation for the total vertical momentum flux

$$M_v = \frac{1}{1+k_n} F_o \frac{x}{U_a} \tag{8.36}$$

where the vertical momentum flux is defined as $M_v = \int\int_{x-section} UW\, dydz$ and k_n = added virtual mass coefficient. The 3D model computations confirm the above equation, giving a result almost identical to that depicted in Fig. 8.11. The added mass coefficient is found to be 1.0 for the thermal defined by the $0.1C_m$ contour. Redefining the thermal by the $0.01C_m$, $0.05C_m$, $0.25C_m$ and $e^{-1}C_m$ contours gives values of $k_n = 0.92, 0.96, 1.04$ and 1.09 respectively. In other words, the advected line thermal itself contains only about half of the total vertical momentum resulting from the buoyancy effect; the other half accounts for the approximately irrotational flow of the surrounding fluid set up by the rising thermal. This result is consistent with those found in experiments and numerical studies of advected line puffs and time-dependent line puffs

vi) *Bifurcation of plume in crossflow* - One normally associates the double maxima structure of the scalar field with a buoyant jet in crossflow. Numerical experiments have confirmed that the bifurcation in concentration structure may be absent for cases with small buoyancy - as supported by the experiment of Wong (1991). For example, an advected line thermal with $Fr = 4.12$ with $\Delta\rho_o/\rho_a = 1.5 \times 10^{-3}$ and $U_a = 0.05\,\mathrm{m/s}$, exhibits only one cross-section concentration maximum centered at the centerline. As the strength of the buoyancy generated circulation can be shown to depend on F_o/U_a, this result seems to be reasonable: for weakly buoyant plumes the circulating flow along and near the thermal centerline is not strong enough to compete with the

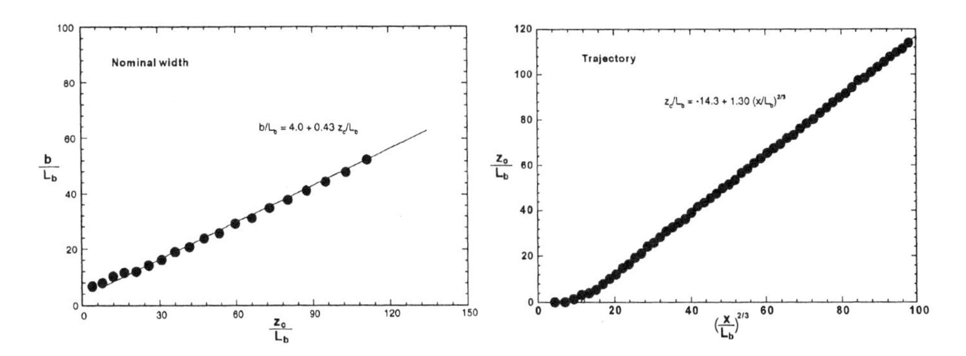

Figure 8.20. Width and trajectory of advected line thermal

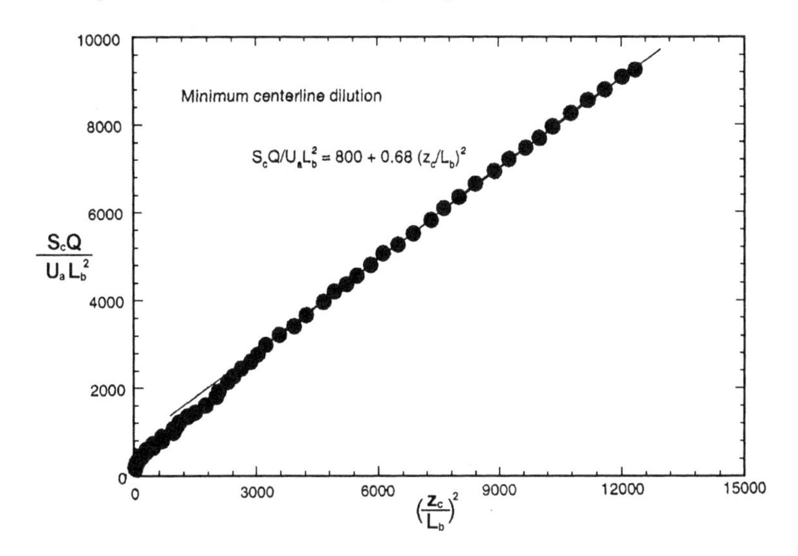

Figure 8.21. Dimensionless centerline dilution of advected line thermal

cross-section turbulent mixing. These cases also result in a smaller aspect ratio of around 1.2.

5. MEASUREMENTS IN LINE THERMALS

5.1 EXPERIMENTS ON ADVECTED LINE THERMALS

Experiments on advected line thermals have been performed by Knudsen (1988), Wong (1991), and Gaskin (1995). All the results show that the cross-section concentration is often double-peaked, with an aspect ratio of around $b_h/b_v \approx 1.2$, where the half-widths b_h, b_v are defined by the $e^{-1}C_m$ concentration contour. Wong's experiments show clearly

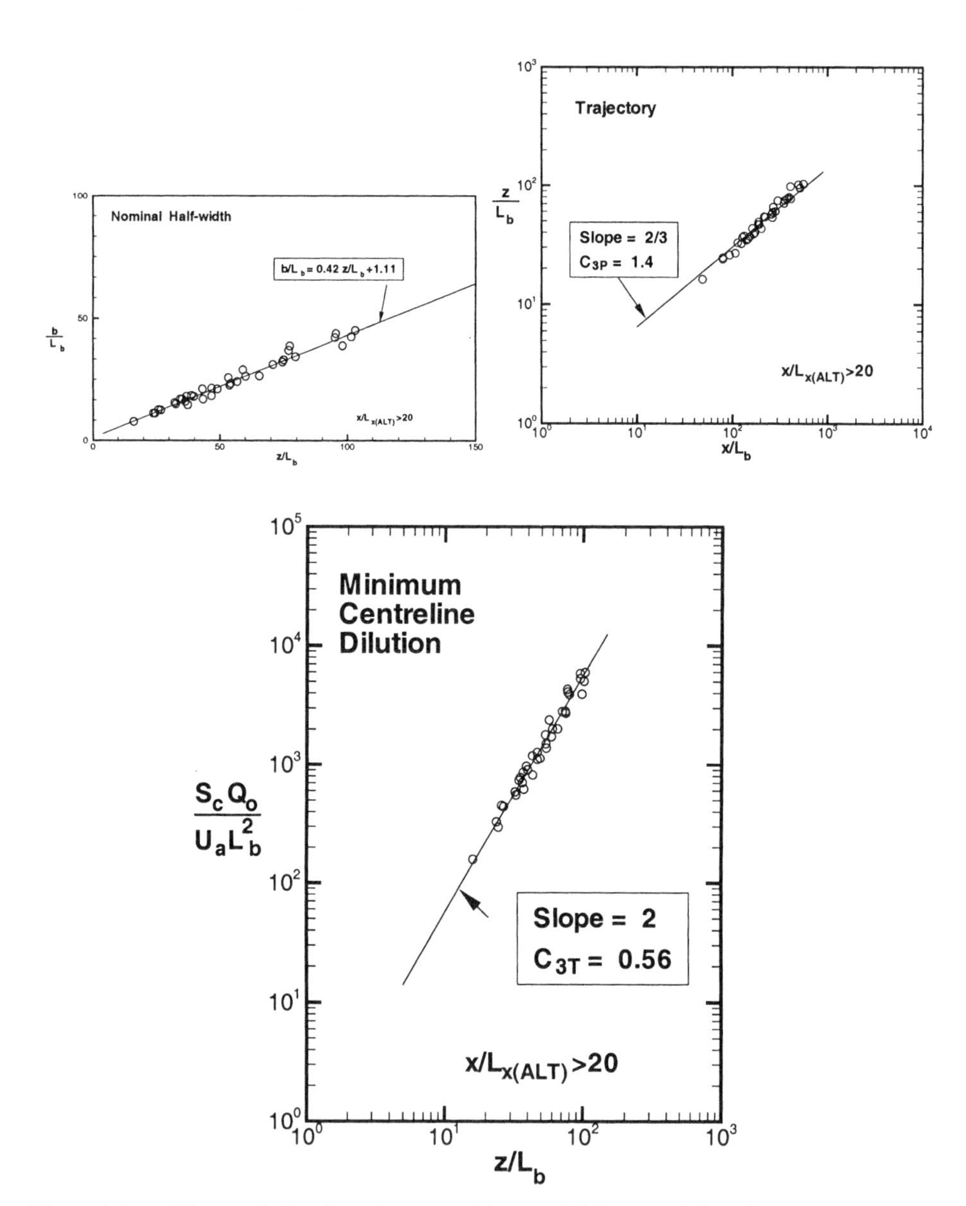

Figure 8.22. Observed visual trajectory, width, and dilution of the advected thermal (Wong 1991)

the validity of the linear spread relation, with a spreading coefficient of $C_{1t} = 0.42$ (Fig. 8.22). The thermal trajectory can be described with the 2/3 power law with a trajectory coefficient of $C_{2t} = 1.40$, although the best fit gives $C_{2t} = 0.8$ and a power variation of about 0.767. This can be related to the use of heat as a source of buoyancy. Compared with the case when buoyancy is created by salinity differences (as in Fan 1967 and Knudsen 1988), both numerical models and experiments show that a heated jet with the same initial $\Delta\rho_o$ would take on a lower (more submerged) jet trajectory due to loss of buoyancy (see also Chapter 3). However, the dilution achieved at the same vertical elevation is approximately the same. The minimum centerline dilution follows well the BDFF relation, with a dilution coefficient of 0.56. The ALT has also been studied using video image processing. In these experiments, 250 instantaneous frames from a one-minute video record of the shadowgraph were used to obtain the ensemble-averaged image. The time-averaged visual boundary is defined by a threshold value selected based on visual judgement of thresholding an *instantaneous* image against its visual boundary. The visual centerline trajectory $z_d(x)$ (defined as the mid-point between the upper and lower boundaries), and the corresponding vertical half-width $R_v(x)$ can then be obtained. The time-averaged image reveals a relatively straight initial portion before the jet is bent over (2/3 power law). Fig. 8.23 shows a plot of the centerline trajectory; the initial trajectory can be described by $z_d/D \approx 0.46x/L_x$, before the transition to the developed stage occurs at $x/L_x \approx 10$. The initial linear trajectory is consistent with the approximately constant vertical velocity computed for the line thermal, with $W_m/W_r \approx 0.53$. We note that $dz/dx = \overline{W}/U_a$, and $\overline{W} \approx W_m$ in the initial phase. Using the definition of W_r and t_c and $U_o = U_a$, it can be shown that $W_m \approx 0.53\ W_r$ would have led to an initial trajectory of $z/D = 0.53\sqrt{\pi/4}\ x/L_x = 0.47x/L_x$.

The trajectory coefficient has a value of $C_{2t} = 1.37$. Fig. 8.24 shows that the vertical half-width varies linearly with elevation, $R_v \sim z$, with $C_{1t} = 0.48$ and 0.39 in the initial and developed phases respectively. Since the upper visual boundary represents the front of the corresponding line thermal, $z_f = z + R_v$, the thermal spreading rate as defined by dz_f/dR_v has a value of 3.11 in the initial phase and 3.57 in the developed phase. As the 3D computation gives an aspect ratio of 1.43, this implies that $z_f \approx 2.2\ R_h$.

Using ($k_n = 1$, $\beta_n = 0.39$), Eq. 8.27 would give $C_{2t} = 1.1$; this compares favorably with experimental values of 1.19 and 1.26, derived from an analysis of the data of Wright (1977b) and Knudsen (1988) respectively. In entrainment-based models of buoyant jet in crossflow (e.g. Hoult *et al.*1969), it can be shown that the entrainment coefficient in the BDFF should be the same as the spreading rate (Prob. 8.5). In previous studies, the entrainment coefficient α_o was typically determined from the observed jet trajectory in the BDFF (2/3 power relation, Eq. 8.27) without accounting for added mass. It can be seen that the entrainment

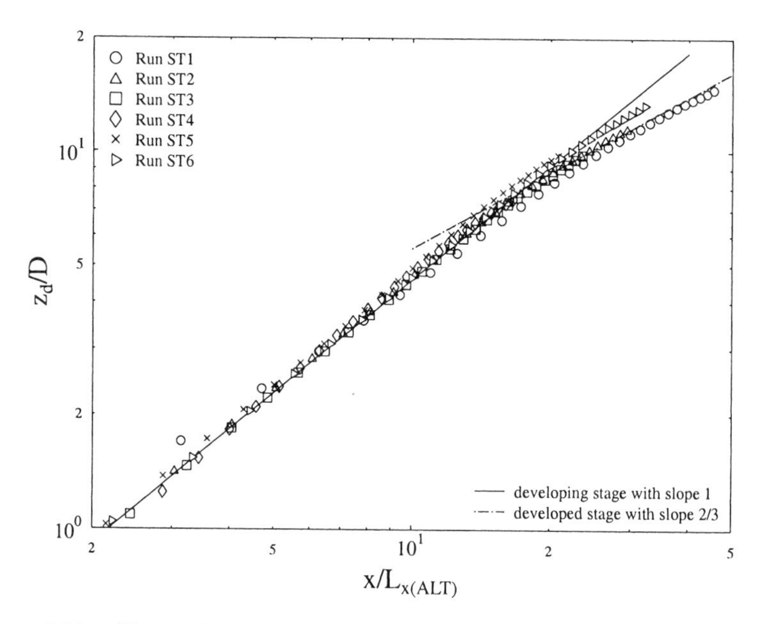

Figure 8.23. Observed visual trajectory of the advected thermal: a) $z \sim x$ for initial stage; b) $(z/L_b) \sim (x/L_b)^{2/3}$ in developed phase

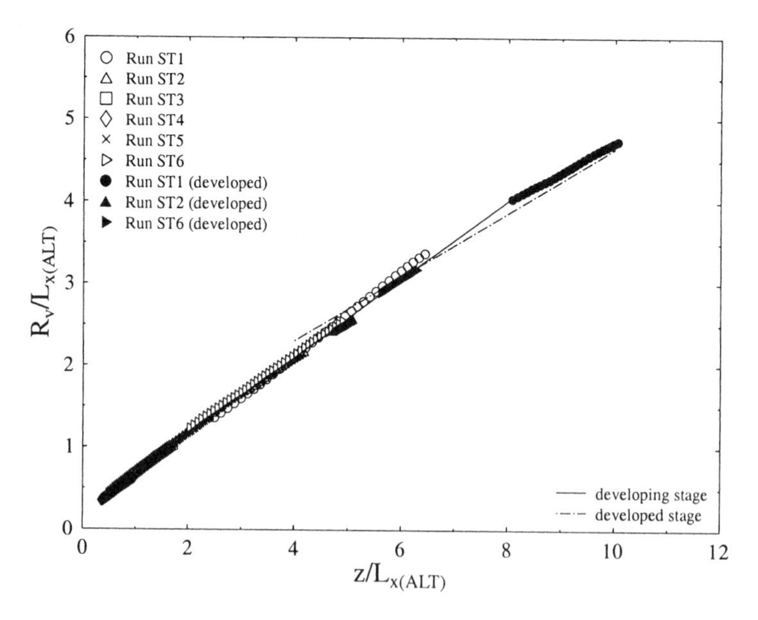

Figure 8.24. Spreading rate of advected line thermal: linear spread of visual vertical half-width vs elevation

coefficient $\alpha_o = \beta\sqrt{1+k}$ so determined will not match the observed spreading coefficient, an anomaly noted earlier (Hoult and Weil 1972). Using the reported entrainment coefficient, the spread rate observed in these studies can be inferred and given in Table 8.2.

5.2 CONCENTRATION MEASUREMENTS

Detailed concentration measurements in the bent-over phase of plumes in crossflow (BDFF) as well as advected line thermals (ALT) have been made in a number of studies. The nature of the measurements are summarized below along with the concentration spread rate C_{1t}, trajectory coefficient C_{2t} and centerline minimum dilution constant C_{3t} in the BDFF so obtained. The key values can also be found in Table 8.2.

Fan 1967 — Vertical buoyant (saline) jet in crossflow; centerline concentration (conductivity) normal to jet axis in the vertical plane of symmetry; concentration and visual trajectory (z_c, z_d and selected cross-section measurements. Momentum-dominated jets ($Fr = 10-80, K = U_o/U_a = 4-16$). Analysis of Fan's data gives $C_{1t} = 0.34$ (visual), $C_{3t} = 0.56$.

Ayoub 1971 — Horizontal buoyant jet in coflow and crossflow; centerline concentration (conductivity) normal to jet axis in the vertical plane of symmetry; trajectory (z_c, z_d) and selected cross-section measurements. Momentum-dominated jets with 3D trajectories ($Fr = 15-90, K = 5-20$).

Wright 1977 — Towed vertical saline jet in uniform and stratified still ambient; centerline concentration (conductivity) along a vertical traverse in the plane of symmetry; visual trajectory. Extends to weak crossflow regime ($K \geq 4$). Analysis of all data with $z/L_b \geq 1$ gives $C_{2t} = 1.19$ (visual) and $C_{3t} = 0.44$. Data with $z/L_b \geq 2$ has an average dilution constant of $C_{3t} = 0.51$,

Knudsen 1988 — Advected line thermal; towed horizontal saline plume in uniform still ambient. Cross-section (conductivity) measurements in six experiments. $C_{1t} = 0.36$, $C_{2t} = 1.39$, $C_{2t} = 1.1$ (visual), $C_{3t} = 0.41$.

Wong 1991 — Advected line thermal; heated horizontal thermal plume in coflow. Cross-section temperature measurements (3-minute average) on a 17×14 grid and visual trajectory. A total of 42 cross-sections of 12 thermals. Digital analysis of video-recorded dyed plume. $C_{1t} = 0.42$, $C_{2t} = 0.8$ (best fit 0.767 power law) , $C_{2t} = 1.4$ (2/3 power law), $C_{3t} = 0.56$. Analysis of video records of dyed thermals gives $C_{1t} = 0.43$ (assuming aspect ratio of 1.2), and $C_{2t} = 1.37$ (z_d).

Cheung 1991 — Vertical and horizontal buoyancy-dominated heated jets in crossflow. ($Fr = 2-10, K \geq 2$). Transverse and normal centerline temperature profiles and visual trajectory particularly in BDNF-BDFF transition. Detailed cross-section measurements for horizontal buoyant jet in crossflow. $C_{3t} = 0.51$.

Gaskin 1995 — Advected line thermal; towed saline jet in uniform still ambient. Cross-section conductivity measurements; Instantaneous and time-averaged LIF measurements of longitudinal centerline section of thermal; instantaneous LIF data on cross-section. PIV measurement of external entrainment flow. $C_{1t} = 0.41$, $C_{2t} = 1.32$ (z_d).

Pun 1999 — Advected jet and plumes; towed saline jet in uniform still ambient. Detailed LIF data in centerline section of vertical buoyant jet in weak crossflow. $C_{2t} = 1.35$ (z_d). Data on MDNF and BDNF.

5.3 SUMMARY OF EXPERIMENTAL DATA

In the experiments described above, different measurement methods have been used; e.g. the concentration half-width has been defined from the visual half-width (between the upper and lower boundaries), from a vertical concentration traverse, or a nominal radius based on the extreme widths of the $e^{-1}C_m$ concentration contour. Different periods of time-averaging have also been used. Nevertheless, the following points can be made:

- The nominal radius based on the $e^{-1}C_m$ contour ranges from 0.36 (Knudsen 1988) to 0.42 (Wong 1991). In Knudsen's experiments, if we define the concentration half-width as λb, and assume (as in puffs) a visual radius of $\sqrt{2}b$, then the theoretical spread rate of the ALT is $C_{1t} = \sqrt{2}/1.2 \times 0.36 = 0.424$ where the concentration to velocity width ratio has been assumed to be $\lambda = 1.2$. Digital image processing of the video records of advected line thermals (Chu 1996) gives a spread rate of $C_{1t} = 0.39$ for the visual vertical half-width. Assuming an aspect ratio of $b_h/b_v = 1.2$, this gives a nominal visual half-width spread rate of $C_{1t} = \sqrt{1.2} \times 0.39 = 0.43$. For the measured ALT concentration half-width of $C_{1t} = 0.42$, the corresponding visual radius can also be interpreted as $C_{1t} = 0.42/1.2 \times \sqrt{2} = 0.50$

- The visual trajectory constant is quite stable, in the range of $C_{2t} = 1.1 - 1.37$; the digital analysis of ALT visual trajectory gives a value of 1.37. These values are generally lower than the trajectory defined by z_c.

- The experiments of buoyant jet in a current suggests a centerline dilution constant of $C_{3t} = 0.51 - 0.56$, while the towed jet experiments give a generally lower value of $C_{3t} = 0.41 - 0.51$. The variability is partly related to the sensitivity of the dilution constant to choice of z (z_d or z_c) and that not all data lie in the fully developed BDFF. The relatively higher dilution constant measured with a stationary jet in a current may be affected by the ambient turbulence.

- In the bent-over phase, the average dilution is given approximately by

$$\overline{S} = U_a \frac{\pi B^2}{Q_o} = 0.80 \frac{U_a z^2}{Q_o} \tag{8.37}$$

where $B = 0.5z$ is adopted. For the simple jet/plume, the ratio of average to centerline dilution is 1.7. For advected line puffs, a ratio of average to centerline dilution constant of $\overline{S}/S_c = 1.45$ has also been measured. Using the turbulence model results for the line thermal and line puff, the ratio $\overline{S}/S_c$ can also be determined to be around 1.7 (Prob. 8.6). A value of 2.08 has also been determined from Fan's (1967) data (Chu and Goldberg 1974). Since the experimental value for the ALT is unknown, adopting a value of $\overline{S}/S_c = 1.7$ results in:

$$S_c = 0.46 \frac{U_a z^2}{Q_o} \tag{8.38}$$

This value is seen to lie within the range of measured dilution constants. It is also consistent with field measurements of initial dilution at sea outfalls after accounting for the surface layer (see Chapter 10).

- The overall numerical and experimental results suggest that a value of $C_{1t} \approx 0.4 - 0.5$, $C_{2t} \approx 1.3$, and $C_{3t} \approx 0.46$ can be reasonably adopted for the BDFF.

The LIF data on instantaneous flow of advected line thermals (Gaskin 1995; Gaskin and Wood 2001) revealed a highly variable and complex three-dimensional structure. The leading (top) edge of the thermal is hydrodynamically unstable and is seen to result in random protrusions into the ambient flow (see Fig. 8.17). This results in much greater concentration fluctuations than the advected line puff. The ratio of concentration fluctuation to the maximum time-mean concentration (C_{rms}/C_m) can have a maximum value of 0.8 towards the leading edge, while the maximum instantaneous concentration can be up to 3.5-4 times C_m maximum time-averaged value. In contrast, the corresponding figures for the advected line puff are approximately 0.4-0.5 and 2.0 respectively (Fig. 7.26 and 7.27). The intermittency factor is however broadly similar for both thermals and puffs. The formation and transport of 3D 'sub-thermal' structures in the far field of the ALT has been discussed by Gaskin and Wood (2001).

Table 8.2. Summary of experimental constants (visual spreading rate, visual and centerline trajectory, centerline and minimum dilution) of ALT

Previous studies	Advected line thermals (BDFF)				
	Spreading rate	Trajectory		Dilution	
	β C_{1t}	z_d C_{2t}	(z_c) (C_{2tc})	S_c C_{3t}	$(S_m$ $(C_{3t}$
Richards (1963)	0.62[a]	0.90[a]			
Tsang (1971)	0.42[a]	1.09[a]			
Hewett, Fay and Hoult (1971)	0.50[b]		(0.98)		
Hoult and Weil (1972)	0.42[b]	1.10			
Chu and Goldberg (1974)	0.35[b]	1.24			
Wright (1977)		0.85 $(L_m/L_b)^{1/6}$	(1.4 $(L_m/L_b)^{1/6}$)	0.41	
Chu (1979)	0.34	1.30		0.68	(0.4
Lee and Neville-Jones (1987a, 1987b)			(0.94)		(0.3
Cheung (1991)				0.51	
Knudsen (1988)	0.42[c]	1.10	(1.39)	0.41	
Wong (1991)	0.51[d]	1.36	(1.40)	0.56	(0.5
Gaskin (1995)	0.49[c]		(1.32)		
JETLAG (Lee and Cheung 1990)	0.45[b]	1.1		0.45[e]	
Chu (1996) - shadowgraphy	0.43	1.37			
1D Lagrangian model ($\beta_n = 0.4$ and $k_n = 1$)	0.4[f]	1.14	(1.20)	0.39	(0.3
				0.43[e]	(0.3

[a] using $R_h/R_v \simeq 1.2$
[b] deduced from entrainment coefficient using $k = 1$
[c] deduced from concentration half-width
[d] average of visual and concentration half-width spreading rate
[e] based on thermal trajectory
[f] selected value

6. BUOYANT JET IN CROSSFLOW

Since in general a buoyant jet has both momentum and buoyancy, whether a discharge is momentum or buoyancy dominated depends on L_m/L_b. If $L_m/L_b \ll 1$, the effect of buoyancy is exerted over a much greater distance than momentum, and the discharge is said to be buoyancy-dominated. Conversely, for $L_m/L_b \gg 1$, the initial momentum is relatively more important than buoyancy before the jet is bent-over, and the discharge is said to be momentum-dominated. Since the initial jet momentum loses its effect within a distance of the order of $L_s = M_o^{3/4}/B_o^{1/2}$ (the momentum length scale in stagnant fluid, Chapter 3), for any discharge the effect of buoyancy will govern asymptotically. In other words, the buoyancy-dominated far field (BDFF) is the ultimate flow regime of any discharge. In Lagrangian terms, given enough time (at large distances), no matter how weak the initial buoyancy is, the buoyancy effect will be felt. It can be demonstrated from analysis of experimental data or numerical model predictions that the behaviour of a momentum-dominated jet in crossflow may be described by a MDNF-MDFF-BDFF sequence of flow regimes (Prob. 8.7); similarly it is likely the behaviour of a buoyancy-dominated jet in crossflow can be described by a MDNF-BDNF-BDFF sequence (Wright 1977; Fischer *et al.* 1979). In general the important flow regimes are the flows very close to and very far from the source; the transition is usually not conveniently dealt with by length scales and is better treated using mathematical models. Note also that the three length scales L_s, L_m and L_b are not independent, with $L_s/L_b = (L_m/L_b)^{3/2}$.

6.1 LINE THERMALS AND PUFFS

Line puffs and thermals are in many ways similar types of free shear flows. Line puffs are generated by an instantaneous source of momentum, while thermals are generated by an instantaneous source of buoyancy. Both are vortex-pair like flows which move through and mix with the surrounding fluid, carrying the ambient fluid with them; the added mass coefficient for both flows is approximately 1. The mixing and turbulence intensity is greatest towards the front of the puff/thermal. The dilution constants and maximum dimensionless eddy viscosities are also comparable. There are however significant differences in the computed scalar concentration field. In the case of line puffs, the motion is dominated by the initial impulse. In the relatively short-lived asymptotic phase, the aspect ratio of the puff is about 1, and (contrary to experiments of jet in crossflow) no concentration peaks are predicted. The line thermal, on the other hand, is a motion that develops gradually from the buoyant impulse, and the flow has a clearly bifurcated double concentration maxima.

The results indicate that the unsteady line thermal is a reasonably good approximation to the bent-over plume in crossflow. The bifurcation of the scalar field in a typical buoyant plume in crossflow is predicted; the computed C_m/C_c and aspect ratio of 1.5 and 1.43 respectively can be compared to measured values of $C_m/C_c = 1.2-1.7$ and $b_h/b_v \approx 1.2-1.4$. It appears that the detailed structure of the scalar field would depend critically on the initial conditions as well as the location of the section (i.e. whether in the BDNF-BDFF transition or in the developed phase). In addition, in experiments the distinct bifurcation is observed only with relatively large initial buoyancy in a weak to moderate crossflow. For experiments with weaker buoyancy and/or stronger current, such a bifurcation is not observed, and $C_m \approx C_c$ (Wong 1991, Cheung and Lee

1991). The absence of double concentration peaks may also be related to the turbulence in the ambient flow (Gaskin 1995). A summary of the experimental results for advected line thermals and 2D thermals is also given in Table 8.2. Additional results on the BDFF concerning laboratory and field data will be dealt with in Chapter 10.

Line puffs and thermals are useful analogs for correlating experimental data of buoyant jet in crossflow. Nevertheless, the 3D ALP numerical computations (Lee, Chen and Kuang 2002) show that the excess x-velocity, $u_e = u - U_a$ is not negligible. Although the overall excess x-momentum is conserved (=zero), the distribution of u_e is highly non-uniform; maximum values of u_{em}/U_a around 0.3 are found even in the bent-over phase. The excess x velocity only becomes negligible for $x^* \sim 100$. Hence the representation of a section of the bent-over jet in crossflow (or advected line puff) by a line puff moving at the ambient velocity has certain limitations. This appears to explain the inability of the line puff to simulate the bifurcation of the scalar field. Such a limitation does not exist with the line thermal. The 3D ALT computations (Chen and Lee 2002) reveal that at $x^* = 8$ the horizontal excess velocity differs negligibly from the ambient velocity, with the maximum excess velocity $u_{em} \approx 0.05 U_a$.

Although line thermals are evidently better mixing agents than line puffs, it may appear strange on first sight that they have similar spreading and dilution laws in terms of the vertical distance z. The similar dilution characteristics seem to suggest that buoyancy is not important - perhaps a surprising result. This apparent anomaly is clarified in an example below (Example 8.3).

6.2 APPLICATION EXAMPLE

For prediction of plume mixing in a crossflow, the numerical model results can be cast in a more convenient form. For completeness the well-known solution for the BDNF is also included.

Buoyancy-dominated Near Field (BDNF)

The classical relations for the characteristic properties of a round plume in stagnant fluid are (Chapter 3):

$$W_m = 4.71\ F_o^{1/3}\ z_c^{-1/3} \tag{8.39}$$

$$b_{vc} = 0.11\ z_c \tag{8.40}$$

$$S_c = 0.096\ \frac{F_o^{1/3}\ z_c^{5/3}}{Q_o} \approx 0.1\ \frac{F_o^{1/3}\ z_c^{5/3}}{Q_o} \tag{8.41}$$

In the BDNF the jet is only weakly advected; mixing is similar to that of a plume in stagnant fluid. Similar to the advected jet (Chapter 7), the following equations can then be obtained for the advected plume:

$z/L_b \ll 1$ **(BDNF):**

$$\frac{z_c}{L_b} = 2.36\ (\frac{x}{L_b})^{3/4} \tag{8.42}$$

$$b_{vc} = 0.11\ z_c \tag{8.43}$$

$$\frac{S_c Q_o}{U_a L_b^2} = 0.1\ (\frac{z_c}{L_b})^{5/3} \tag{8.44}$$

The above simple advected plume model for the BDNF is verified by the detailed LIF experiments of a vertical plume in weak crossflow (Pun and Davidson 1999) which confirms the validity of the $x^{3/4}$ trajectory law and give a trajectory coefficient of 2.57. This supports the use of the average velocity in defining the plume trajectory (see also related discussion on coflow jet formulation in Chapter 6 and MDNF in Chapter 7).

According to theory, the BDNF asymptotic flow regime is characterized by $z/L_b \ll 1$; for many engineering application estimates. Concentration measurements have shown the transition occurs in the range of $z/L_b = 0.1 - 1$ (Cheung 1991); the LIF trajectory data show a transition at $z/L_b \approx 0.5$. For practical purposes, a transition at around $z/L_b \approx 1$ can often be assumed; the transition issue will be further discussed in Chapter 10.

Buoyancy-dominated Far Field (BDFF)

Based on the ALT numerical model and experimental results, the following equations for the advected line thermal are recommended:

$z/L_b > 1$ **(BDFF):**

$$\frac{z_c}{L_b} = 1.3\ (\frac{x}{L_b})^{2/3} \tag{8.45}$$

$$b_{vc} = 0.4\ z_c \tag{8.46}$$

$$\frac{S_c Q_o}{U_a L_b^2} = 0.46\ (\frac{z_c}{L_b})^2 \qquad \text{or}$$

$$S_c = 0.46 \frac{U_a\ z_c^2}{Q_o} \tag{8.47}$$

Note that the dilution constant in the vortex flow in the bent-over phase is the same for both the MDFF and the BDFF.

EXAMPLE 8.1 *Vertical buoyant jet in crossflow - BDFF*
Consider a vertical buoyant jet in crossflow in the laboratory as shown in Fig. 8.1 , with the following discharge and ambient flow data: $W_o =$ 0.25 m/s, $U_a = 0.063$ m/s, $D = 0.01$ m, $\Delta\rho_o/\rho_a = 0.0257$. *Estimate the elevation, width, and dilution of the buoyant jet at a distance of* $x = 0.12$ m *downstream.*

Solution: The densimetric Froude number of the jet can be determined to be $Fr = \underline{5}$. First compute the jet volume, momentum, and buoyancy fluxes as:

$$Q_o = 0.25 \times \pi \times 0.01^2/4 = 0.196 \times 10^{-4}\ m^3/s$$

$$M_o = Q_oW_o = 0.491 \times 10^{-5}\ m^4/s^2$$

$$F_o = Q_og_o' = 0.494 \times 10^{-5}\ m^4/s^3$$

The crossflow momentum and buoyancy length scales are then: $L_m = L_{mv} = \frac{M_o^{1/2}}{U_a} = \underline{0.0352}$ m, and $L_b = \frac{F_o}{U_a^3} = \underline{0.0198}$ m. Since $L_m/L_b =$ 1.78, the two crossflow length scales are comparable to each other. At $x = 0.12$ m, the jet trajectory is located in the BDFF, with

$$\frac{z}{L_b} = 1.3\ (\frac{x}{L_b})^{2/3}$$

or $z = 1.3 \times 0.12^{2/3} \times (0.0198)^{1/3} = \underline{0.086}$ m. Since $z/L_b > 1$ and $z/L_m > 1$, use the BDFF equation 8.46 to estimate the width, $w = 2B =$ $2\times0.4\times0.085 = 0.068$ m. The centerline dilution at this elevation is given by Eq. 8.47, with $S_c = 0.046 \times 0.063 \times 0.085^2/(0.196 \times 10^{-4}) = \underline{10.7}$. The minimum dilution in the cross-section is probably less by 20-30 percent. Since $B = \sqrt{b_h b_v}$, assuming an aspect ratio of $b_h/b_v = 1.2$, the vertical and horizontal half-widths can be estimated as: $b_v = B/\sqrt{1.2} = \underline{0.062}$ m, and $b_h = 1.2\times b_v = \underline{0.075}$ m. Note that the above calculations assume a buoyant jet in an unbounded crossflow. The effect of the free surface on the dilution is treated in Chapter 10.

EXAMPLE 8.2 *Plume in weak crossflow - BDNF/BDFF transition*
Consider a laboratory vertical heated plume in crossflow with the following characteristics: $W_o = 0.1075$ m/s, $U_a = 0.0175$ m/s, $D = 0.0075$ m, $\Delta\rho_o/\rho_a = 0.012$ *(corresponding to a jet discharge temperature of* $T_o =$ 56° C, *and* $\Delta T_o = 21.8^o$ C*). Estimate the dilution at a depth of* $z = 0.05$ *m* $z = 0.136$ *m above the source.*

Solution: The densimetric Froude number of the jet is $Fr = 3.62$. The jet volume, momentum, and buoyancy fluxes can be determined as: $Q_o = 0.475 \times 10^{-5}$ m^3/s, $M_o = 0.511 \times 10^{-6}$ m^4/s^2, and $F_o = 0.5592 \times 10^{-6}$ m^4/s^3. The jet momentum length scale is $l_s = M_o^{3/4}/F_o^{1/2} = 0.026$ m. The crossflow momentum and buoyancy length scale are respectively $L_m = 0.0408$ m, and $L_b = 0.104$ m.

This is a case of $L_m \sim L_b$, with $L_m/L_b = 0.39$. At $z = 0.05$ m, since $z/L_b \approx 0.5$, it is natural to use the BDNF relation Eq. 8.41 to compute the dilution of the advected plume, with $S_c = 0.1 \times (0.5592 \times 10^{-6})^{1/3} \times 0.05^{5/3}/(0.475 \times 10^{-5}) = \underline{1.07}$. At $z/D = 0.05/0.0075 = 6.7$, the buoyant jet is just beyond the zone of flow establishment, and hence the center of the plume is not much diluted. In this case, however, the prediction based on such a reasoning would under-estimate the actual observed value by a factor of 6!

What is at fault? We note that at $z = 0.05$ m, $z/L_b < 1$, and $z/l_s = 1.9$, suggesting that the flow is buoyancy-dominated at this depth. At the same time, $z > L_m$ and hence the momentum-induced velocity has already decayed to less than the ambient velocity value, although the flow is still influenced by the discharge buoyancy. If we assume a vortex flow, the dilution (for either the MDFF or BDFF) would be given by $S_c = 0.46 \times 0.0175 \times 0.05^2/(0.475 \times 10^{-5}) = \underline{4.7}$, much closer to the measured dilution of 6.0. In situations when we are dealing with transitions, it is often difficult to estimate dilutions based purely on length scales. Alternatively the width and dilution in the transition can be predicted by a 1D Lagrangian model for the case of a general buoyant jet with both discharge momentum and buoyancy (Chapter 9 and 10). Similarly, application of the BDFF equation gives the centerline dilution at $z = 0.136$ m as $S_c = \underline{31.4}$.

EXAMPLE 8.3 *Advected line thermals vs advected line puffs*
Consider a vertical discharge with the following characteristics: $W_o = 1.0$ m/s, $U_a = 0.2$ m/s, $D = 0.1$ m, $\Delta\rho_o/\rho_a = 0.025$. *Use Eq. 8.27 to determine the BDFF trajectory. Similarly, use Eq. 7.18 to determine the MDFF trajectory for a corresponding nonbuoyant discharge. Estimate the half-width and dilution at* $z = 5, 10$ *m, and the corresponding travel times.*

Solution: The densimetric Froude number of the buoyant jet is $Fr = 6.5$. The jet volume, momentum, and buoyancy fluxes can be determined as: $Q_o = 7.85 \times 10^{-3}$ m^3/s, $M_o = 7.85 \times 10^{-3}$ m^4/s^2, and $F_o = 1.926 \times 10^{-3}$ m^4/s^3. Noting that $F_{lo} = F_o/U_a$, and $M_{lo} = M_o/U_a$, Eq. 8.27 (with $k = 1, \beta = 0.4$) then gives $z = (1.492 F_{lo})^{1/3} t^{2/3}$ for the line thermal, and similarly Eq. 7.18 gives $z = (2.98M_{lo})^{1/3}t^{1/3}$ for the line puff. The trajectories are sketched in the attached figure (ignoring the near field).

The figure shows clearly that although the width and dilution achieved at the same z vertical distance above the jet are the same for the thermal and the puff, the time taken to rise to 5-10 m is 10-30 times longer for the puff. The buoyant thermal is clearly a much more vigorous mixing agent than the momentum puff; the ratio of the half-width at a given time for the thermal and puff is $B_r = R_{thermal}/R_{puff} = (0.5\frac{F_o t}{M_o})^{1/3}$. On simplification, we have $B_r = (0.5\frac{g'_o t}{W_o})^{1/3}$. The relative mixing efficiency of thermal vs puff is determined by the ratio $g'_o t/W_o$ i.e. the buoyant velocity to the intial velocity. Since this ratio quickly exceeds unity, the thermal can be seen to be a much more efficient mixing device than puffs when interpreted in Lagrangian manner.

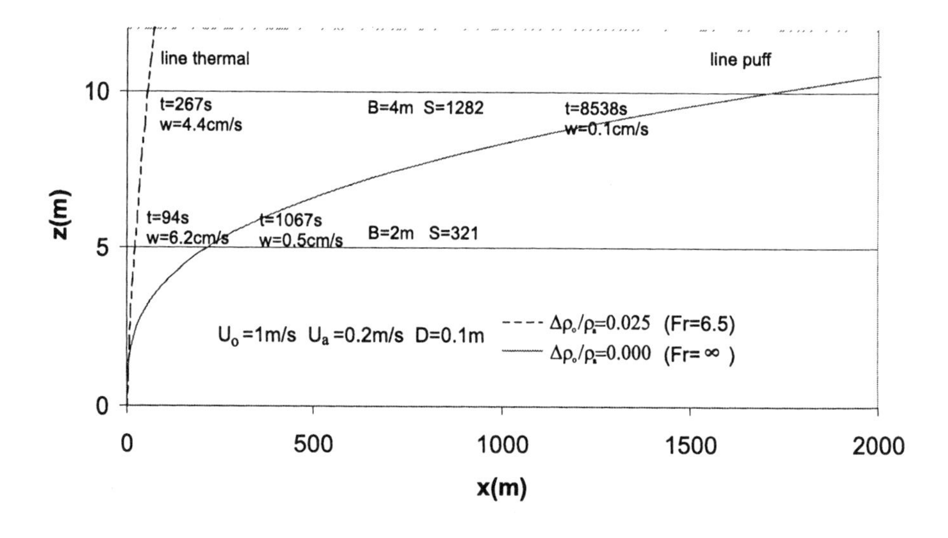

Figure 8.25. Comparison of line thermal and line puff characteristics for plume/jet in crossflow

7. SUMMARY

To an observer moving with the ambient flow, a section of a bent-over plume in crossflow rises like a line thermal. Turbulence model computations of unsteady 2D line thermals and steady 3D advected line thermals show that the flow is self-similar beyond a dimensionless time of around $t^* = 10$. The thermal is characterized by a double-vortex flow, and typically a kidney-shaped cross-section scalar field with off-centered double concentration maxima. The aspect ratio of the thermal is around 1.2-1.4. The cross-sectional maximum concentration can be up to 40-60 percent greater than the centerline maximum concentration. Both the predictions and the experiments show that the line thermal is a very good analogy of the bent-over plume (BDFF). The existence of added mass is confirmed. A review of measurements of thermals is presented. The predicted relations for time-averaged trajectory, width, and dilution are well-supported by experimental data. It is demonstrated that the mixing is governed by the large scale flow structure, and the path is defined by the average rather than the centerline velocity.

The instantaneous flow of the advected line thermal is highly variable and much more complex than the time-averaged structure. Thermals exhibit much greater concentration fluctuation than jets, plumes, and puffs. Thermals are by nature much more efficient mixers than puffs. Surprisingly, the Eulerian spreading and dilution relations are very similar. For both the advected thermal (BDFF) and the advected puff (MDFF), numerical computation and direct measurements have confirmed the linear spread with vertical distance above source. The recommended values for the added mass and spreading coefficients are:

$$k_n = 1 \quad \text{and} \quad \beta_n = 0.4 \quad \text{for line thermals and puffs} \tag{8.48}$$

PROBLEMS

8.1 Consider the waste gas jet discharge in Example 3.6 in Chapter 3, with $w_o = 5$ m/s, $\Delta\rho_o/\rho = 0.034$, and $D = 0.2$ m. Estimate the dilution of the waste plume at $z = 10$ m in a cross wind of a) $U_a = 1$ m/s, and b) $U_a = 3$ m/s. Compare with the plume dilution in a stagnant atmosphere. What is the downstream location of the plume at this height?

8.2 *Concentration vs visual trajectory of jet in crossflow*
Consider the predictions of the concentration trajectory z_c and maximum vertical velocity W_m of the line puff derived from the $k-\epsilon$ model (Table 7.1):

$$\begin{aligned} z_c &= 1.7 M_{lo}^{1/3} t^{1/3} \\ W_m &= 1.61 M_{lo}^{1/3} t^{-2/3} \end{aligned}$$

a) The visual trajectory (center of mass) z_d is given by $dz_d/dt = \overline{W}$, where $\overline{W}$ is the average velocity. Noting that $W_m/\overline{W} = 3.18$, determine the ratio of z_c/z_d. Is your result in accord with observations?

b) Using the numerical results for the line thermal in Table 8.1, obtain z_c/z_d for a plume in crossflow.

8.3 *Added mass of line puff/thermal*

The virtual added mass of a line puff or thermal can be likened to that of a 2D cylinder. Consider a circular cylinder of radius a and mass per unit length M, moving at a velocity $U(t)$ in the x-direction through a fluid of density ρ_a. The otherwise stagnant fluid will be displaced by the cylinder and a flow pattern is formed. The stream function $\psi(r,\theta)$ of the instantaneous irrotational flow outside of the cylinder at any time is (see e.g. Lamb 1932):

$$\psi = -\frac{Ua^2}{r}\sin\theta$$

where a=radius of cylinder, and (r,θ) are the radial and angular co-ordinates (with respect to the centre of the sphere at time t, and the $x-$ direction respectively). The radial and tangential velocities at any point in the flow are then:

$$\begin{aligned} V_r &= -\frac{1}{r}\frac{\partial\psi}{\partial\theta} = U\frac{a^2}{r^2}\cos\theta \\ V_\theta &= \frac{\partial\psi}{\partial r} = U\frac{a^2}{r^2}\sin\theta \end{aligned}$$

Show that the total kinetic energy of the external flow is:

$$E_{flow} = \int_a^\infty \frac{1}{2}\rho_a q^2 2\pi r dr = \frac{1}{2}M_a U^2$$

where $q^2 = V_r^2 + V_\theta^2$, and $M_a = \rho_a(\pi a^2)$ is the virtual added mass of the moving cylinder. The total kinetic energy required to move the cylinder is then:

$$E = \frac{1}{2}(M + M_a)U^2$$

If F is the external force acting on the cylinder in the $x-$ direction, then the rate at which F does work must equal to the rate of increase of the total kinetic energy, $FU = \frac{dE}{dt}$. Hence show that:

$$F = (M + M_a)\frac{dU}{dt}$$

i.e. the cylinder experiences a resistance of $-M_a\frac{dU}{dt}$ due to the presence of the fluid.

For line puffs and thermals, with $\rho \approx \rho_a$, and $M \approx M_a$, if the above equation is assumed, we obtain:

$$F = (M + M_a)\frac{dU}{dt} = (1+k)M\frac{dU}{dt}$$

where $k = 1$ is the added mass coefficient. However, unlike a circular cylinder, the interface between the puff/thermal element and the ambient fluid is not a solid boundary; it is a moving fluid interface across which there is significant entrainment (particulary at the rear). The puff/thermal is also kidney-shaped and the flow around it is not separated. In view of these complexities the added mass coefficient must be determined by a full numerical calculation of the time-dependent flow and passive scalar field (Fig. 7.11 or 8.11) or by direct measurement (Fig. 7.25).

8.4 *Entrainment coefficient of line thermal*

Consider the following typical entrainment model for a line thermal or puff: the rate of change of volume of the thermal is proportional to the inflow due to turbulent entrainment:

$$\frac{d}{dt}(\pi B^2) = 2\pi B(\alpha\overline{W})$$

where $B = 0.44\ z_c$ is the characteristic radius of the thermal and $\overline{W}$ = average vertical velocity. The entrainment velocity is assumed to be proportional to the thermal velocity, with α=entrainment coefficient.

a) Using the turbulence model predictions of the concentration trajectory z_c, maximum vertical velocity W_m, and $W_m/\overline{W}$ for the line thermal (Table 8.1), show that $\alpha \approx 0.46$. Obtain a similar value for the entrainment coefficient of the line puff using Table 7.1.

b) If we assume the line thermal to be an analog for the bent-over plume in crossflow, the BDFF, then $dB/dt = U_a dB/dx$. Noting that $dz/dx = \overline{W}/U_a$, show that $B = \alpha z$ - i.e. in this case the entrainment coefficient is the same as the spreading rate.

8.5 *Entrainment-based model for bent-over buoyant jets*

An entrainment-based model for the line thermal (BDFF) or line puff (MDFF) of a bent-over buoyant jet in crossflow can be formulated in Eulerian form.

a) For a buoyant jet in crossflow, the entrainment relation is often stated as the sum of two parts, one due to shear entrainment, and one due to the entrainment of the crossflow:

$$\frac{dQ}{ds} = 2\pi B[\alpha_s(U - U_a\cos\phi) + \beta' U_a \sin\phi]$$

where $Q = V\pi B^2$ = jet volume flux, s=natural streamwise co-ordinate, α_s = shear entrainment coefficient, and β' = crossflow entrainment coefficient, and ϕ=local jet inclination to x-axis. In the bent-over phase, $V \approx U_a$, and the second crossflow term dominates. Noting that $\sin\phi = dz/ds$, show that $B = \beta' z$. The use of the this model together with momentum conservation leads to the 2/3 power trajectory law, Eq. 8.27.

b) In experiments, the BDFF trajectory coefficient of a plume in crossflow has been determined to be $C_{2t} = 1.1$, and the observed vertical half-width is $B_v = 0.35\ z$, where C_{2t} is given by Eqs. 8.27 and 8.28. If added mass is neglected, determine the spreading rate from the trajectory observation. Compare your result with the observation and give an interpretation of the data.

8.6 *Average to centerline minimum dilution for line thermal*

For the line thermal, the volume (per unit length) of the thermal (defined within the $0.01C_m$ concentration contour) is given by L^2. The average dilution is hence given by $\overline{S} = L^2/Q_{lo}$, where the terms have their usual meaning. On the other hand, the centerline minimum dilution is given by $S_c = 1.14z_c^2/Q_{lo}$. Using the turbulence model predictions of z_c and L given in Table 8.1 show that $\overline{S}/S_c \approx 1.7$ for the line thermal. Obtain a similar value for the line puff using Table 8.1.

8.7 *MDFF-BDFF trajectory*

Consider a vertical buoyant jet discharge with $w_o = 4$ m/s, $U_a = 0.2$ m/s, $D = 0.1$ m, and $\Delta\rho_o/\rho_o = 0.005$. Using the model VISJET or equivalent, predict the buoyant jet trajectory. By plotting the predicted trajectory on a log-log scale, identify the slopes for the asymptotic flow regimes (if any) for the MDFF (1/3) and BDFF (2/3). Comment on your results. How would your results change if $\Delta\rho_o/\rho_o = 0.025$?

8.8 *Line plume in stratified crossflow*

The behaviour of a line plume in a stratified crossflow can be effectively studied using a Lagrangian model using the principles outlined in Chapters 3, 5, and 8 (Tate, P.M. and Middleton, J.H. , "The height of rise and dilution of a two dimensional plume in a stratified crossflow", Sydney Water Board internal report, 1990, unpublished). The discharge from an ocean outfall of length L is sometimes modelled as a line source of buoyancy. The discharge and buoyancy flux per unit diffuser length of this line plume are $Q_o = Q_s/L$, $F_o = Q_s(\rho_a(0) - \rho_o)g/L$ respectively, where Q_s=total source flow from diffuser.

Consider a line plume in a perpendicular stratified crossflow of velocity U_a and density profile $\rho_a(z)$. In a moderate to strong current, when the plume velocity ($\sim (F_o/\rho)^{1/3}$) is less than the ambient current (or $F = U_a^3/(F_o/\rho) \leq 1$), it is expected the line plume is quickly bent over by the crossflow and the $x-$velocity in the curtain of mixed sewage is approximately U_a. Fig. 8.26 shows an element of this line plume, characterized by length dx, vertical half-width B, volume $(dx2B)$, buoyancy $\Delta\rho g = (\rho_a - \rho)g$, and vertical velocity W.

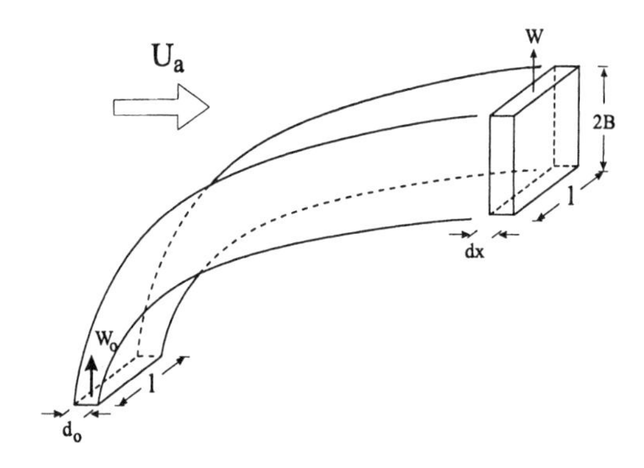

Figure 8.26. A line plume in stratified crossflow

a) By using the Lagrangian spreading hypothesis, and applying momentum and buoyancy conservation, show that the line plume satisfies the following equations:

$$\begin{aligned}
\frac{dB}{dt} &= \beta W \\
\frac{dM}{dt} &= \frac{d}{dt}(\rho 2WB) = (\Delta\rho g 2B) = F \\
\frac{dF}{dt} &= \frac{d}{dt}(\Delta\rho g 2B) = (\rho 2WB)(\frac{g}{\rho}\frac{d\rho_a}{dz}) \\
&= -N^2 M
\end{aligned}$$

where M, F = vertical momentum of and buoyancy force acting on the plume element, N=stratification frequency, and β=spreading or entrainment coefficient (Hint: see Eq. 3.60 - 3.61, and Eq. 5.8).

b) For a linear stratification, N=constant, show that the buoyancy force $F(t)$, and the trajectory $z(t)$ of the line plume element are given by:

$$F(t) = \frac{F_o}{U_a}\cos(Nt)$$
$$z^2 = -(\frac{F_o}{\rho\beta N^2 U_a})\cos(Nt)$$

c) At the equilibrium height of rise z_e, the vertical momentum $M = 0$ ($Nt = \pi$). The average dilution is given by $\overline{S} = U_a(2B)/Q_o$. Show that z_e and $\overline{S}$ are given by:

$$z_e = \frac{1}{\sqrt{\beta}}(\frac{F_o}{\rho U_a})^{1/2}N^{-1} = 1.29(\frac{F_o}{\rho U_a})^{1/2}N^{-1}$$
$$\overline{S} = 2\sqrt{\beta}U_a^{1/2}\frac{(F_o/\rho)^{1/2}}{Q_o}N^{-1} = 1.55U_a^{1/2}\frac{(F_o/\rho)^{1/2}}{Q_o}N^{-1}$$

where $\beta = 0.6$ has been used. As a comparison, the experiments of a line plume in a stratified crossflow (Roberts *et al.*1989) give for the equilibrium level of the line plume:

$$z_e = 1.5(\frac{F_o}{\rho U_a})^{1/2}N^{-1}$$

for $0.1 < F < 100$, where $F = U_a^3/(F_o/\rho)$ is a crossflow parameter. The minimum dilution in the spreading layer is given by:

$$S_m = 0.67\frac{F_o^{1/3}z_e}{Q_o}F^{1/6}(2.19F^{1/6} - 0.52)$$

It should be noted that the first term in the Roberts dilution equation has the same form as that given by the Lagrangian model, with a coefficient of 1.46.

8.9 *Ocean outfall in stratified current*

Consider an ocean outfall with total discharge flow $Q_s = 10$ m^3/s, diffuser length $L = 800$ m, and initial relative density difference of 0.025. Estimate the equilibrium level and dilution S in the waste field in a stratified current of $U_a = 0.2$ m/s, and stratification frequency $N = 0.01$ s^{-1}. Compare the estimates obtained from the Lagrangian model with that using the Roberts formula.

Chapter 9

GENERAL LAGRANGIAN FORMULATION

The development of jets, plumes, puffs or thermals (JPPT) in coflow and crossflow, and in a surrounding environment of non-uniform density, have been the topics treated at length in previous chapters. The interaction of the JPPT with the density stratification in the environment was considered in chapter 5. The effects of uniform coflow and uniform crossflowwere examined in chapters 6, 7 and 8. In the present chapter, the combination of all the effects that have been examined as limiting cases are included in a general formulation. The concept of dominant eddies, the method of excesses, and the hypotheses of entrainment and spreading are revisited in a general context. The general formulation, for non-uniform coflow and crossflow and for non-uniform density stratification, is carried out here following the Lagrangian method by Lee and Cheung (1990) and Chu (1994). The equivalent formulation using the Eulerian method is given in a more recent work by Chu and Lee (1996).

1. ELEMENTAL VOLUME

The Lagrangian formulation is based on mass balance in an *elemental volume* and the concept of *dominant eddies.* In puffs and thermals, the elemental volume is the dominant eddy occupying the entire puffs/thermals volume. The formulation for the puffs and the thermals is naturally a Lagrangian problem. Jets and plumes can also be formulated using the Lagrangian method if an elemental volume is correctly defined for the problem. Figure 9.1 shows the elemental volume in the general problem of a buoyant jet in a non-unform crossflow of changing direction. The elemental volume is defined by the dominant eddies that are extending across the full width of the turbulent region and moving along the path of the buoyant jet. The volume is defined along the path

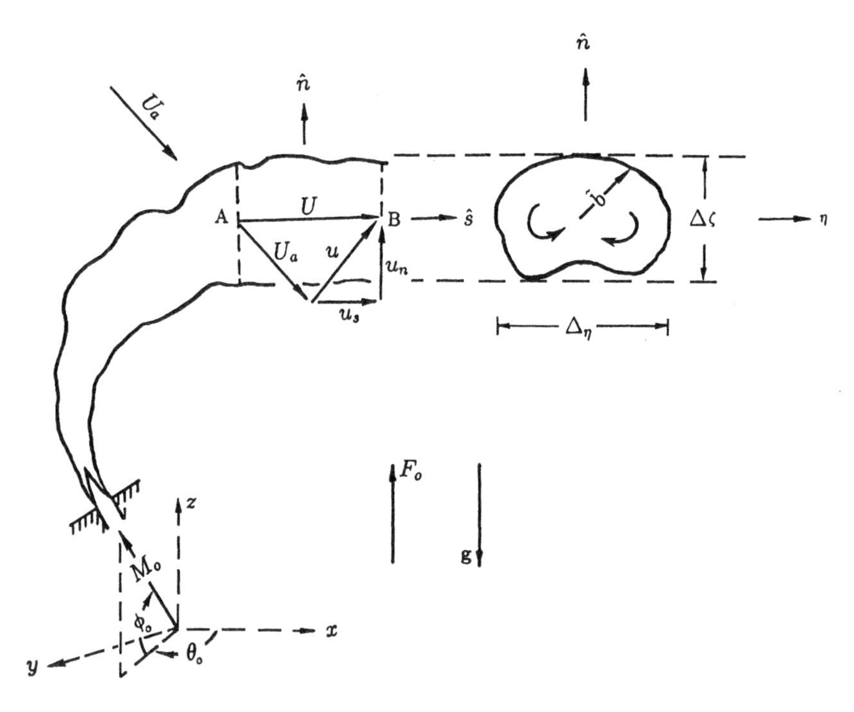

Figure 9.1. Buoyant jet in a non-uniform crossflow of changing direction. The three dimensional trajectory is produced as buoyancy force acts in a general direction relative to the cross flow. The elemental volume, AB, is defined by the discharge of momentum and buoyancy at the source over a period of one time unit.

over the length AB as marked in the figure. The orientation of AB is along the path of motion. The magnitude and direction of AB is determined by the velocity vector, $\tilde{\mathrm{U}}$, tangential to the path of dominant eddies. The dominant eddies over the length AB are produced by momentum and buoyancy injection at the source over a period of one time unit. With such definition of the elemental volume, the momentum and buoyancy initially associated with AB are equal in magnitude of the momentum and buoyancy fluxes at the source, M_o and F_o, respectively.

As the dominant eddies move through their surrounding, materials are drawn into the eddies from the surrounding. The result is a continuous increase of materials in the elemental volume. This increase of materials in the elemental volume does not change the *excess* value in the volume if the excess is defined as the concentration of the materials relative to its value in the environment. The following section explains how the problem is conveniently formulated by the *method of excess.*

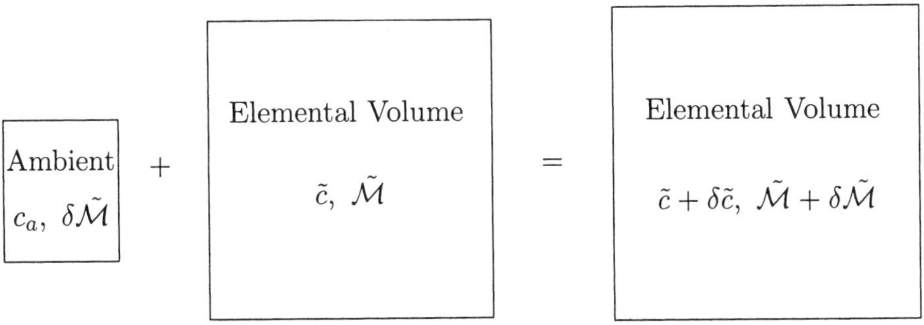

Ambient $c_a,\ \delta\tilde{\mathcal{M}}$ + Elemental Volume $\tilde{c},\ \tilde{\mathcal{M}}$ = Elemental Volume $\tilde{c}+\delta\tilde{c},\ \tilde{\mathcal{M}}+\delta\tilde{\mathcal{M}}$

Figure 9.2. Mixing of fluid in an elemental volume of carrying mass $\tilde{\mathcal{M}}$ and concentration $\tilde{c}$ with a small carrying mass $\delta\tilde{\mathcal{M}}$ of the ambient fluid of concentration c_a

2. METHOD OF EXCESSES

Due to entrainment, the 'mass' and volume in an elemental volume increase continuously. The volume is not fixed in space but moves with the velocity of the dominant eddies. The conventional fluid-mechanic methods such as the method of the control volume and the method of the material volume are not suitable for the present formulation. We overcome the difficulty associated with the entrainment using the method of excess. The concentration relative to its value in the ambient environment is the concentration *excess* $(c - c_a)$, where c_a is the ambient concentration. The mass-balance equation for the excess is derived by considering the mixing between the fluid in the elemental volume and the entrained ambient fluid.

The block diagram in Figure 9.2 shows the mass-balance. The fluid in the elemental volume of carrying mass $\tilde{\mathcal{M}}$ and concentration $\tilde{c}$ is mixed with a small carrying mass $\delta\tilde{\mathcal{M}}$ of the ambient fluid of concentration c_a. As the fluid in the environment is drawn into the dominant eddies in the elemental volume, tracer 'mass' is exchanged between the eddies and its environment. For the present formulation, the concentration $\tilde{c}$ and c_a are intensive variables and are defined as the tracer 'mass' per unit mass of the carrying fluid. With this definition, the tracer 'mass' in the elemental volume is $\tilde{c}\tilde{\mathcal{M}}$. The tracer 'mass' of the entrained fluid entering the elemental volume is $c_a\delta\tilde{\mathcal{M}}$. After the entrain mixing, the mass concentration becomes $\tilde{c} + \delta\tilde{c}$ and the carrying mass becomes $\tilde{\mathcal{M}} + \delta\tilde{\mathcal{M}}$. Equating the tracer 'mass' of the system before and after the entrainment,

$$c_a\delta\tilde{\mathcal{M}} + c\tilde{\mathcal{M}} = (\tilde{c} + \delta\tilde{c})(\tilde{\mathcal{M}} + \delta\tilde{\mathcal{M}}), \tag{9.1}$$

which gives the mass-balance equation

$$\delta[(\tilde{c} - c_a)\tilde{\mathcal{M}}] = -\tilde{\mathcal{M}}\,(\delta c_a). \tag{9.2}$$

The tracer 'mass' concentration in this equation can represent either the concentration of scalars, such as salinity and heat, or concentration of vectors such as linear and angular momentum.

Momentum Equation If the tracer 'mass' is linear momentum, the concentration of the momentum will be the velocity vector, which is momentum per unit mass of the carrying fluid. With this usage, the Equation 9.2 becomes the momentum equation, which is

$$\delta[(\tilde{\mathbf{U}} - \mathbf{U}_a)\tilde{\mathcal{M}}] = -\tilde{\mathcal{M}}(\delta \mathbf{U}_a) \tag{9.3}$$

if momentum is conserved. However, the momentum of the system may change due to the buoyancy force. The equation for momentum changed by the impulse of the buoyancy force is

$$\delta[(\tilde{\mathbf{U}} - \mathbf{U}_a)\tilde{\mathcal{M}}] = -\tilde{\mathcal{M}}(\delta \mathbf{U}_a) + \tilde{\mathbf{B}}\delta t \tag{9.4}$$

that, in a more compact form, is as follows:

$$\delta\tilde{\mathbf{m}} = -\tilde{\rho}\tilde{V}\delta\tilde{\mathbf{U}}_a + \tilde{\mathbf{B}}\delta t \tag{9.5}$$

where $\tilde{\mathbf{m}} = \tilde{\rho}\tilde{V}\tilde{\mathbf{u}}$ is the momentum *excess* and $\tilde{\mathbf{B}}$ is the buoyancy force vector. The velocity excess, $\tilde{\mathbf{u}} = \tilde{\mathbf{U}} - \mathbf{U}_a$, is the velocity of the dominant eddies relative to the ambient. The density of the carrying fluid in the elemental volume is $\tilde{\rho}$. The mass of the carrying fluid in the elemental volume of volume $\tilde{V}$ is $\tilde{\mathcal{M}} = \tilde{\rho}\tilde{V}$. The tilde symbol, $\tilde{}$, denotes the averages over the dominant eddies. The excess momentum $\tilde{\mathbf{m}}$ is modified at a rate proportional to the buoyancy force, $\tilde{\mathbf{B}}$. In the absence of buoyancy force ($\tilde{\mathbf{B}} = 0$), the excess momentum would be constant if the ambient velocity is constant ($\delta\tilde{\mathbf{U}}_a = 0$).

Buoyancy Equation The buoyancy force is not a constant if the fluid in the environment is density stratified. The basic equation for the variation of the buoyancy is

$$\delta\tilde{\mathbf{B}} \simeq g\hat{\mathbf{k}}\tilde{V}\delta\rho_a. \tag{9.6}$$

where $\tilde{\mathbf{B}} = B\hat{\mathbf{k}}$ and $B = g(\rho_a - \rho)\tilde{V}$; g is the gravity constant, $\hat{\mathbf{k}}$ is the base vector pointing in the opposite direction to the force of gravity, and $(\rho - \rho_a)$ is the density of the fluid relative to the reference density ρ_a. This is the same buoyancy-variation equation given by Eq. 5.8. Note that $\tilde{B}$ is positive if $\rho < \rho_a$ and $\delta\tilde{B}$ is positive if $\delta\rho_a > 0$.

General Form of the Equation for the Excess Following similar procedure, it is possible to obtain excess equations for any tracer mass in the flow (e.g. nutrient, suspended sediments, fuel, toxic fume, smoke, heat and moisture). These equations are similar in form and all can be expressed in terms of the excesses as follows:

$$\delta[(\tilde{c} - c_a)\tilde{\rho}\tilde{V}] = -\tilde{\rho}\tilde{V}\delta(c_a) - \mathcal{K}\delta t \tag{9.7}$$

where $[(\tilde{c} - c_a)\tilde{\rho}\tilde{V}]$ is the concentration excess of the tracer mass, $\tilde{\rho}\tilde{V}$ is the mass of the carrying fluid; $\tilde{c}$ is the tracer concentration (tracer mass per unit mass of the eddies), c_a the tracer concentration of the ambient fluid, and $\mathcal{K}$ the 'mass' dissipation rate. By definition, excesses are zero in the environment. Entrainment of materials from the environment do not change the values of the excesses in the elemental volume. Without the source term, the excesses are constant if the ambient concentration is constant; i.e.,

$$\delta[(\tilde{c} - c_a)\tilde{\rho}\tilde{V}] = 0 \quad \text{if} \quad \delta(c_a) = 0 \quad \text{and} \quad \mathcal{K} = 0. \tag{9.8}$$

Analogous relations may be obtained for momentum and buoyancy excesses.

3. SPREADING HYPOTHESIS

The formulation is completed with a turbulent closure equation. In a one-dimensional formulation, this closure equation is based on an entrainment or spreading hypothesis. A number of variations in the hypothesis has been proposed. The great majority of the reported models are derived from an Eulerian point of view; many are not entirely consistent with all the experimental data. In the present formulation, the radius of the dominant eddies, $\tilde{b}$ or R, is assumed to increase at a rate in proportional to the relative speed, $|\tilde{\mathbf{u}}|$, as follows:

$$\frac{D\tilde{b}}{Dt} = \beta|\tilde{\mathbf{u}}| \tag{9.9}$$

where $\tilde{\mathbf{u}} = \tilde{\mathbf{U}} - \mathbf{U_a}$ is the velocity excess. In this formulation, D/Dt is a Lagrangian operator for time differentiation in a Lagrangian reference frame moving with the dominant eddies. Since $|\tilde{\mathbf{u}}| = Ds/Dt$,

$$\frac{D\tilde{b}}{Dt} = \beta\frac{Ds}{Dt}. \tag{9.10}$$

where s is the migration distance of the eddies relative to the surrounding fluid. The radius of the dominant eddies, $\tilde{b}$, increases at a rate in proportional to this migration distance, s, according to this hypothesis. The spreading rate is

$$\beta = \frac{D\tilde{b}}{Ds}. \tag{9.11}$$

In the general formulation for buoyant jet in a non-uniform crossflow (see Figure 9.1), the excess velocity vector $\tilde{\mathbf{u}}$ is resolved into a tangential component u_s in the direction parallel to $\tilde{\mathbf{U}}$ and into a normal component u_n in the direction perpendicular to $\tilde{\mathbf{U}}$. The entrainment of fluid into the dominant eddies is known to dependent on the orientation of elemental volume. Lee and Cheung (1990) and Lee *et al.*(2000) in their Lagrangian formulation of the buoyant jet problems have introduced the terms *shear-induced* entrainment and *vortex* entrainment to associate the process with the tangential and normal components of the excess velocity. The shear-induced entrainment is the dominant mechanism in the near field while vortex entrainment is dominant in the far field. There is yet no satisfactory model to handle the transition from the shear-induced entrainment to the vortex entrainment. The formulation of the entrainment hypothesis can be either based on an additive hypothesis, or a maximum hypothesis, as follows:

$$\frac{D\tilde{b}}{Dt} = \beta_s|u_s| + \beta_n|u_n|, \tag{9.12}$$

$$\frac{D\tilde{b}}{Dt} = Max[\beta_s|u_s|, \beta_n|u_n|] \tag{9.13}$$

Lee and Cheung (1990) found the maximum hypothesis gives slightly better results. Chu (1994) and Dehghani (1996) however were able to obtained satisfactory model agreement with experimental data using the additive model. A heuristic theory to account for the transition from the shear entrainment to the vortex entrainment regime is given in Equation 10.36 (Lee *et al.*2000). Chu (1994) has recommended a value of the shear-induced coefficient $\beta_s \simeq 0.17$ based on the data from the limiting case of the jets and plumes in stagnant fluid. When used together with an additive spreading hypothesis, the value recommended for the vortex coefficient is $\beta_n \simeq 0.34$. Analysis of data obtained for advective line puff and advective thermal in the previous chapters has lead to a higher votex coefficient of $\beta_n \simeq 0.4$. The value of the spreading coefficients are critically dependent on the definitions of $\tilde{b}$ and $\tilde{\mathbf{u}}$. Our recommendation is to define the boundary of the dominant eddies at the location where the turbulence intermittency $\gamma = 50$ %, so that $\tilde{b}$ (or R) is the radius of the region defined by this boundary and $\tilde{\mathbf{u}}$ is the average of the velocity

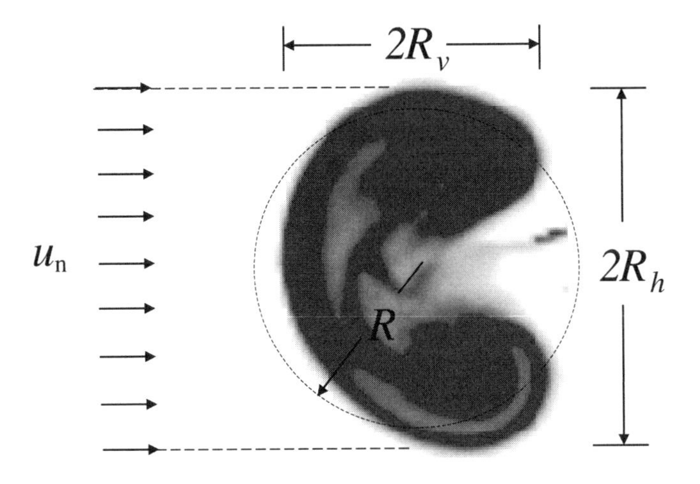

Figure 9.3. Projection Area Entrainment (PAE) hypothsis and Surface Area Entrainment (SAE) hypothesis. R (or $\tilde{b}$) is the radius of an equivalent cylinderical or spherical volume; u_n is component of the velocity relative to the dominant eddy.

in this region. As noted in the previous chapters, unlike the Eulerian coefficient, the Lagrangian coefficient β_s is approximately constant for straight jets and plumes as well as coflowing jets. The following sections show how the Lagrangian coefficient is obtained from the entrainment hypothesis introduced in a reference frame moving with the dominant eddies of the turbulent flow.

3.1 PROJECTED AREA ENTRAINMENT

Besides the spreading hypothesis (Equations 9.12 and 9.13), an alternate model is the Project Area Entrainment (PAE) hypothesis (Lee and Cheung 1990; Lee *et al.* 2000). The PAE model was introduced to explain the process of the vortex entrainment (Figure 9.3). The entrainment into the elemental volume is assumed in this model to be equal to the ambient-flow intercepted by the 'windward' face of the volume.

For a line puff or a line thermal, the volume per unit length of the line element is πR^2; the frontal projection area per unit length is $\mathcal{A}_p = 2R_h$; the flow into the frontal projection area on the wind-ward face is

$\mathcal{A}_p u_n = 2R_h u_n$. Equating the rate of volume change to the wind-ward interception,

$$\frac{D}{Dt}(\pi R^2) = 2R_h u_n \tag{9.14}$$

which gives the spreading rate

$$\frac{DR}{Dt} = \frac{R_h}{\pi R} u_n = \beta_n u_n, \tag{9.15}$$

and the normal component of the coefficient

$$\beta_n = \frac{R_h}{\pi R}. \tag{9.16}$$

Note that R is the radius of the equivalent cross section while $2R_h$ and $2R_v$ are the widths of the dominant eddy normal to and along the ambient flow direction, as shown in Figure 9.3. Hence, $\pi R^2 \simeq \pi R_h R_v$ (i.e., $R = \sqrt{R_h R_v}$) and the value of β_n is determined by Equation 9.16:

$$\beta_n = \frac{1}{\pi}\sqrt{\frac{R_h}{R_v}} \simeq 0.35. \tag{9.17}$$

This value of the spreading coefficient, obtained directly from PAE hypothesis, is remarkably close to the value recommended for line puffs/thermals by Chu (1994), and close to the experimental value of around 0.4 for advected line puffs and thermals (Chapter 7 and 8). The above calculation was based on the experimental and theoretical value of R_h/R_v = 1.2 (Chapter 8).

Similar calculation can be made for the asymmetric puffs/thermals. For this case the volume is $\mathcal{V} = \frac{4}{3}\pi R^3$; the frontal projection area is $\mathcal{A}_p = \pi R_h^2$. The mass balance for the elemental volume is

$$\frac{D}{Dt}(\frac{4\pi}{3}R^3) = (\pi R_h^2)u_n \tag{9.18}$$

This PAE model gives a spreading rate

$$\frac{DR}{Dt} = \beta_n u_n = \frac{R_h^2}{4R_v^2} u_n \tag{9.19}$$

for the axisymmetric volume and a spreading coefficient

$$\beta_n = \frac{1}{4}(\frac{R_h}{R_v})^2 \simeq 0.36 \tag{9.20}$$

if $R_h/R_v \simeq 1.2$. This spreading coefficient obtained by PAE model for the round puffs/thermals is comparable to the value of 0.34 for the line puffs/thermals. The round-puff data obtained by Richards (1965) has a spreading rate varying from $\beta_n = 0.25$ to 0.8.

3.2 SURFACE AREA ENTRAINMENT

Alternatively, the entrainment rate can be assumed to be proportional the 'surface area' instead of the projected area. This SAE (Surface Area Entrainment) model gives identical spreading coefficient for round and line puffs/thermals that is in better agreement with experimental data. The mass balance equation of the SAE model is

$$\frac{D}{Dt}(\mathcal{V}) = \beta_n(\mathcal{A})u_n. \tag{9.21}$$

The entrainment rate is assumed to be proportional to the surface area $\mathcal{A}$ and velocity excess u_n. For line puffs/thermals, the surface area is $\mathcal{A} = 2\pi R$, and the volume is $\mathcal{V} = \pi R^2$, per unit length of the line element. Equation 9.21 becomes

$$\frac{D}{Dt}(\pi R^2) = \beta_n(2\pi R)u_n. \tag{9.22}$$

For round puffs/thermals, the volume is $\mathcal{V} = \frac{4}{3}\pi R^3$, the surface area is $\mathcal{A} = 4\pi R^2$, and the equation is

$$\frac{D}{Dt}(\frac{4\pi}{3}R^3) = \beta_n(4\pi R^2)u_n. \tag{9.23}$$

Both Equation 9.22 for the line puffs/thermals, and Equation 9.23 for the round puffs/thermals, give the same spreading rate

$$\frac{DR}{Dt} = \beta_n u_n. \tag{9.24}$$

The coefficient β_n now represents the effective fraction of the surface area (or effective fraction of velocity) for turbulent entrainment. The values of the spreading coefficients for both the line puffs/thermals and round puffs/thermals are the same according to SAE hypothesis/model. The SAE model is conceptually simple and therefore most adaptable to complex flows encountered in practical application. Since the entrainment rate is directly proportional to the surface area, the volume of turbulent fluid with greater the surface area will have greater rate of entrainment into the volume.

The SAE hypothesis also leads to a naturally simple rule for the treatment of *'merging'*. Merging occurs as multiple parts of the jets and plumes are occupying the same space and in paths of the jets and plumes with large curvature. In such complex situation encountered in practical engineering applications, the natural approach would be to simply adapt the SAE model assuming that the entrainment rate is proportional to the surface area. Computation algorithms already have been developed to find the surface area in highly complex geometry (see Chapter 10).

The entrainment rate in the overlapping region is reduced as effective surface area is reduced due to merging.

3.3 SHEAR ENTRAINMENT

The shear entrainment hypothesis is also consistent with the SAE model as can be demonstrated by examining jets/plumes in stagnant fluid. In a Lagrangian reference frame moving with the dominant eddies, the cross sectional area of the jets/plumes increases at a rate proportional to the surface area and the shear velocity. For round jets/plumes, the volume per unit length of the jet/plume element is $\mathcal{V} = \pi\tilde{b}^2$ and the surface area per the same unit length is $\mathcal{A} = 2\pi\tilde{b}$. The mass balance between volume increase in the element and rate of entrainment is

$$\frac{D}{Dt}(\pi\tilde{b}^2) = \beta_s(2\pi\tilde{b})u_s \qquad (9.25)$$

For plane jet, the volume per unit length is $\mathcal{V} = 2\tilde{b}$ and the surface area per unit length is $\mathcal{A} = 2$. Hence,

$$\frac{D}{Dt}(2\tilde{b}) = \beta_s(2)u_s \qquad (9.26)$$

Both Equations 9.25 and 9.26 give the same equation

$$\frac{D\tilde{b}}{Dt} = \frac{\partial\tilde{b}}{\partial t} + u_s\frac{\partial\tilde{b}}{\partial s} = \beta_s u_s. \qquad (9.27)$$

which becomes

$$\frac{d\tilde{b}}{ds} = \beta_s \qquad (9.28)$$

in steady flow. Data of both plane and round jets/plumes are consistent with this spreading rate (Equation 9.28) derived from SAE model. The shear-entrainment coefficient for *steady flow* is $d\tilde{b}/ds = \beta_s = 0.17$ (Chu, 1994), if $\tilde{b}$ is defined at the laminar and turbulent interfaces where the intermittency factor is 50%.

3.4 SUMMARY

To summarize, our recommendation is to use the SAE model in the general formulation. For steady flow, and the additive hypothesis adopted, the values of the β-coefficients are $\beta_s = 0.17$ and $\beta_n = 0.34$ for shear-entrainment and vortex-entrainment, respectively.

4. PUFFS AND THERMALS

The formulation for the puffs and the thermals is carried out as a prelude to the general formulation of the more difficult problem to be presented in the next section. The problem of the puffs and thermals is simple because a single dominant eddy occupies the entire volume of a puff or a thermal. The round thermal is considered here as the first example using the method of excess.

EXAMPLE 9.1 *A 1.0 m³ of dredge spoil contaminated with 10.0 kg of cyanide is accidentally released from a ship into a river. The density of the dredge spoil is 1100 kg/m³. The river velocity relative to the ship is 8 m/s. An axisymmetrical thermal is formed as the cloud of the spoil mixes with the river water. (a) Find the momentum excess of the thermal upon entering the water surface. (b) Find the dilution and cyanide concentration after the thermal has reached a depth of 5 m below the water surface.*

SOLUTION: Example 5.2 is considered again with the effect of the crossflow included. (a) Upon entering the water surface, the velocity of the spoil (1100 kg) is increased suddenly from zero to 8 m/s. The momentum excess also changes suddenly. At time $t = 0$, the x-component of the momentum excess is $m_x(0) = 1100\text{kg} \times (0-8)\text{m/s} = \underline{-8800}$ kg-m/s; the z-component $m_z(0) = 0$. (b) The subsequent variation of the excess momentum is determined by the momentum equations, Equation 9.5:

$$\delta \tilde{m}_x = 0 \tag{9.29}$$

$$\delta \tilde{m}_z = B\delta t \tag{9.30}$$

The horizontal momentum excess stays constant. The vertical momentum excess changes at a rate proportional to the downward buoyancy force, which is $B = (1100 - 1000) \times 9.81 = 981$ N. Integrating these excess momentum equations once with time, gives

$$\tilde{m}_x = (1+k)\rho(\frac{4\pi R^3}{3})\frac{d\xi}{dt} = -8800, \tag{9.31}$$

$$\tilde{m}_z = (1+k)\rho(\frac{4\pi R^3}{3})\frac{d\zeta}{dt} = 981\, t. \tag{9.32}$$

where $(d\xi/dt, d\zeta/dt)$ are the velocity excess. The added-mass coefficient is $k = 0.5$ if the shape of the thermal is assumed to be spherical. The absolute displacements (dx, dz) are the relative displacements $(d\xi, d\zeta)$ plus the contribution from the crossflow. If the crossflow is in the x-direction,

$$dx = U_a\, dt + d\xi, \;\; dz = d\zeta. \tag{9.33}$$

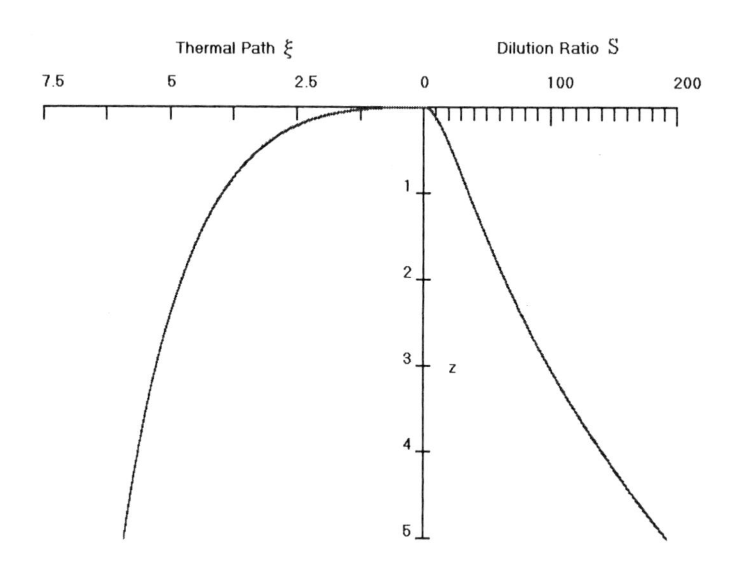

Figure 9.4. Path and dilution ratio of the round thermal in a coordinate system moving with the crossflow (Example 9.1)

The radius of the thermal is assumed to increase at a rate proportional to the velocity excess

$$\frac{dR}{dt} = \beta_n|\tilde{\mathbf{u}}| = \beta_n\sqrt{(\frac{d\xi}{dt})^2 + (\frac{d\eta}{dt})^2}. \tag{9.34}$$

Hence, the radius increases linearly with distance along the path of the thermal as follows:

$$dR = \beta_n\sqrt{(d\xi)^2 + (d\zeta)^2}. \tag{9.35}$$

The spreading coefficient $\beta_n = 0.34$. Equations 9.31, 9.32 and 9.35 are numerically integrated to give R, ξ, and ζ. Figure 9.4 shows dilution ratio, and the path of the thermal, in a coordinate system moving with the ambient cross flow. The thermal moves backwards relative to the ship due to the negative excess momentum of - 8800 Kg-m/s. At 5 m depth, the dilution ratio is $S = \mathcal{V}_o/\mathcal{V} = \underline{192}$, which may be compared with the dilution ratio $S = 21$ for the same thermal at the same depth without the crossflow. The effect of the crossflow is significant. It enhances the dilution by ninefold. The volume of the thermal at this depth is 192 m^3 and the cyanide concentration is $\tilde{c} = 10/192 = \underline{0.0521}$ kg/m^3 (i.e, 52.1 mg/L).

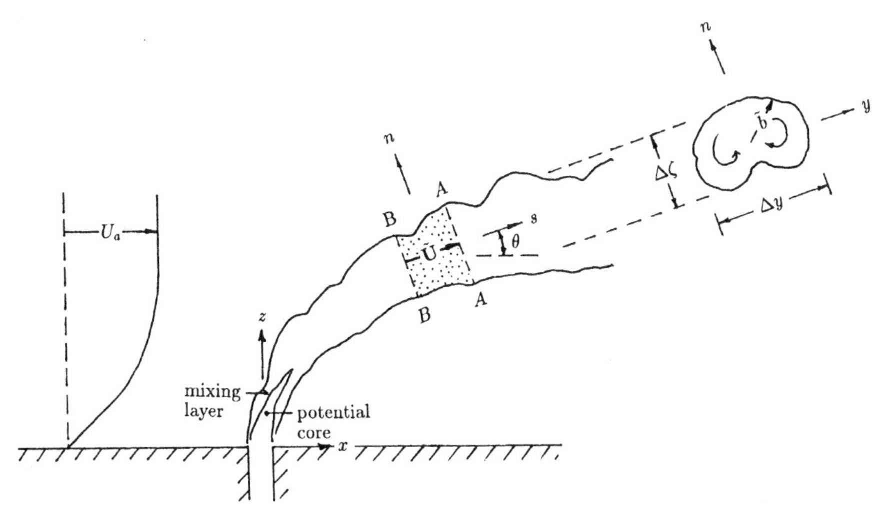

Figure 9.5. Elemental volume AB defined by the velocity of the dominant eddies, **U**, along the path of a buoyant jet in a non-uniform cross flow. The momentum flux and and the buoyancy flux at the source are M_o and F_o, respectively; the crossflow velocity is $\mathbf{U_a}$

5. BUOYANT JET IN CROSSFLOW

The formulation for jets and plumes in crossflow follows similar procedure as for the puffs and thermals except that the *elemental volume* must be defined so that the initial condition can be correctly related to the momentum flux and the buoyancy flux at the source. The sketch in Figure 9.5 shows how such an elemental volume is defined. By definition, the elemental volume AB is the part of the jet and plume that has a length equal to speed $|\tilde{U}|$. Such an elemental volume is produced by injection of momentum and buoyancy at the source for a period of one time unit. The dominant eddy at one end of the element, A, are released from the source a unit of time earlier than the time when the dominant eddy at other end of the element, B, are released from the source. By defining the elemental volume this way, the momentum and buoyancy associated with the elemental volume is equal to the momentum flux M_o and the buoyancy flux F_o at the source. The interaction between the adjacent elements may be ignored so the formulation for the jets and plumes produced by the continuous source is similar to the method as described in the previous section for puffs and thermals.

The formulation for jets and plumes in a non-uniform crossflowis by far the most complex problem being considered in this monograph. Due to the possible change in the direction of the cross flow, the buoyant jet follows in general a three-dimensional trajectory. Although the governing equations are the same, much details regarding the geometry is required before the actual numerical computations can be carried out for properties such as the jet trajectory and dilution ratio. Figure 9.1 shows the buoyant jet and the elemental volume AB produced by the momentum and buoyancy flux, M_o and F_o, at a general orientation relative to the crossflow and direction of the buoyancy force. The direction of the absolute velocity vector $\tilde{\mathbf{U}}$ is in a direction along the path of the buoyant jet. This velocity is equal to the sum of the velocity of the crossflow $\mathbf{U_a}$ and the relative velocity $\tilde{\mathbf{u}}$ as shown in the figure. The geometry of the flow in the zone of flow estabilishment (ZFE) is affected by the development of the potential core and the formulation for the ZFE near the source is different from the formulation for ZEF far away from the source. Therefore the problems for the two regions are considered separately. The flow in the ZEF will be considered first. The problem of the ZFE will be examined in a subsequent section.

5.1 ZONE OF ESTABLISHED FLOW

The calculation in the ZEF is conducted using a Lagrangian coordinate moving with the cross flow. A plume elemental volume AB is selected so that the initial momentum and initial buoyancy associated with the volume is equal to momentum flux M_o and buoyancy flux F_o. This is the elemental volume due to the discharge of momentum and buoyancy at the source for a period of one time unit. The length of this element is equal to the absolute velocity $\tilde{U}$ along the path of the buoyant jet as shown in Figure 9.1. The vector tangential to the path of the jet is

$$\tilde{\mathbf{U}} = \tilde{U}\hat{\mathbf{s}}. \tag{9.36}$$

The excess velocity vector is $\tilde{\mathbf{u}} = \tilde{\mathbf{U}} - \mathbf{U_a}$. It is resolved into two components, u_s and u_n, as follows:

$$\tilde{\mathbf{u}} = \tilde{u}_s\hat{\mathbf{s}} + \tilde{u}_n\hat{\mathbf{n}} \tag{9.37}$$

where $\hat{\mathbf{s}}$ and $\hat{\mathbf{n}}$ are unit vectors, and $u_s = \tilde{\mathbf{u}} \cdot \hat{\mathbf{s}}$ and $u_n = \tilde{\mathbf{u}} \cdot \hat{\mathbf{n}}$ are the velocity components, in the tangential and normal directions to the path, respectively. The tangential unit vector is calculated using the formula

$$\hat{\mathbf{s}} = \frac{\tilde{\mathbf{U}}}{|\tilde{\mathbf{U}}|} \tag{9.38}$$

while the normal vector using

$$\hat{\mathbf{n}} = \frac{\tilde{\mathbf{u}}_\mathbf{n}}{|\tilde{\mathbf{u}}_\mathbf{n}|} = \frac{\tilde{\mathbf{u}} - \tilde{\mathbf{u}}_\mathbf{s}}{|\tilde{\mathbf{u}} - \tilde{\mathbf{u}}_\mathbf{s}|} \tag{9.39}$$

The vector $\tilde{u}$ is resolved into components so that the radius of the elemental volume, $\tilde{b}$, is assumed to spread at a rate dependent on the orientation of the element as follows:

$$\frac{\delta \tilde{b}}{\delta t} = \beta_s |\tilde{u}_s| + \beta_n |\tilde{u}_n|. \tag{9.40}$$

An additive assumption is employed for the two basic entrainment mechanisms. The spreading coefficient β_s is the rate associated with the shear entrainment and the coefficient β_n is the rate associated with vortex entrainment as previously described. The two coefficients are known to have very different values. The spreading coefficients in the tangential shearing,

$$\beta_s = 0.17, \tag{9.41}$$

is assumed to be the same as the straight jets and plumes in stagnant environment. The spreading coefficient associated with the vortex pair in the direction perpendicular to the path is selected to fit with the experimental data. The best-fit value is

$$\beta_n = 0.34. \tag{9.42}$$

This spreading coefficient of the vortex entrainment is the same value as the coefficient recommended previously for the line thermal.

The incremental changes in buoyancy and momentum are governed by the buoyancy and momentum equations:

$$\delta \tilde{F} = \tilde{g} \tilde{V} \delta \rho_a, \tag{9.43}$$

$$\delta \tilde{\mathbf{m}} = -\rho_a \tilde{V} \delta \mathbf{U}_a + \tilde{\mathbf{F}} \delta t \tag{9.44}$$

where $\tilde{\mathbf{F}} = \tilde{F} \hat{\mathbf{k}}$ is the buoyancy flux vector. The buoyancy terms $\tilde{B}$ in Equations 9.6 and 9.5 becomes buoyancy flux $\tilde{F}$, since the initial buoyancy in the elemental volume AB is produced by the flux released at the source over a period of one time unit.

Since the elemental volume is produced over a period of one time unit, the length of the volume is $\tilde{U}$ and the volume $\tilde{V} = \pi \tilde{b}^2 \tilde{U}_s$ where U is velocity of the dominant eddies along the path of the jet. The momentum associated with the elemental volume, $\tilde{V}$, has two parts. Part of the momentum comes from the turbulent part of the jet and part of it from the induced irrotational motion surrounding the turbulent jet. The part with the turbulent fluid is $\tilde{V} = \pi \tilde{b}^2 |\mathbf{U}|$. The added-mass effect due to the irrotational motion depends on the orientation of the jet relative to the cross flow. A pair of added-mass coefficients is introduced to account for the relative motion in the tangential and normal directions perpendicular to the path of the crossflowas follows:

$$m_s = \rho(1+k_s)\tilde{V}u_s, \quad \text{and} \quad m_n = \rho(1+k_n)\tilde{V}u_n \tag{9.45}$$

where (m_s, m_n) are the components of the excess momentum vector $\tilde{m}$ and (u_s, u_n) are the components of the excess velocity in the tangential and normal directions, respectively.

The solutions for the five unknowns, $\tilde{b}, \tilde{u}_s, \tilde{u}_n, \tilde{U}_s, \theta$, are obtained by numerical integration of Equations 9.40, 9.43, and 9.44 using the algebraic relations imposed by Equation 9.45. The added mass coefficient selected for the tangential shearing is the same as the one for the free jet, i.e.,

$$k_s = 0.18. \tag{9.46}$$

The added-mass coefficient for relative motion normal to the path is

$$k_n = 1.00. \tag{9.47}$$

which is assumed to be the same as the value for acceleration of a circular cylinder normal to the flow.

5.2 POTENTIAL CORE IN THE ZEF

The entrainment process in the potential core region near the source is determined by the size and velocity of the dominant eddies in the mixing layer. The size of the eddies in the mixing layer is defined by the outer and inner radius, R and r, as shown in Figure 9.6. A mixing layer develops around the core due to the exchange of momentum between the fluid in the core and the cross flow. The inner radius, r, decreases while the outer radius, R, increases with distance, $\tilde{s}$, from the source. The inner radius reduces to zero at the end of the potential core. The calculation for the flow in the potential core follows a procedure as given below.

The fluid in the core is lighter than the fluid in the cross flow. The increase in velocity due to buoyancy is determined by the Bernoulli equation:

$$W^2 - g_o' z = W_o^2 \tag{9.48}$$

where W_o and $g_o' = g(\rho_a - \rho)/\rho_a$ are the velocity and the reduced gravity at the source, respectively.

The dominant eddies in the mixing layer move along the shear layer in the tangential direction. The velocity of the eddies is the average of the velocities in the core and the cross flow: $\tilde{U}_c = \frac{1}{2}(W + U_{as})$.

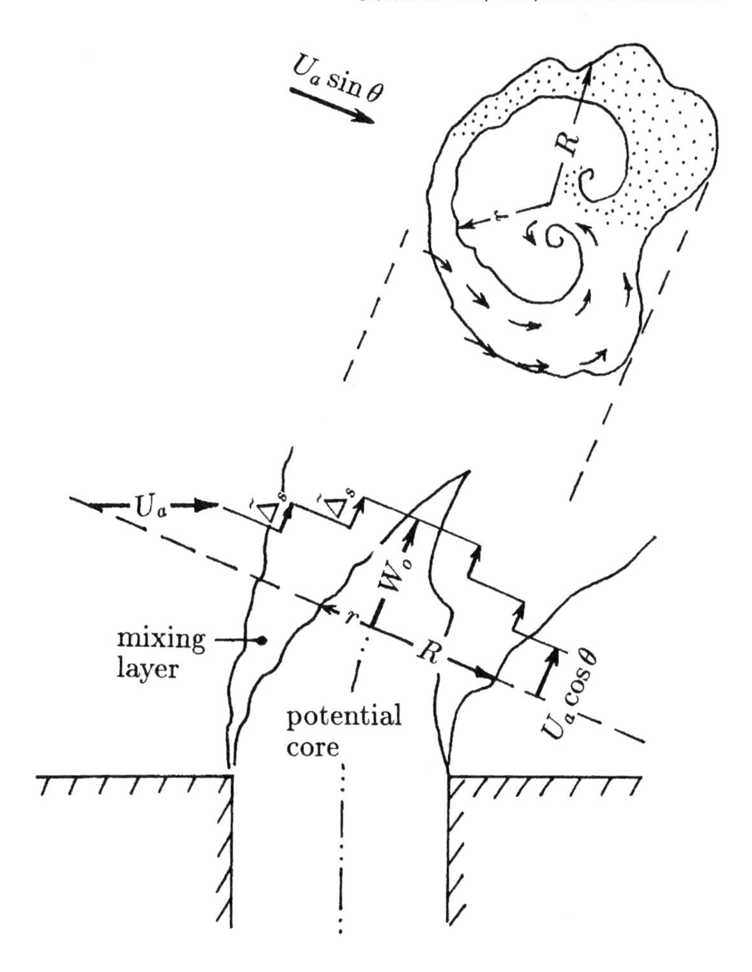

Figure 9.6. The development of the mixing layer in the potential core.

In the tangential direction, the difference in velocity across the mixing layer is $\tilde{\Delta}_s = \frac{1}{2}(W - U_{as})$. In normal direction, the difference is $\tilde{\Delta}_n = U_{an}$. The components of the crossflowvelocity in tangential and normal directions are U_{as} and U_{an}, respectively. The part of the fluid entrained into the mixing layer due to the tangential shearing is $\beta_s \tilde{\Delta}_s 2\pi(R + r)$ while the part due to shearing in the normal direction is $\beta_n \tilde{\Delta}_n 2\pi(R+r)$. The total entrained volume divided by the circumference, $2\pi(R + r)$, gives the entrainment velocity $v_e = \beta_s \tilde{\Delta}_s + \beta_n \tilde{\Delta}_n$. The thickness of the mixing layer, $(R - r)$, is assumed to increase at a rate proportional to the entrainment velocity:

$$\frac{\delta(R-r)}{\delta t} = \tilde{U}_c \frac{\delta(R-r)}{\delta s} = \beta_s \tilde{\Delta}_s + \beta_n \tilde{\Delta}_n. \tag{9.49}$$

The change in momentum in the potential core follows a relation

$$\delta\tilde{m} = -\rho\tilde{Q}\delta\mathbf{U}_a + \tilde{B}\delta t \tag{9.50}$$

that is similar to the one in the ZEF, except that the volume flux and the body force terms due to buoyancy now have slightly different expressions:

$$\tilde{Q} = \pi r^2 W + \pi(R^2 - r^2)\tilde{U}_c, \tag{9.51}$$

$$\tilde{B} = \pi r^2 W \mathbf{g}' + \pi(R^2 - r^2)\tilde{U}_c \frac{\mathbf{g}'}{2}. \tag{9.52}$$

The excess momentum, $\tilde{m}$, is again resolved into s- and n-components as before. The s-component is

$$\tilde{m}_s = 2\rho\pi r^2 W\tilde{\Delta}_s + \rho(1 + k_s)\pi(R^2 - r^2)\tilde{U}_c\tilde{\Delta}_s. \tag{9.53}$$

The calculation for the potential core ends when $r = 0$. The variation of the velocity in the potential core, δW, is determined by

$$W\delta W - g_o'\delta z = 0, \tag{9.54}$$

which is obtained by differentiation Equation 9.40. The variation δr is

$$\delta\tilde{m}_s = f_1\delta r + f_2, \tag{9.55}$$

which is obtained by the differentiation of Equation 9.50. Hence,

$$\delta r = \frac{\delta\tilde{m}_s - f_2}{f_1}, \tag{9.56}$$

where

$$f_1 = 4\pi\rho r W\tilde{\Delta}_s + 2\pi\rho(1 + k_s)(R - r)U_c\tilde{\Delta}_s,$$

$$f_2 = -\pi\rho[2r^2 W\delta W + r^2 W\delta U_{as} - r^2 U_{as}\delta W]$$

$$-\frac{1}{4}\pi\rho(1 + k_s)[8RU_c\tilde{\Delta}_s + (R^2 - r^2)(2W\delta W - 2U_{as}\delta U_{as})]. \tag{9.57}$$

The deflection of the jet is affected by entrainment of ambient fluid into the mixing layer surrounding the core. The process in mixing layer is particularly important for jets and plume with low exit-to-crossflow velocity ratio. The calculation for the mixing layer follows essentially the same procedure as in the ZEF. The variation $\delta\tilde{m}$ is determined by Equation 9.55, the variation δW by Equation 9.54, the variation $\delta(R-r)$ by Equation 9.49, and δr by Equation 9.56.

Numerical calculations using the above formulation for jets and plumes of non-unform crossflowwere conducted by Dehghani (1996). The results

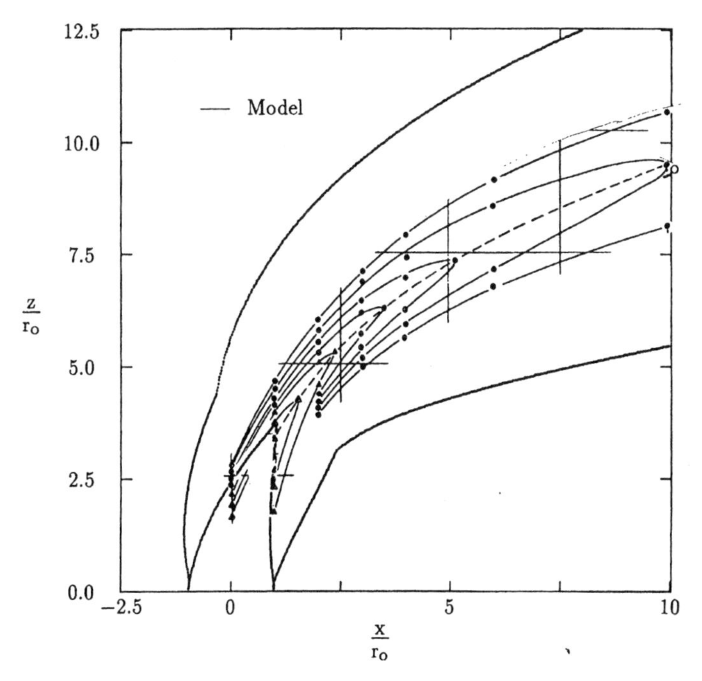

Figure 9.7. Jet in the region near the potential core; $W_o/U_a = 2.37$. The contours of constant velocity are reproduced from Chassaing *et al.* (1974).

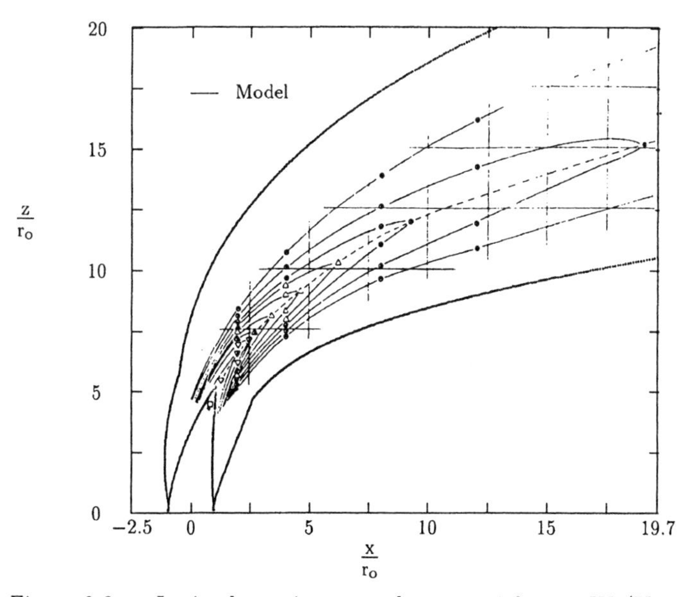

Figure 9.8. Jet in the region near the potential core; $W_o/U_a = 3.95$.

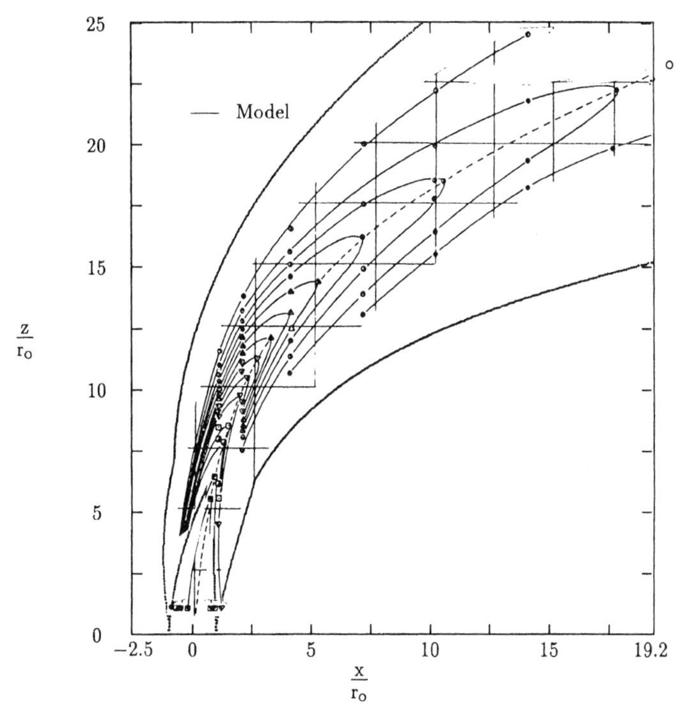

Figure 9.9. Jet in the region near the potential core; $W_o/U_a = 6.35$.

obtained using the standard coefficients (Equations 9.41, 9.42, 9.46 and 9.47) were very consistent with experimental observations for a wide range of crossflowconditions. Some of the numerical results obtained for non-buoyant jets discharged in a direction normal into a crossflow of constant density is presented in this chapter. Figures 9.7, 9.8 and 9.9 show the boundary of the deflecting jet in the potential core region, obtained for exit-to-crossflow velocity ratio $W_o/U_a = 2.37, 3.95$, and 6.35, respectively. The contour lines of constant velocity obtained for the same exit-to-crossflow velocity ratios by Chassaing *et al.*(1974) are superimposed on the same figure for comparison with the numerical results obtained by the Lagrangian model. The development of the potential core can be seen in these figures quite accurately predicted by the model. At the end of the potential core, $r = 0, R = \tilde{b}, \tilde{\Delta}_s = \tilde{u}_s$ and $\tilde{\Delta}_n = \tilde{u}_n$. Figure 9.10 plotted the length of the potential z_c (normalized by the diameter of the pipe D) for a range of exit-to-crossflowvelocity ratio varying from $W_o/U_a = 2$ to 45. The data by Pratte and Baines (1967) was obtained by detecting the change of velocity at the end of the core using a hot-film anemometer. The observations follow the same trend as the model, with generally somewhat lower values than predicted. The asymptotic value

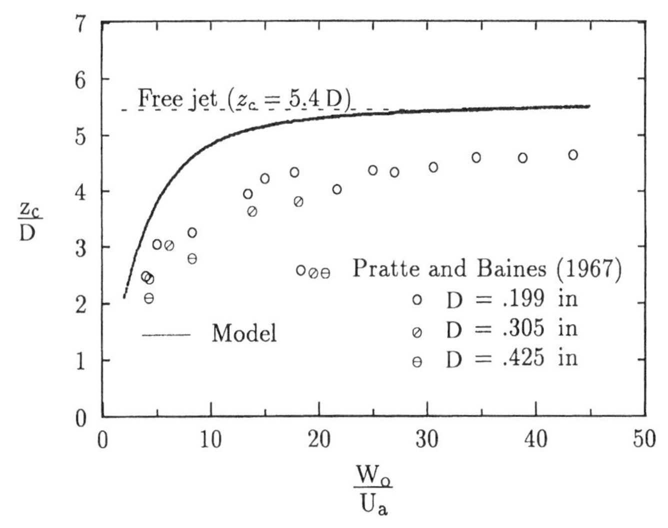

Figure 9.10. The length of the potential core obtained from the model and comparison with experimental observation.

$(z_c = 5.4D)$ for the high exit-to-crossflow velocity ratio is in agreement with the observation in straight jets (see the data of Albertson *et al.*in chapter 2).

Figure 9.11 shows the top and bottom boundaries of the turbulent jet in cross flow. The data are obtained by flow visualization of smoke jets in a wind tunnel (Pratte and Baines, 1967). The lines are the model predictions. Obtained in the ZEF far away from the source, these predictions by the Lagrangian model are also in close agreement with the experimental observation.

6. SUMMARY

The Lagrangian method of the jets, plumes, puffs or thermals (JPPT) is based on two sets of parameters – the spreading coefficients (β_s, β_n) and the added mass coefficients (k_s, k_n). The spreading coefficients are the parameters that characterize the turbulent motion within the dominant eddies. The added mass coefficients on the other hand are to account for the effect of the irrotational motion surrounding the turbulent flow. These parameters are introduced as required by the one-dimensional (along the path) formulation of the problem.

The formulation of the general problem is described in this chapter primarily based on the works of Chu (1994) and Dehghani (1996) which adopts a Lagrangian spreading hypothesis for turbulent closure. A parallel development of the Lagrangian method was made earlier by Cheung

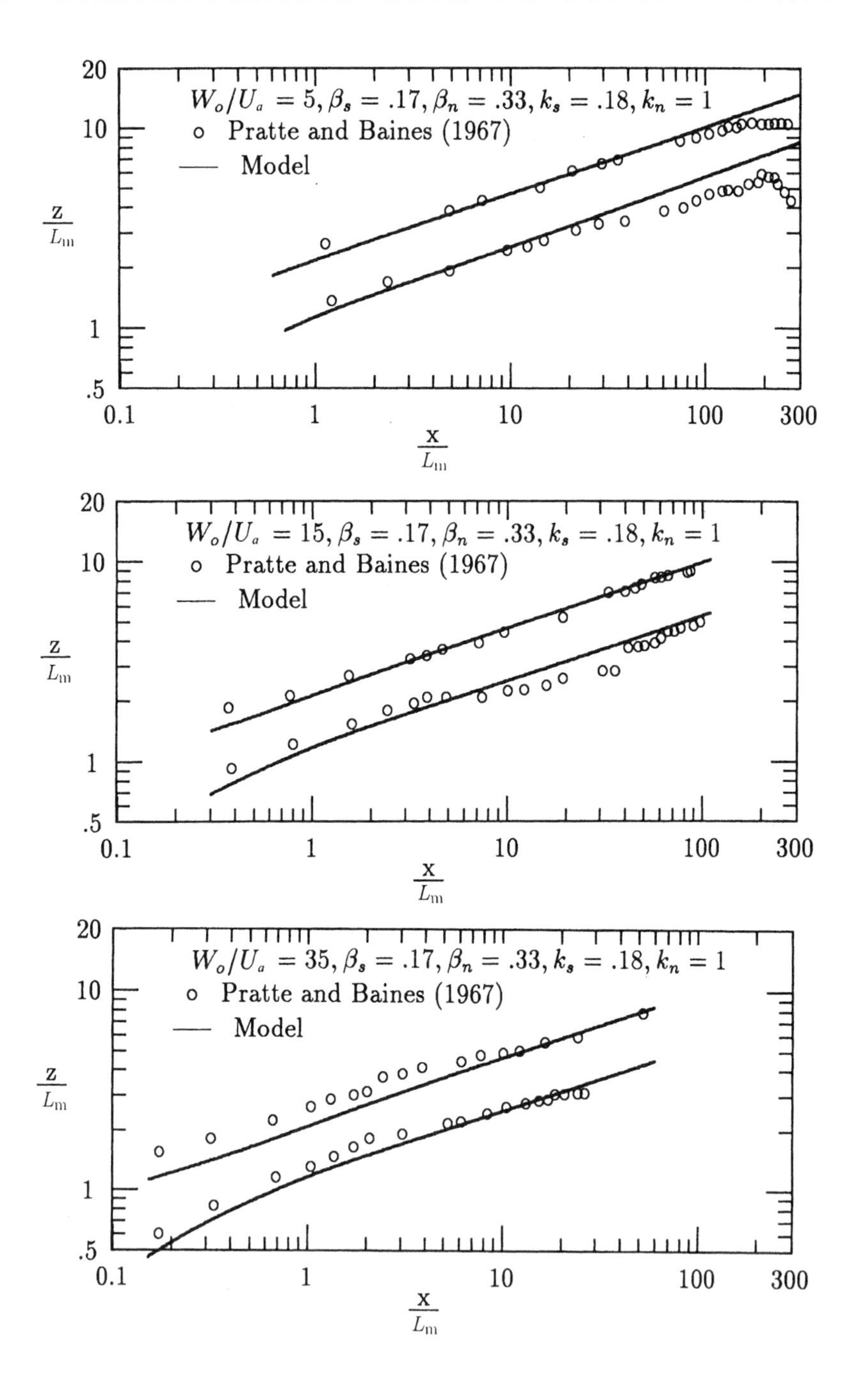

Figure 9.11. The development of the jet in the far field region; (a) $W_o/U_a = 5$, (b) $W_o/U_a = 15$, (c) $W_o/U_a = 35$. The set of curves in each figure denote the upper and lower boundaries of the turbulent jet. The experimental data of Pratte and Baines (1967) were obtained by flow visualization of smoke jets in a wind tunnel. $L_m = \sqrt{M_o/\rho}/U_a$ = momentum length scale.

(1991) and Lee and Cheung (1990). The Lagrangian jet model JETLAG by Lee and Cheung (1990) and Lee *et al.*(2000) is based on the project-area entrainment (PAE) hypothesis. A heuristic theory is adopted to treat the shear to vortex entrainment transition. The model has been extensively validated against basic laboratory data and applied to a variety of practical engineering problems. An interactive three-dimensional virtual reality model (VISJET) using JETLAG as the model engine will be given in the next and final chapter of this monograph.

With the advance of computational method and turbulence modelling, turbulence in jets, plumes, puffs and thermals has increasingly been simulated by two- and three-dimensional models. Techniques such as k-ϵ model and LES model have provided greater details of the local process. The fine scales of the turbulence often are simulated to match directly with the exeperimental observation. (See chapters 5, 7 and 8 for examples of k-ϵ modelling and LES modelling results.) Nevertheless, the one-dimensional simulation method described in this monograph will continue to be significant in engineering analysis for its simplicity and its ability to focus on the overall physical process and the dominant structure of the turbulent flow.

PROBLEMS

9.1 A jet is discharged into a river with a flow rate $Q = 0.0138\ m^3/s$ and a direction perpendicular to the crossflowof the river. The jet diameter is $D = 0.1$ m. The river velocity is $U_a = 0.2\ m/s$. Find a) the excess momentum $\tilde{\mathbf{m}}$ and b) the tangential and normal components of the excess velocity, u_s and u_n, of the jet immediately after the jet enters the river. See Equation 9.45 for the definitions of the excess momentum and excess velocity. Use $\beta_s = 0.17$ and $\beta_n = 0.34$ in your calculation.

9.2 The momentum of a buoyant plume in a cross wind is changed by the buoyancy force in the vertical direction and by the variable wind in the horizontal direction. Approximate the wind profile by the following formula

$$U_a = U_\infty[1 - \exp(-\frac{z}{\delta})].$$

in which U_∞ is the velocity at the edge, and δ the displacement thickness, of the atmospheric boundary layer; z is the elevation above ground.

a) Write down the momentum, buoyancy and spreading equations for an elemental volume moving with the dominant eddies. Explain how the variable wind enters as an effect to the equations.

b) Specify the initial condition for a point source of buoyancy flux F_o in the cross wind. Describe how the dilution and the path of the plume are calculated in a computer program. List the steps in the numerical calculation.

9.3 A computer program is developed to calculate the buoyant jet in a crossflow of non-uniform velocity and non-uniform density stratification. The numerical calculations are carried out using an elemental volume following the motion of the dominant eddies. In a Cartesian coordinate system, the absolute and relative positions of the elemental volume are x_i and ξ_i, respectively. The incremental displacement of the volume over a period of time dt is

$$dx_i = \tilde{U}_i dt = (U_{ai} + \tilde{u}_i)dt; \quad i = 1, 2, 3$$

where U_{ai} is the velocity of the cross flow; $\tilde{U}_i = dx_i/dt$ and $\tilde{u}_i = d\xi_i/dt$ are the abosolute and relative velocities of the dominant eddies, respectively.

a) Show that the components of the unit vector in the tangential and normal direction to the path of the motion are, respectively,

$$s_i = \frac{U_i}{\sqrt{U_\ell U_\ell}}; \quad i = 1, 2, 3$$

$$n_i = \frac{u_{ni}}{\sqrt{u_{n\ell} u_{n\ell}}}; \quad i = 1, 2, 3$$

where $u_{ni} = u_i - (u_\ell s_\ell)s_i$. The repeated index denotes the inner product, which is sum over all three components in the Cartesian coordinates.

b) If the momentum deficit vector is m_i, show that the components of the momentum deficit vector in the tangential and normal directions are $m_s = m_\ell s_\ell$ and $m_n = m_\ell n_\ell$.

c) The components of the velocity in the normal and tangential directions are

$$u_s = \frac{m_s}{\rho(1+k_s)\tilde{V}} \quad \text{and} \quad u_n = \frac{m_n}{\rho(1+k_n)\tilde{V}},$$

respectively. Explain why these velocity components, u_s and u_n, are calculated in the computer program.

Chapter 10

NUMERICAL MODELLING AND FIELD APPLICATION

Jets and plumes are often encountered in environmental engineering applications. Wastewater disposal into coastal waters is an example. In this final and last chapter, we examine a number of data sets obtained in field investigations, and show how numerical model and asymptotic formulae are developed to guide the interpretation of these data.

The problems of wastewater disposal are often analyzed by dimensional consideration in asymptotic flow regimes. Simple formulae for the jet trajectory and dilution in each of the asymptotic flow regimes are obtained by length scale analysis, and flow classification furnishes a means of correlating laboratory and field data. Basic asymptotic flow regimes of a turbulent buoyant jet in a stratified current have been examined: straight jets and plumes, effect of buoyancy and ambient density stratification, strong and weak coflowing jet, puffs and thermals. Relations derived from the length scale analysis are useful in the limiting cases. However, for many practical problems we are faced with the prediction of mixing in the transition between regimes. For example, a strongly buoyant jet in a weak crossflow (e.g. the 95 percentile value of the ambient current speed) is often a design condition. Dilution predictions in the BDNF-BDFF transition, a difficult problem (see later section), are hence needed. As a second example, prediction of the trapping level of a submerged sewage field (which may determine whether the swimmers in a nearby beach will be protected) may be based on a measured vertical ambient density stratification that is neither linear nor two-layered. Another commonly encountered situation is when the direction of momentum is not aligned with the direction of buoyancy, and the jet trajectory is three-dimensional - e.g. in shallow coastal waters the effluent is often discharged horizontally into a perpendicular alongshore tidal current (horizontal jet in crossflow). In these and similar situations, effective use of length scales is much more difficult; the results can sometimes be unreliable or misleading. For engineering application, a general mathematical model for predicting the characteristic jet properties is required (e.g. Muellenhoff *et al.*1985). As pointed out by Wood (1993), such a model should be validated in the asymptotic flow regimes. A discussion of representative mathematical models and their pertinent features can be found in Lee *et al.*(1987c) and Chu and Lee (1996).

We begin in this chapter by examining first the characteristics of the field data sets in their asymptotic regimes. This will be followed by discussing the transition problem between the regimes. We will examine the so called BDNF-BDFF regimes and compare the predictions with both laboratory and field data. The problem of mixing in a weak current and the necessity for the solution by a Lagrangian numerical model (JETLAG) is discussed. The JETLAG computational model embodies many of the concepts and shares the spirit of the Lagrangian approach outlined in the previous chapter. However, JETLAG uses a projected area entrainment hypothesis (shown to be consistent with the spreading hypothesis in the last chapter) and a heuristic theory to treat the shear to vortex entrainment transition. The model has been extensively validated against laboratory and field data, and successfully applied in a wide range of practical problems. The implementation of JETLAG and the visualization of the plume geometry (using the supplied VISJET software) are illustrated by a series of examples.

1. INITIAL DILUTION OF BUOYANT PLUMES IN A CURRENT: THE BDNF AND BDFF

Fig. 10.1 shows the general case of a horizontal buoyant jet discharge; without loss of generality the ambient current U_a is assumed to flow in the x-direction. The trajectory is shown both for the vertical $(x - z)$ buoyancy plane, and the horizontal $(x - y)$ momentum plane. Based on the concepts outlined in Chapter 8, it is possible to delineate the Buoyancy-dominated Near Field (BDNF) and the Buoyancy-dominated Far Field (BDFF) regimes for a buoyancy jet discharge (e.g. coastal sewage effluent). The relevant equations can be summarized as:

Buoyancy-dominated Near Field (BDNF) $z/L_b = zU_a^3/F_o \ll 1$

$$\textit{Dilution}: \quad \frac{SQ_o}{U_a L_b^2} = C_1 \left(\frac{z}{L_b}\right)^{\frac{5}{3}} \quad \text{or} \quad \frac{SQ_o}{F_o^{\frac{1}{3}} z^{\frac{5}{3}}} = C_1 \qquad (10.1)$$

$$Trajectory: \quad \frac{z}{L_b} = C_2\,(\frac{x}{L_b})^{\frac{3}{4}} \tag{10.2}$$

<u>Buoyancy-dominated Far Field (BDFF)</u> $z/L_b = zU_a^3/F_o \gg 1$

$$Dilution: \quad \frac{SQ_o}{U_a L_b^2} = C_3\,(\frac{z}{L_b})^2 \quad \text{or} \quad \frac{SQ_o}{F_o^{\frac{1}{3}} z^{\frac{5}{3}}} = C_3(\frac{z}{L_b})^{\frac{1}{3}} \tag{10.3}$$

$$Trajectory: \quad \frac{z}{L_b} = C_4\,(\frac{x}{L_b})^{\frac{2}{3}} \tag{10.4}$$

where $L_b = F_o/U_a^3$ is the crossflow buoyancy length scale. Here z and S are the characteristic vertical location (defined e.g. by location of maximum concentration, or by the center of mass) and dilution. Q_o, F_o are the discharge volume and specific buoyancy fluxes respectively.

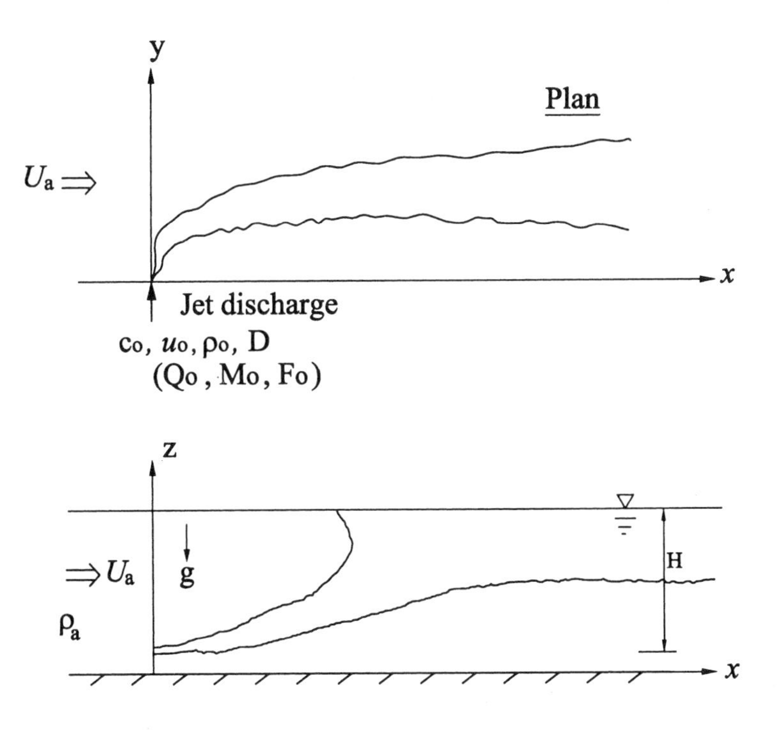

Figure 10.1. A horizontal buoyant jet in a crossflow

According to this classification, jet behaviour is determined by the parameter $H/L_b = HU_a^3/F_o$. Whether a given elevation H is in the BDFF or BDNF depends on the ambient current U_a and the buoyancy flux F_o. The weaker the ambient current and the stronger the buoyancy

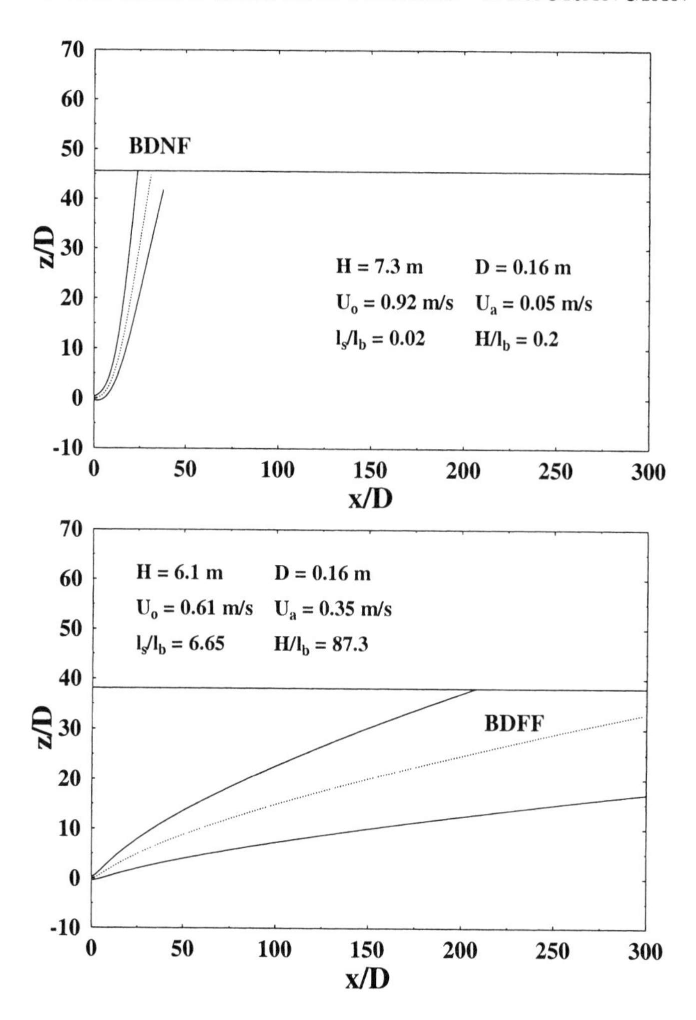

Figure 10.2. Example of an actual field condition for BDNF and BDFF (Hastings outfall, UK)

flux, the more likely a given vertical location on the jet trajectory will lie in the near field. Thus in a coastal situation, as the tidal current varies, the free surface can alternately lie in the BDNF and BDFF. Fig. 10.2 shows an example of an actual field condition for each case (Hastings outfall, UK); the plume geometry is generated by a Lagrangian model (to be discussed in the next section). The buoyant plume is portrayed for a situation i) around tidal slack (surfacing plume in BDNF), and ii) when the current gains strength (surfacing plume in BDFF). (Note: The reader should be cautioned that the terminology of near and far field in this context is different from the customary usage of near/far field to distinguish the region close to the effluent (and dynamically affected by the buoyant discharge) from the region further away and dominated by ambient advective and decay processes.)

The above length scale analysis has been verified in the laboratory (e.g. Wright 1977a; Lee and Cheung 1991; Knudsen 1988; Wood 1993). Fig. 10.3 shows the correlation of measured centerlinedilution of a vertical buoyancy-dominated heated jet as a function of z/L_b. In these experiments the tracer concentration was measured in the centerline plane, either along a horizontal traverse (for near vertical jets) or a vertical traverse (bent-over jets). The location z_c of the maximum concentration C_m in such a profile can be obtained, from which the centerline minimum dilution, $S_m = C_o/C_m$, can be determined. It can be seen that the dilution relations for both the limiting cases of a BDNF (weakly advected plume, with a Gaussian cross-section concentration), and the BDFF (advected thermal, with a vortex cross-section) are well-supported by the experiments, with a BDNF/BDFF dilution constant of 0.10 and 0.51 respectively. The data also shows that the BDNF/BDFF transition occurs at $z/L_b \approx 0.1 - 1$.

Fig. 10.4 shows the results of length scale analysis of all available field data of initial dilution collected at a number of sea outfalls in the United Kingdom and in Southeast Florida. In each of these field experiments, the initial dilution refers to the minimum dilution (inferred from the concentration of a tracer) in the "sewage boil" of a surfacing plume. The UK field data are measured for single port discharges (Sidmouth, Bridport, Gosport) or widely spaced ports along a multiport diffuser (Hastings); the discharge can be characterized as a horizontal buoyant jet in a perpendicular unstratified crossflow (Lee and Neville-Jones 1987a). The *in-situ* dilution measurements are based on radioactivity counts or Rhodamine WT dye concentrations. The field data for the Hastings Outfall (Bennett 1981, 1983) and the Southeast Florida ocean outfalls (SEFLOE experiments, Proni *et al.*1994) are derived from a large number of dye injection experiments. The surface dye concentration was tracked with a continuous flow fluorimeter, and the minimum dilution was inferred from the maximum dye concentration in the crossing of the boil or from grab samples taken from the observed boil. In addition, part of the Florida dilution data was based on salinity derived from CTD (conductivity-temperature-depth) measurements. Ambient stratification was present in some of the Florida experiments; but apparently the dilution was too low as surface boils were always observed. Together with the measured port discharge, initial density difference, ambient velocity, and water depth above the jet discharge, the dilution constant can be derived for each field condition. The Florida ocean outfall data complement the UK data very well; the discharges are horizontal buoyancy-dominated jets located in deeper waters and characterized by $H/L_b \leq 3$. Most of the data are from single-port outfalls (Hollywood, Broward); the data for multiport outfalls (Miami-Central and Miami-East) are shown only for comparison but not used.

In Fig. 10.4, the dimensionless initial dilution (minimum surface dilution) in moving water is plotted against the dimensionless discharge depth H/L_b. It is seen that the measured dimensionless dilution (which spans eight decades) for $H/L_b \gg 1$ is well-supported by the BDFF relation (z^2 dilution law in advected thermal), with an average dilution constant of $C = 0.32$ for $H/L_b \geq 1$. The statistical variabilities of the BDFF dilutions have also been studied (Lee and Neville-Jones 1987b); the median value of the BDFF constant is 0.27. Note that the correlation embodies the effect of the surface layer thickness; the dilution constant is hence somewhat lower than that obtained from the laboratory correlations. Although the field data were mainly for horizontal jets in a perpendicular crossflow, the difference of vertical vs horizontal discharges could not be distinguished against the background scatter in the field data.

For the buoyancy-dominated limit, the relatively few field data for $H/L_b \leq 0.1$ can be described by the BDNF equation with a dilution constant of $C = 0.14$. Even without accounting for the surface layer thickness in a near-stagnant water, the BDNF dilution theoretically should only have a value of 0.091. However, the observed dilution is much greater than the stillwater value. This may reflect additional mixing in the boil, and effects not accounted for by the theory (e.g. wave effects, and current non-uniformity). The Florida data is seen to be consistent with the Hastings data. In the BDNF-BDFF transition, $H/L_b = 0.1 - 1$, the BDNF dilution constant has an average value of 0.26. The collective data for $H/L_b \leq 0.5$ gives a dilution constant of 0.24. The measured dilution for the Miami outfalls are in general lower; this has been attributed to plume merging (Proni *et al.*1994) and the data have not been included in the analysis.

The results of the data analysis can be summarized in the following initial dilution equations which can be used to estimate in general initial dilution of buoyant wastewater discharges in a current:

<u>Buoyancy-dominated Near Field (BDNF)</u> $H/L_b = HU_a^3/F_o \leq 0.5$

$$Dilution: \quad \frac{SQ_o}{U_a L_b^2} = 0.26\ (\frac{H}{L_b})^{\frac{5}{3}} \quad \text{or} \quad S = 0.26\ \frac{F_o^{\frac{1}{3}} H^{\frac{5}{3}}}{Q_o} \qquad (10.5)$$

Buoyancy-dominated Far Field (BDFF) $H/L_b = HU_a^3/F_o \geq 0.5$

$$\textit{Dilution}: \quad \frac{SQ_o}{U_a L_b^2} = 0.32 \left(\frac{H}{L_b}\right)^2 \quad \text{or} \quad S = 0.32 \frac{U_a H^2}{Q_o} \qquad (10.6)$$

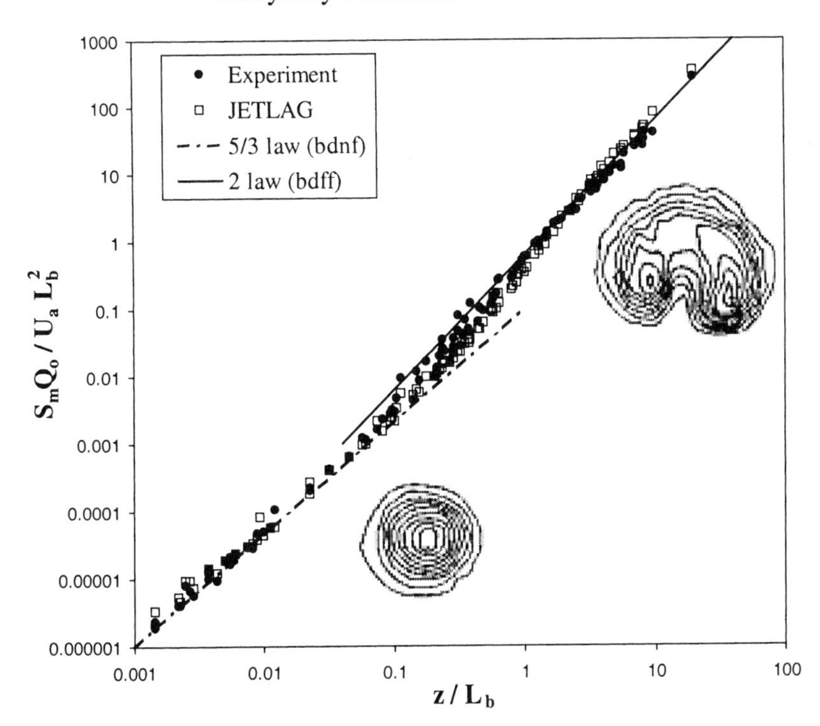

Figure 10.3. Comparison of predicted and measured dimensionless centerline dilution as a function of z/L_b for vertical heated jet in crossflow (Cheung 1991)

EXAMPLE 10.1 *For a vertical sewage discharge with jet velocity $u_o = 1.2$ m/s, D=0.125 m, H=14 m, estimate the initial dilution in moving water for a design ambient current of $U_a = 0.05$ m/s and 0.2 m/s respectively. Assume a relative density difference of 0.025 at the discharge point.*

Solution: To find the dilution in a current, first compute the volume flux $Q_o = 1.2 \times \frac{\pi}{4} \times 0.125^2 = 0.01473\ m^3/s$, and the buoyancy flux $F_o = 1.2 \times \frac{\pi}{4} \times 0.125^2 \times 9.81 \times 0.025 = 3.612 \times 10^{-3}\ m^4/s^3$. For $U_a = 0.05$ m/s, $HU_a^3/F_o = 0.48$, therefore Eq. 10.5 with an average dilution constant of 0.26 can be used to compute the BDNF dilution $S_m = 0.26 \times (3.612 \times 10^{-3})^{1/3} \times 14^{5/3}/0.01473 = \underline{221}$.

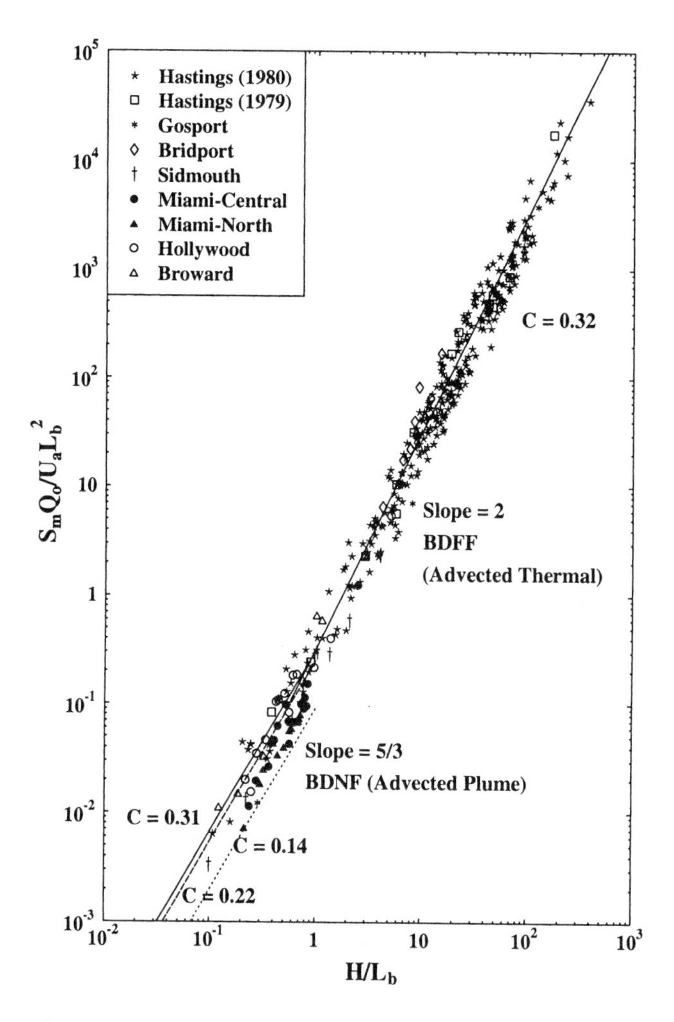

Figure 10.4. Comparison of initial dilution prediction equation with field data

For $U_a = 0.2\ m/s$, $HU_a^3/F_o = 30.7$, the plume lies in the BDFF when it surfaces. Use Eq. 10.6 to obtain $S_m = 0.32 \times 0.2 \times 14^2 / 0.01473 = \underline{852}$.

Compared to a stillwater dilution of 85 (with BDNF dilution coefficient of ≈ 0.1), it can be seen the initial dilution is increased considerably even in the presence of a weak current. In coastal locations with adequate currents, Eq. 10.5 and 10.6 give much more realistic prediction of dilution than the stillwater dilution. In typical situations, the travel time of a buoyant plume to the free surface is of the order of a minute; while the time scale of tidal variation is of the order of an hour (Prob. 3.1). Thus in a tidal situation, the initial mixing problem can often be treated as a quasi-steady problem. Fig. 10.5 illustrates the typical variation of initial dilution over a tidal cycle at a sea outfall.

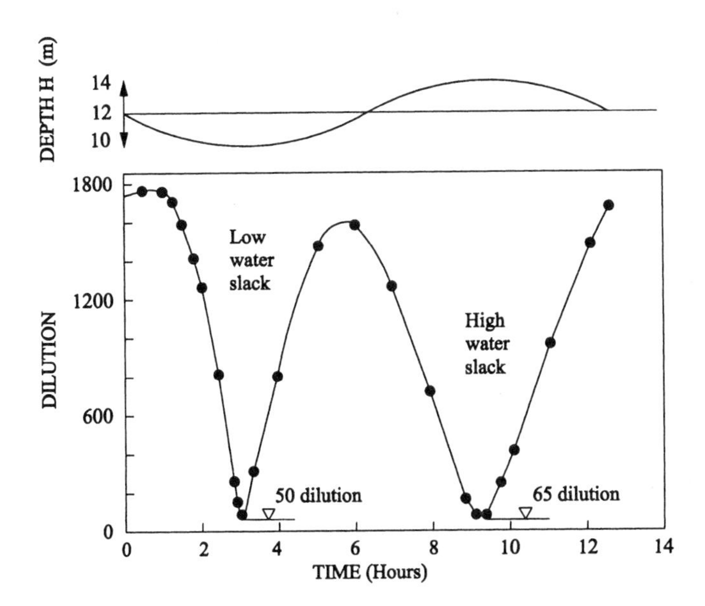

Figure 10.5. Huge variation of the initial dilution over a tidal cycle. Dilution ratio changes from 50 at low water slack to nearly 1800 at flood tide.

1.1 THE BDNF-BDFF TRANSITION

The above field data suggests that even in a weak current the dilution can be about 3 times greater than the still water value (with a dilution constant of approx. 0.1). At first sight this seems to be paradoxical as one would expect the BDNF dilution to be asymptotically the same as that of a stillwater plume. In the field, in the great majority of time, there is an ambient current. In most cases, the dilution departs from the stillwater value at a z/L_b of the order of 0.1. This can be noted in the centerline dilution measurements of laboratory experiments (Wright 1977; Lee and Cheung 1991) and is apparently related to the change in jet structure from the Gaussian to the vortex type as the ambient current is increased. Fig. 10.6 shows the measured centerline minimum dilution relative to the stillwater dilution for a vertical buoyancy-dominated jet in a weak crossflow (Lee and Cheung 1990; Cheung 1991). In this figure it is clearly seen that the dilution can significantly increase even in a weak current. However for $z/L_b \leq 0.2$, the stagnant plume dilution value is recovered. The prediction of mixing in this BDNF-BDFF transiton is a difficult problem; it will be further discussed in the next section.

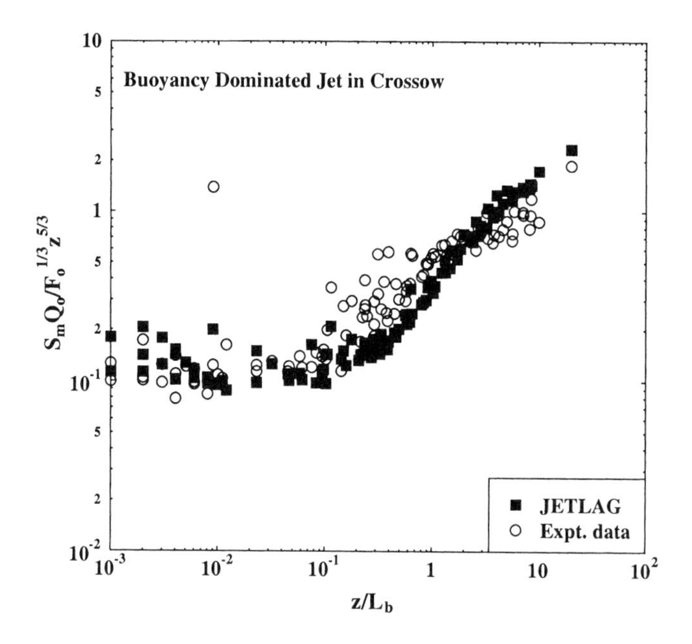

Figure 10.6. Dimensionless dilution of buoyant plume in weak current. JETLAG predictions (solid symbol) are compared with experimental data (open symbol).

2. JETLAG - A LAGRANGIAN BUOYANT JET MODEL

For environmental impact assessment, outfall design studies, and post-operation monitoring, it is desirable to have a computer model that is able to predict the initial mixing of buoyant wastewater discharges in a current. This is needed, for example, to help define mixing zones outside of which water quality objectives are legally enforced. For discharges into shallow coastal waters, water quality objectives simply cannot be achieved if the effect of a tidal current (which is present for most of the time) is not taken into account. In order to carry out satisfactory risk assessment, the model must give reliable predictions over a wide range of conditons (jet orientation, ambient current and stratification). On the other hand, few models can treat satisfactorily jets with three-dimensional trajectories, such as oblique buoyant jets or rosette-shaped jet groups discharging from modern ocean outfall risers. This section describes a Lagrangian model for buoyant jets with three-dimensional trajectories. The model combines the current version of the Lagrangian model JETLAG with computer graphics techniques to give virtual reality simulations of merging jets issuing from an ocean outfall. In the following the Lagrangian entrainment formulation is first presented. Representative comparisons of model predictions with experimental data are then presented for flow situations that are difficult to model but often encountered in practice. The use of computer graphics for visualizing and assisting the model predictions is then highlighted.

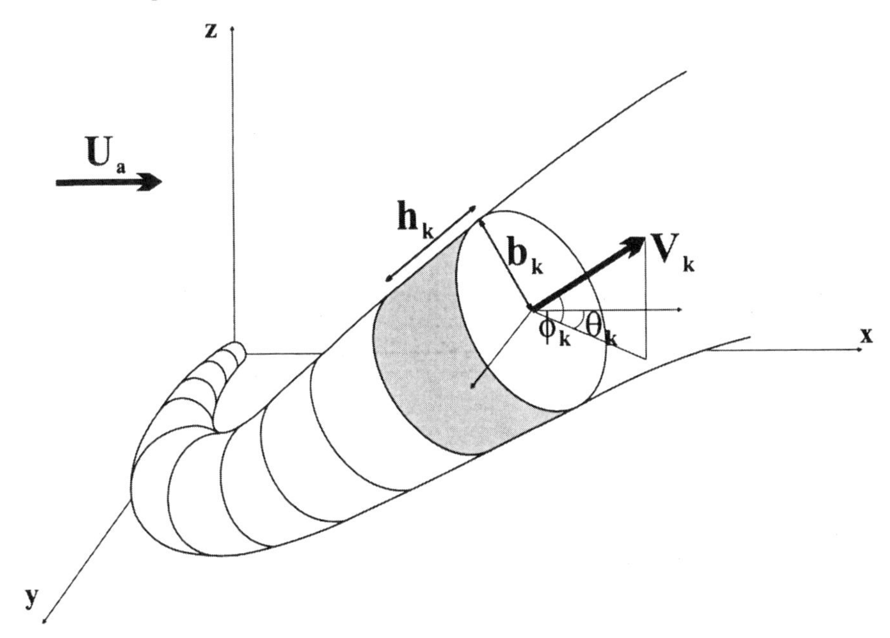

Figure 10.7. General Lagrangian model for buoyant jet with three-dimensional trajectories; schematic diagram of jet trajectory traced out by Lagrangian plume elements

2.1 OVERVIEW OF MODEL

The Lagrangian model JETLAG (Lee and Cheung 1990) predicts the mixing of buoyant jets with three-dimensional trajectories. Conceptually, the present JETLAG model is a generalization of a previous treatment of jets with a two-dimensional trajectory (Winiarski and Frick 1976; Frick 1984), and is developed after extensive testing against laboratory data. The model does not, strictly speaking, solve the usual Eulerian governing differential equations of fluid motion and mass transport. Instead, the model simulates the key physical processes expressed by the governing equations. The unknown jet trajectory is viewed as a sequential series of "plume-elements" which increase in mass as a result of shear-induced entrainment (due to the jet discharge) and vortex-entrainment (due to the crossflow) - while rising by buoyant acceleration and being sheared over by the crossflow. The model tracks the evolution of the *average* properties of a plume element at each step by conservation of horizontal and vertical momentum, conservation of mass accounting for entrainment, and conservation of tracer mass/heat. The vortex entrainment is determined by a heuristic Projected Area Entrainment (PAE) hypothesis originally proposed by Frick (1984) for buoyant jets with 2D trajectories, while pressure drag is ignored. A justification for the PAE hypothesis based on fundamental principles has been advanced in the previous chapter. A discussion of Frick's (1984) model can be found in Cheung and Lee (1996).

Fig. 10.7 shows a round buoyant jet directed into a uniform horizontal cross flow of velocity U_a. Without loss of generality, the ambient current is assumed to flow along the x-axis. The major parameters governing the flow are the jet volume flux $Q_o = V_o \pi D^2/4$, the kinematic momentum flux $M_o = Q_o V_o$, and the specific buoyancy flux $F_o = Q_o g'_o = Q[(\rho_a - \rho_o)/\rho_a]g$, where V_o, ρ_o are jet discharge velocity and density, respectively, and g is gravitational acceleration. The initial jet discharge angle is (ϕ_o, θ_o), where ϕ_o is vertical discharge angle with respect to the horizontal $x-y$ plane, and θ_o is horizontal discharge angle with respect to x-axis. The objective is to predict the unknown plume geometry and the characteristic dilution and velocity in the jet cross section.

2.2 BASIC CONCEPTS OF LAGRANGIAN MODEL

The unknown plume trajectory is viewed as consisting of a sequential series of plume elements (Fig. 10.7). Each plume element, which can be thought of as a section of a bent cone, is characterized by its location, average velocity, pollutant concentration, temperature and salinity, width, and thickness. We examine the changes in the properties of a plume element at discrete time steps of Δt.

The turbulent entrainment of the ambient fluid into the plume element is calculated at each step. Based on the increase in elemental mass, the momentum, energy, and tracer mass conservation equations can be solved in their integral form to give the velocity, density, and concentration at the next step. The kinematic and geometrical relations are then used to give the trajectory and relevant dimensions. The evolution of a single plume element is then equivalent to a solution of the properties along the unknown jet trajectory in the three-dimensional steady flow. By defining the thickness of a plume element as proportional to the local jet velocity, the Lagrangian method is similar to a time integration along the jet trajectory. The formulation, however, allows a certain flexibility in accounting for the detailed jet geometry in the determination of the vortex entrainment. The initial characteristics of the element at the source can be clearly related to the discharge parameters.

2.3 MODEL FORMULATION

At the kth step consider a plume element located at (x_k, y_k, z_k) with velocity (u_k, v_k, w_k) where (u, v) is the horizontal velocity, and w is the vertical velocity; $V_k = \sqrt{u_k^2 + v_k^2 + w_k^2}$ is the magnitude of the velocity (Fig. 10.7). The temperature, salinity, and density are denoted by T_k, S_k and ρ_k. The jet axis makes an angle of ϕ_k with the horizontal plane, and θ_k is the angle between the x-axis and the projection of the jet axis on the horizontal plane. The half-width or radius of the plume element is b_k; h_k is the thickness/length, defined as proportional to the magnitude of the local jet velocity, $h_k \propto V_k$. The mass of the plume element is then given by $M_k = \rho_k \pi b_k^2 h_k$.

Given the increase in mass due to turbulent entrainment, ΔM_k, the plume element characteristics at the next step are obtained by applying conservation of mass, horizontal and vertical momentum, energy, and tracer mass to the discrete element.

- Mass:

$$M_{k+1} = M_k + \Delta M_k \tag{10.7}$$

$$M_{k+1} = \rho_{k+1} \pi b_{k+1}^2 h_{k+1} \tag{10.8}$$

- Concentration and density:

$$S_{k+1} = \frac{M_k S_k + \Delta M_k S_a}{M_{k+1}} \tag{10.9}$$

$$T_{k+1} = \frac{M_k T_k + \Delta M_k T_a}{M_{k+1}} \tag{10.10}$$

$$\rho_{k+1} = \rho(S_{k+1}, T_{k+1}) \tag{10.11}$$

$$c_{k+1} = \frac{M_k c_k + \Delta M_k c_a}{M_{k+1}} \tag{10.12}$$

- Horizontal momentum:

$$u_{k+1} = \frac{M_k u_k + (\Delta M_k) U_a}{M_{k+1}} \tag{10.13}$$

$$v_{k+1} = \frac{M_k v_k}{M_{k+1}} \tag{10.14}$$

- Vertical momentum:

$$w_{k+1} = \frac{M_k w_k + M_{k+1}\left(\frac{\Delta\rho}{\rho}\right)_{k+1} \mathrm{g}\Delta t}{M_{k+1}} \tag{10.15}$$

$$(HVEL)_{k+1} = (u_{k+1}^2 + v_{k+1}^2)^{1/2} \tag{10.16}$$

$$V_{k+1} = (u_{k+1}^2 + v_{k+1}^2 + w_{k+1}^2)^{1/2} \tag{10.17}$$

- Thickness/radius:

$$h_{k+1} = \frac{V_{k+1}}{V_k} h_k \tag{10.18}$$

$$b_{k+1} = \left(\frac{M_{k+1}}{\rho_{k+1}\pi h_{k+1}}\right)^{1/2} \tag{10.19}$$

- Jet orientation:

$$\sin\phi_{k+1} = \left(\frac{w}{V}\right)_{k+1} \tag{10.20}$$

$$\cos\phi_{k+1} = \left(\frac{HVEL}{V}\right)_{k+1} \tag{10.21}$$

$$\sin\theta_{k+1} = \left(\frac{v}{HVEL}\right)_{k+1} \tag{10.22}$$

$$\cos\theta_{k+1} = \left(\frac{u}{HVEL}\right)_{k+1} \tag{10.23}$$

- Location:

$$x_{k+1} = x_k + u_{k+1}\Delta t \tag{10.24}$$

$$y_{k+1} = y_k + v_{k+1}\Delta t \tag{10.25}$$

$$z_{k+1} = z_k + w_{k+1}\Delta t \tag{10.26}$$

$$\Delta s_{k+1} = V_{k+1}\Delta t \tag{10.27}$$

- Initial condition:

$$(u, v, w)_o = (V_o cos\phi_o cos\theta_o, V_o cos\phi_o sin\theta_o, V_o sin\phi_o) \tag{10.28}$$

$$(b, h)_o = (0.5D, 0.5D) \tag{10.29}$$

$$(S, T, \rho, c)_o = (S_o, T_o, \rho_o, c_o) \tag{10.30}$$

$$\Delta t_o = 0.1 \times \frac{h_o}{V_o} \tag{10.31}$$

In these equations, pressure drag is tacitly neglected, and all of the entrained ambient flow is assumed to have velocity $U_a\hat{i}$. The initial location of the plume element is set at the discharge location. It can be readily shown that a consistent formulation can be obtained if the initial plume element velocity, radius, density, salinity, and pollutant concentration, are set equal to the jet discharge values (see Prob. 10.7 for a heuristic proof of the Lagrangian formulation).

By the nature of the formulation, there is no distinction between the zone of flow establishment (the potential core) and the zone of established flow of the buoyant jet; only the average properties in the jet cross section are represented. In effect, the model postulates a top-hat profile for both the velocity and the concentration. The formulation, however, can be readily extended to include the potential core in accordance with the concepts laid out in Chapter 4; such a model also gives satisfactory predictions similar to those in Figure 4.7 and 4.9.

2.4 SHEAR AND VORTEX ENTRAINMENT

The crux of the plume model is the computation of turbulent entrainment for the general situation of an arbitrarily-inclined plume element in a crossflow. The increase in mass of the plume element is due to turbulent entrainment of the ambient flow. Close to the discharge point, or in a very weak currrent, shear-induced entrainment dominates. In the bent-over phase, however, the vortex entrainment of the crossflow dominates. The increase in mass of the plume element at each step, ΔM, is computed as a function of two components: the shear entrainment due to the relative velocity between the plume element and the ambient velocity in the direction of the jet axis, ΔM_s, and the vortex entrainment ("forced" entrainment) due to the ambient crossflow, ΔM_f. The formulation of the shear and vortex entrainment is first separately presented below. The computation of the total entrainment from these components is discussed in a later section.

Shear entrainment :
In the current model, the shear entrainment ($\Delta M_s = E_s$) at each time step k is computed as (Fig. 10.7):

$$E_s = 2\pi\alpha_s b_k h_k \Delta U \Delta t \tag{10.32}$$

$$\alpha_s = \sqrt{2}\ (0.057 + 0.554 \sin\phi_k / F_l^2)\ (\frac{2V_k}{\Delta U + V_k}) \tag{10.33}$$

where V_k = jet velocity, $\Delta U = |V_k - U_a \cos\phi_k \cos\theta_k|$ is the relative jet velocity in the direction of the jet axis, and b_k, h_k are the radius and thickness of the plume element (Fig. 10.7); α_s and F_l are the entrainment coefficient and the local jet densimetric Froude number respectively. This shear entrainment expression can be shown to predict correctly the limiting case of straight jets and plumes, and a jet in coflow (Eq. 3.45 and Eq. 6.44).

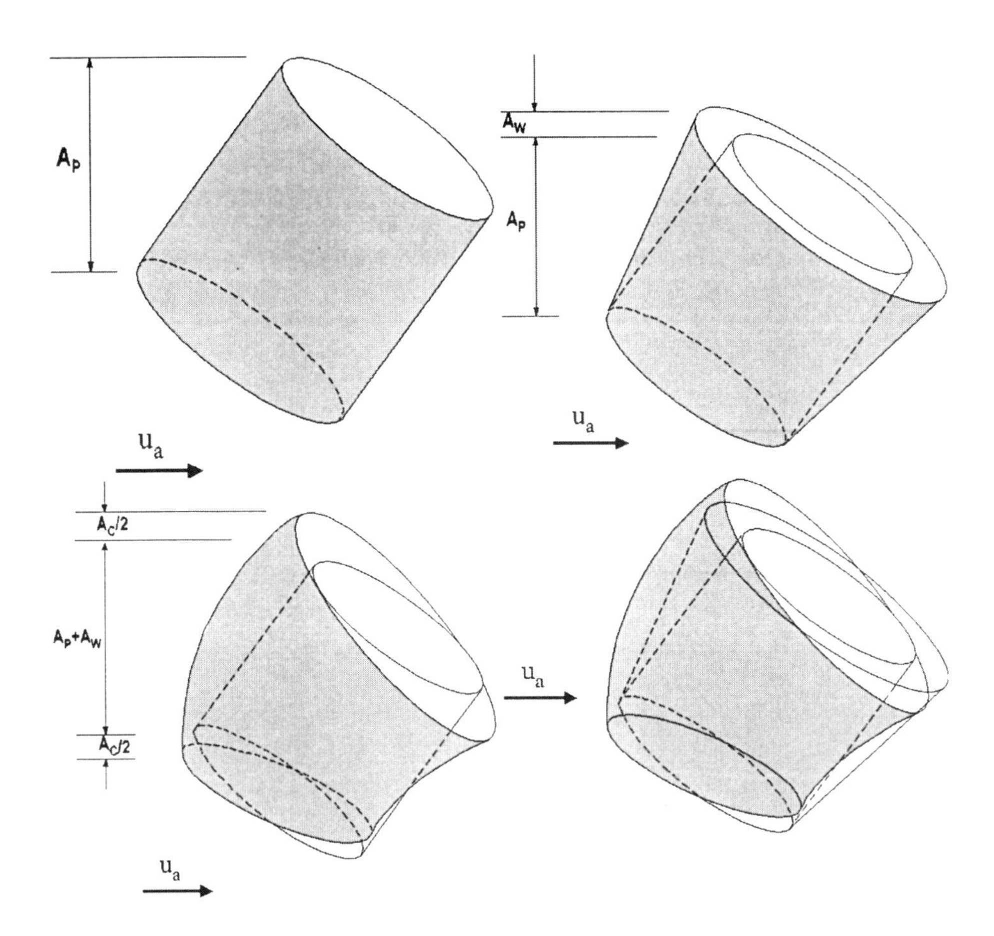

Figure 10.8. Illustration of the three contributions of the Projected Area Entrainment (PAE): projected area term A_p, increase in area due to plume growth A_w, and correction in area due to plume curvature A_c, and the sum of all three terms.

Projected Area Entrainment (PAE)

The vortex entrainment due to the crossflow is modelled using the Projected Area Entrainment (PAE) hypothesis (Frick 1984, Lee and Cheung 1990, Cheung and Lee 1996). Experimental observations (Chu and Goldberg 1974, Chu 1977) have shown that the transfer of horizontal momentum is complete beyond a few jet diameters. We assume that all the ambient flow on the windward side of the plume is entrained into the plume element - i.e. the entrainment due to the crossflow (the vortex pair entrainment in the far field) is equal to the ambient flow intercepted by the 'windward' face of the plume element. The vortex entrainment of the ambient flow into an arbitrarily inclined plume element, the incremental increase in mass at each step $\Delta M_f (= E_f)$, can be formulated and results in the following PAE expression (Lee and Cheung 1990):

$$\begin{aligned}\Delta M_f &= \rho_a U_a \; [2 b_k h_k \sqrt{1 - \cos^2\theta_k \cos^2\phi_k} \\ &\quad + \pi b_k \Delta b_k \cos\phi_k \cos\theta_k \\ &\quad + \frac{\pi b_k^2}{2} \Delta(\cos\phi_k \cos\theta_k)] \; \Delta t \\ &= E_p(\text{projection}) + E_w(\text{growth}) + E_c(\text{curvature}) \\ &= \rho_a U_a (A_p \; + A_w \; + A_c) \; \Delta t \end{aligned} \tag{10.34}$$

An initial estimate of ΔM_f can be obtained as:

$$\begin{aligned}\Delta M_f &= \rho_a U_a b_k h_k \; [2\sqrt{\sin^2\phi_k + \sin^2\theta_k - (\sin\phi_k \sin\theta_k)^2} \\ &\quad + \pi \frac{\Delta b_k}{\Delta s_k} \cos\phi_k \cos\theta_k \\ &\quad + \frac{\pi b_k}{2} \frac{(\cos\phi_k \cos\theta_k - \cos\phi_{k-1} \cos\theta_{k-1})}{\Delta s_k}] \; \Delta t \end{aligned} \tag{10.35}$$

Based on the initial estimate of ΔM_f, the properties of the plume element at the next step $k + 1$ can be computed (see Eqs.10.7 - 10.27). A revised estimate of $\Delta b/\Delta s$ and $\Delta(\cos\phi \cos\theta)/\Delta s$ can be calculated using the values at step k and $k + 1$ [e.g., $\Delta b/\Delta s = (b_{k+1} - b_k)/\Delta s_k$]. An improved estimate $(\Delta M_f)_k$ can then be obtained from Eqs.10.34 and 10.35, and the procedure repeated until convergence is achieved. In all the results presented here, typically two iterations are required for convergence of the key variables (with a relative tolerance of 10^{-4}) at each step. The time step Δt can be fixed or variable; it is chosen via a "predict-correct" procedure to attain a prescribed fractional change in mass (typically of the order of 1%) at each step. It should be emphasized that the ambient density ρ_a takes on local values; thus ambient stratification can be readily handled.

In Eq. 10.34 and 10.35, there are three contributing terms to the projected area: they are respectively the entrainment due to the projected plume area normal to the cross flow - the projection or cylinder term E_p, a correction term due to the increase in plume width E_w, and a correction term due to plume curvature E_c; the total projected area entrainment is $E_f = E_p + E_w + E_c$. Fig. 10.8 illustrates the meaning of the respective projected area terms pictorially for normal situations for a current flowing from the left. It can be shown that each term has its importance over different parts of the trajectory, so that none of the terms can be neglected. It is noted that the first term has been rewritten to minimize roundoff error, and derivatives are backward differenced from values at the present and previous steps. For jets with two-dimensional trajectories, $\theta_k = 0$, Eq. 10.34 reduces to the PAE expression given by Frick (1984).

In developing this general vortex entrainment formulation, a wide variety of flow situations have been considered. Eq. 10.34 is generally valid without having to make special provisions - i.e. the signs and angles take care of themselves properly; it is not necessary (in fact erroneous) to force each term to be positive. The projection term A_p is always positive. However, there are situations when the growth term A_w or curvature term A_c can be negative. For example, $A_w \leq 0$ for $\phi_k \geq 90^o$, when a plume element is directed against the ambient current (oblique jets or horizontal jets with $\theta_o \geq 90^o$). A detailed discussion of the physical interpretation of the individual terms for different cases can be found in Cheung and Lee (1999).

The total entrainment

The total entrainment can be obtained from a maximum hypothesis, $\Delta M = Max(\Delta M_s, \Delta M_f)$; alternatively an additive hypothesis, $\Delta M = \Delta M_s + \Delta M_f$ can be used. Comparison with basic data shows that the maximum hypothesis in general gives better results. However, the use of this hypothesis may give unreasonable predictions for a weak current. When U_a is small, e.g. around tidal slack, or when Q_o is large, it is expected that shear entrainment dominates. However, the relative jet velocity decreases as the ambient current increases - thus leading to a decrease in entrainment with increasing crossflow, a result that is not borne out by experiments of plumes in weak crossflow (Lee and Cheung 1991). Neither of these two approaches can reproduce satisfactorily the initial mixing data in the bdnf-bdff transition (Fig. 10.3). A general modelling framework is needed to handle the transition from the shear entrainment (advected jet/plume) regime to the vortex entrainment (advected puff/thermal) regime.

2.5 FORMULATION FOR NEAR-FAR FIELD TRANSITION

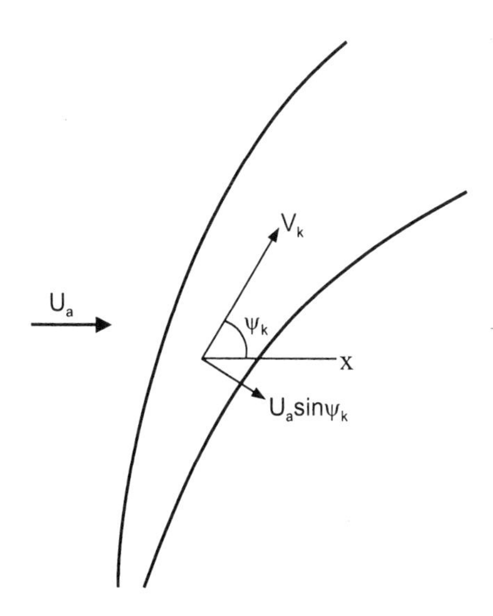

Figure 10.9. Weakly bent-over buoyant jet in crossflow

Consider the general situation of a jet/plume in a weak crossflow, with the jet axis making an angle of ψ_k with the crossflow (Fig. 10.9); the ambient current in the plane of the jet cross-section is then $U_a \sin\psi_k$. The general entrainment formulation, which models the near-far field transition (or weak to strong current), is as follows:

$$\Delta M = E_s \frac{(\pi - \varphi_k)}{\pi} + E_f \ \sin\varphi_k \tag{10.36}$$

where φ_k is a "separation angle" which delineates the relative importance of shear and vortex entrainment. φ_k is computed from the maximum radial shear entrainment velocity and the ambient velocity according to:

$$\cos(\varphi_k) = Min(\frac{V_r(max)}{U_a \sin\psi_k}, 1) \tag{10.37}$$

For a round jet or plume in stagnant fluid, the longitudinal jet velocity is self-similar and approximately Gaussian; the radial entrainment velocity field V_r can be obtained from continuity as a function of the local centerline velocity; since the spreading rate of straight jets and plumes

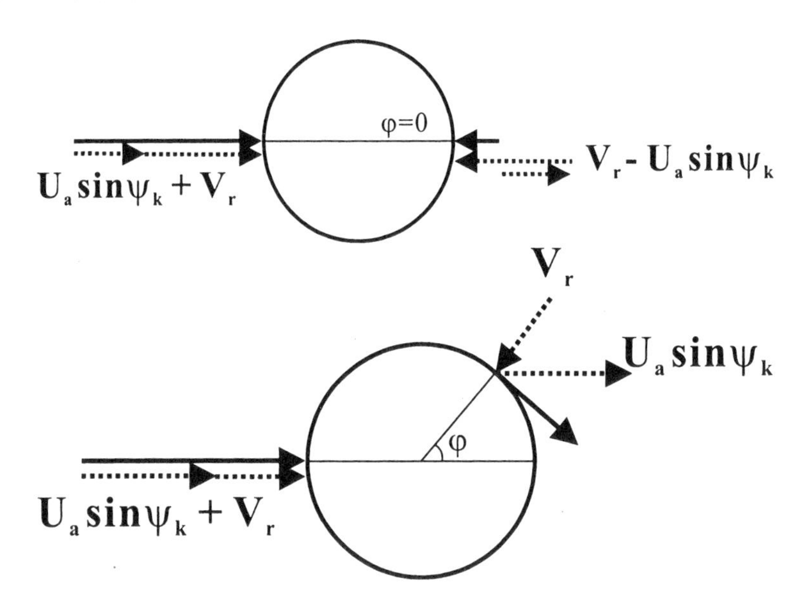

Figure 10.10. Net radial velocity at jet boundary (entrainment flow in jet cross-section); a) total entrainment = shear entrainment, $E = E_s$ (top); b) $E \geq E_s$, with stagnation point at angle φ.

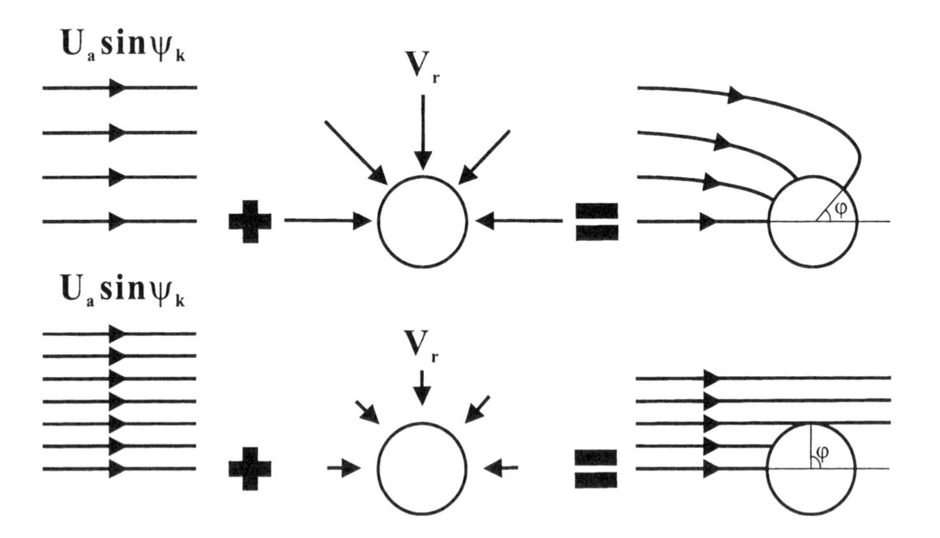

Figure 10.11. Entrainment flow pattern for a) jet in weak crossflow (top) ; and b) jet in strong crossflow (bottom)

is approximately the same $db/ds \approx \sqrt{2}(0.109)$, it can be shown that $V_r(max) = 0.421\alpha_s \Delta U_g$, and occurs at a radial position of $r/b_k \approx \sqrt{2}$ (Problem 10.8). where $\Delta U_g = 2\Delta U$ is the centerline excess velocity.

The above formulation is motivated by the work of Gaskin (1995) and Gaskin *et al.*(1995). In a weak crossflow, both theory and experiments suggest the irrotational entrainment flow in the plane of the jet cross-section can be modelled as the sum of the radial entrainment flow of a straight jet/plume in stagnant fluid, V_r, and the uniform ambient flow defined by $U_a \sin\psi_k$. This provides a heuristic general framework to piece together E_s and E_f. Figure 10.10 shows the external flow outside of the jet obtained in this way (for convenience we define the jet/plume boundary by the location of maximum entrainment velocity, $r = \sqrt{2}b_k$). Consider the radial velocity *into* the plume element along the centerline axis; after superposition the velocity on the windward side of the plume element will be reinforced, $V_r + U_a \sin\psi_k$, while that on the leeward side will be decreased, $V_r - U_a \sin\psi_k$. Note that the turbulent entrainment which accounts for mixing is the inflow into the plume element. If the net radial velocity all around the element boundary is directed inwards, the total entrainment will be exactly equal to the shear entrainment E_s, since the ambient velocity U_a contributes no net entrainment into the element when integrated around the circumference. The critical transition point, when total entrainment exceeds shear entrainment, is given by $U_a \sin\psi_k = V_r(max)$, where $V_r(max)$ can be computed locally in a general way. To see the possibility of a zero radial inflow on the plume boundary, we take the maximum radial shear entrainment velocity and equate to the radial component of the ambient velocity (Fig. 10.10):

$$\frac{U_a \sin\psi_k}{\alpha_s \Delta U_g} \cos\varphi_k = 0.421 \tag{10.38}$$

The above equation shows that, for a given ΔU_g the minimum ambient velocity U_a^* to give a 'stagnation point' (point of zero normal velocity) on the plume boundary is $\frac{U_a^* \sin\psi_k}{\alpha_s \Delta U_g} = 0.421$. For $U_a \geq U_a^*$, Eq. 10.38 can be used to locate the "stagnation point", i.e. the angle φ_k (Fig. 10.10). The corresponding stream function for such a flow is illustrated in Fig. 10.11. Eq.10.37 and Eq.10.36 show that as $U_a \rightarrow 0$, $\varphi_k \rightarrow 0$, $\Delta M \rightarrow E_s$; i.e. the entrainment is entirely due to shear entrainment, while as $U_a \rightarrow \infty$, $\varphi \rightarrow \pi/2$ - i.e. entrainment is almost entirely due to crossflow (E_f typically much greater than E_s). The two symmetrical limiting streamlines that pass through the stagnation points give the width of the upstream flow that gets entrained into the plume element and hence the total entrainment flow.

To summarize, the main consequence of this hypothesis is that for an ambient current up to a value defined by ψ_k and ΔU_g, the total entrainment is entirely due to the shear entrainment E_s. Beyond this critical limit, the entrainment increases to the vortex entrainment value

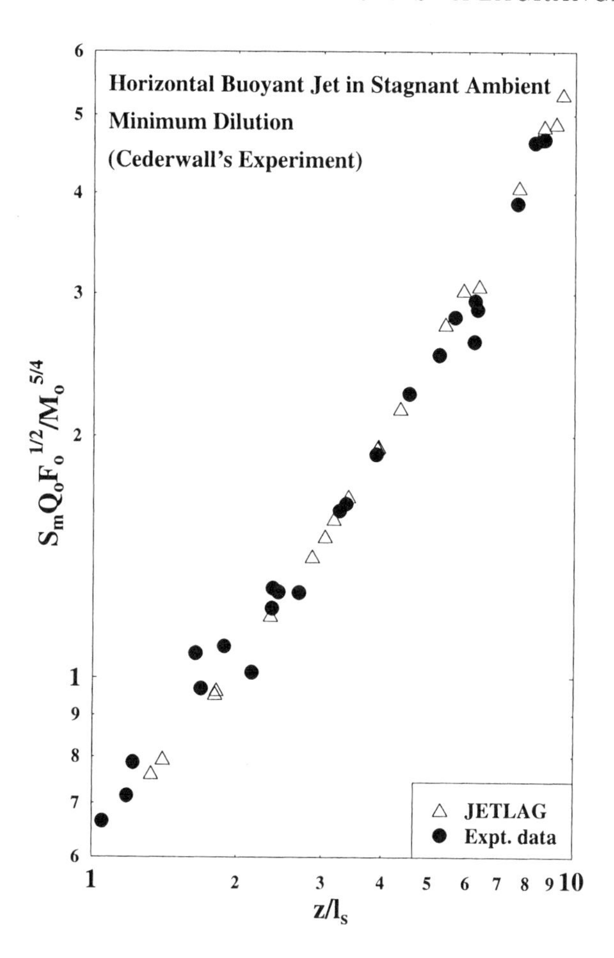

Figure 10.12. Comparison of predicted dilution with measurements of a horizontal buoyant jet in still fluid

E_f. It can be shown that (Problem 10.9) the entrainment flow q_e can be written as (Gaskin 1995):

$$\frac{q_e}{2\pi\alpha\Delta U_g b_g} = \frac{\pi - \varphi_k}{\pi} + \frac{2U_a \sin\psi_k r^*}{2\pi\alpha\Delta U_g b_g} \sin\varphi_k \tag{10.39}$$

where $r^* \approx \sqrt{2}b_k$ is the radial location of the plume boundary. Since $E_s = 2\pi\alpha\Delta U_g b_g$, and the PAE entrainment is $E_f = 2U_a \sin\psi_k r^*$, the above equation can be replaced by the equivalent form, Eq. 10.36, which is amenable to general calculation. Note that Eq.10.36 and 10.37 are generally applicable to jets with three-dimensional trajectories. It should be emphasized that the ultimate test of the heuristic theory outlined is validation against experimental data.

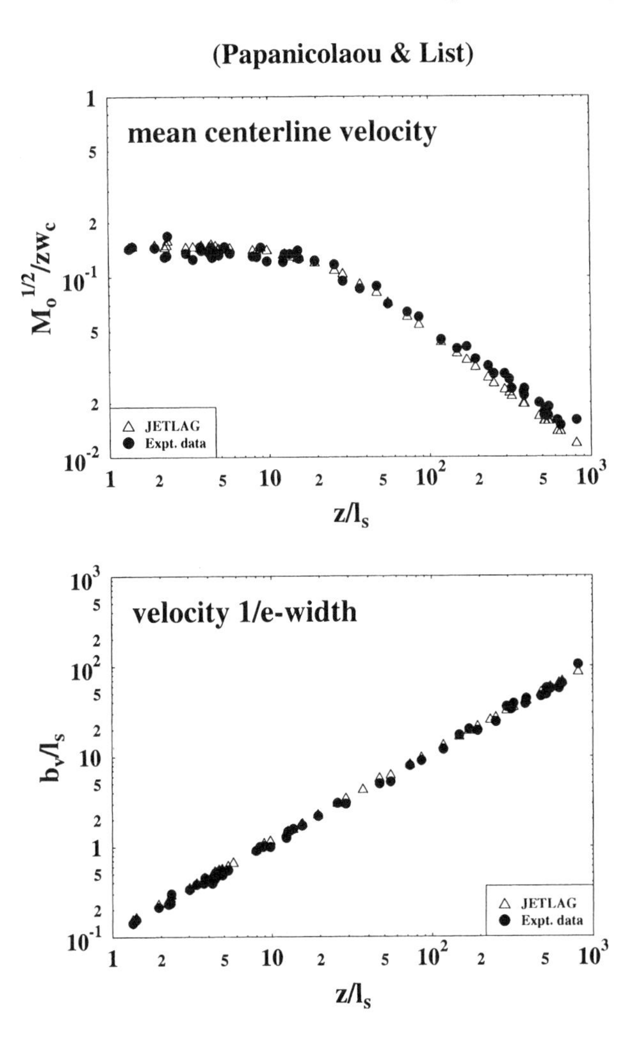

Figure 10.13. Comparison of predicted and measured velocity and concentration for a vertical buoyant jet in stagnant fluid

2.6 COMPARISON OF MODEL PREDICTIONS WITH LABORATORY DATA

Using the general formulation, the turbulent entrainment into the Lagrangian plume element can be computed at each step. The change in plume properties and the plume trajectory can then be obtained. JETLAG reproduces the correct behaviour of i) a round buoyant jet in stagnant or near stagnant fluid, and ii) an advected line puff/thermal in a bent-over momentum/buoyancy dominated jet. The model has been val-

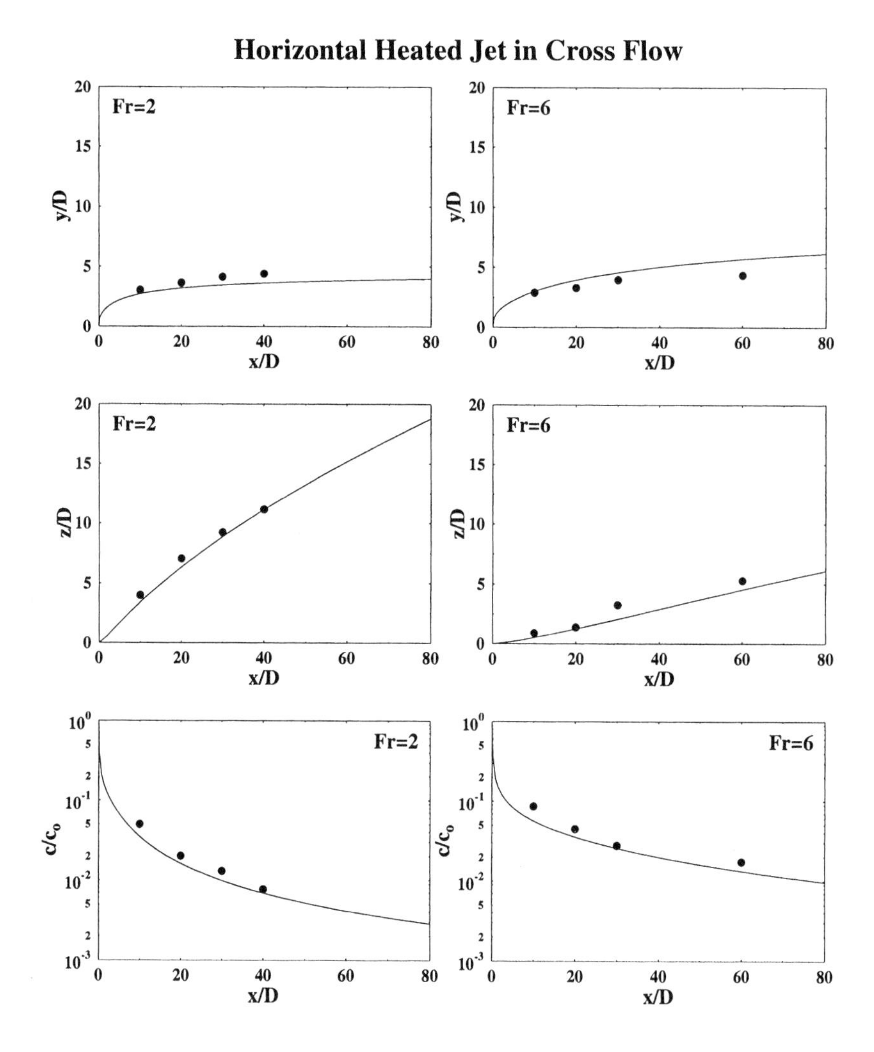

Figure 10.14. Comparison of predicted (solid line) jet trajectory (horizontal $[x, y]$ and vertical $[x, z]$ planes), and tracer concentration with data of horizontal buoyant jets in crossflow (Cheung 1991)

idated against experimental data by different investigators for: straight jets and plumes, vertical buoyant jet and dense plume in crossflow, oblique momentum jet in crossflow, horizontal buoyant jet in coflow; horizontal buoyant jet in crossflow; vertical buoyant jet in stratified crossflow; coflow and counterflowing momentum jets; buoyant plumes in weak current. We present herein only model-data comparisons for representative cases, including some that have proved difficult for integral models in general (e.g. Muellenhoff *et al.*1985; Wood 1993).

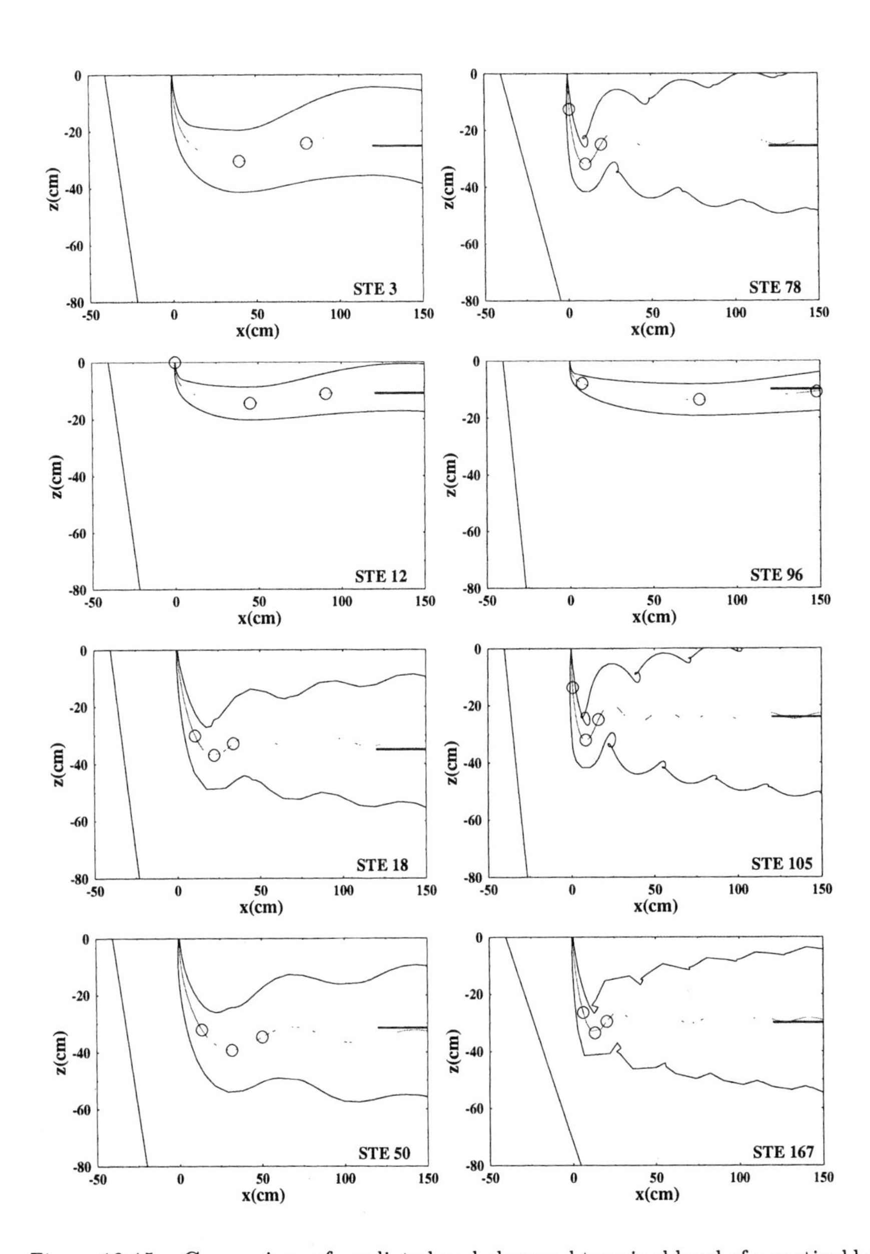

Figure 10.15. Comparison of predicted and observed terminal level of a vertical buoyant jet in stratified crossflow (Experiment of Wright 1977; observed level indicated by horizontal bar)

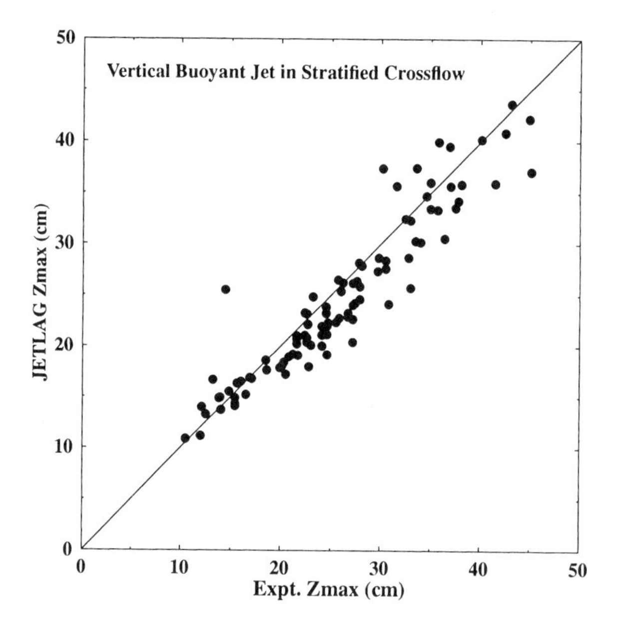

Figure 10.16. Comparison of model predictions and data of maximum rise height of a vertical buoyant jet in stratified crossflow (Wright 1977)

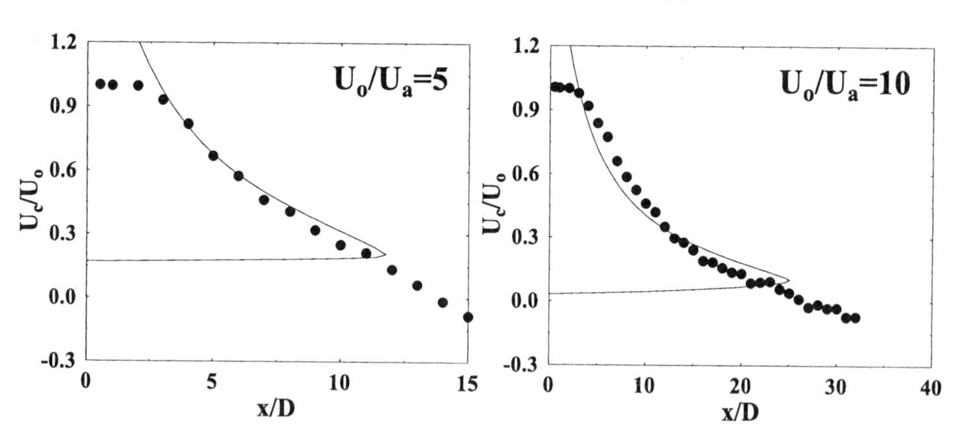

Figure 10.17. Comparison of model predictions (solid line) and data (symbols) of centerline velocity of counterflowing jet (Chan and Lam 1998)

Fig. 10.12 shows the comparison of predicted dilution of a horizontal buoyant jet in still fluid with the classic data of Cederwall (1968); the model is well-supported by the data. Predictions of velocity and concentration for a vertical buoyant jet in stagnant fluid (Fig. 10.13) are also in good agreement with the data of Papanicolaou and List (1988). Fig. 10.14 shows the comparison of predicted and observed trajectories and concentration of a horizontal buoyant jet in crossflow (experimental data from Cheung 1991); both the jet paths in the horizontal and vertical planes and tracer concentration are well-predicted. Fig. 10.3 shows the predicted centerlinedilution for a plume as a function of $z/l_b = zU_a^3/F_o$, where z= elevation above source, and F_o=jet buoyancy flux. In a very weak current, $z/l_b \ll 1$, it is seen that the dilution is given by that of a straight plume, while significant increases in dilution can be achieved even in a weak current, $z/l_b \approx 0.1 - 1$. The present prediction represents a significant improvement over alternative formulations which do not consider the transition problem. Fig. 10.15 shows representative comparison of predictions of jets in stratified crossflow with the experiments of Wright (1977b). A stratified crossflow is often encountered in the field situation. It can be seen the model predictions of final equilibrium level (trap level indicated by circles) compare very well with the experimental data (solid horizontal bar at right of figure). Finally, Fig. 10.16 shows a comparison of predicted maximum plume rise height with data of Wright (1977b), and Fig. 10.17 compares predicted centerline jet velocity in a counterflow with data of Lam and Chan (1997); the model predictions are in reasonably good agreement with experimental data. Other cases of buoyant jet in still water, a coflowing jet, vertical jet in crossflow, dense plumes, and oblique jets have been reported (Lee and Cheung 1990; Lee *et al.*2000). In addition to comparison of model predictions with data of individual experiments, numerical experiments simulating a momentum-dominated jet and a buoyancy-dominated jet have been reported (Lee *et al.*1987c). The model predictions of jet trajectory and dilution reproduce the power laws as given by Fischer *et al.*(1979) for the limiting flow regimes; the value of the trajectory and dilution constants derived from the numerical calculations compare favorably with experiments. The reader is reminded that in all the calculations reported herein, none of the model parameters is changed.

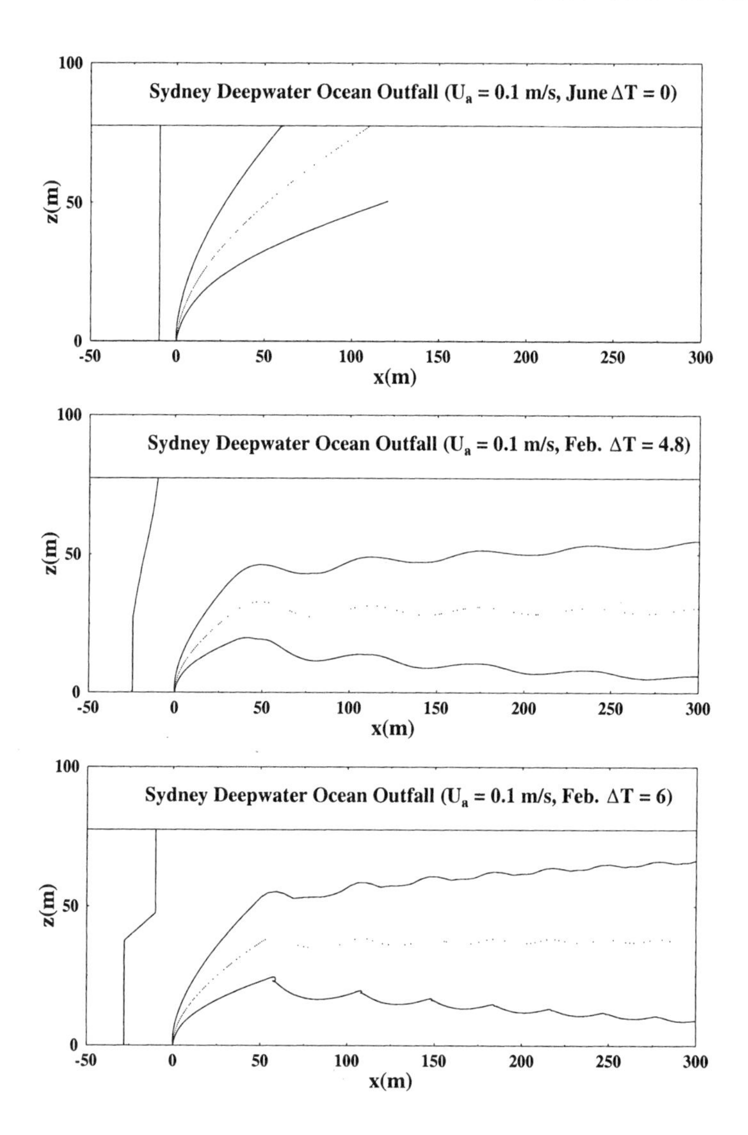

Figure 10.18. Computed ocean outfall plume behaviour: Sydney Ocean Outfall, Australia

3. FIELD APPLICATION AND VERIFICATION

3.1 POST-OPERATION MONITORING OF SYDNEY OUTFALL

Sydney, Australia is serviced by three major ocean outfalls. Since around 1991, primary-treated domestic effluent is discharged into the coastal water through three offshore outfalls at North Head, Bondi, and Malabar (Gordon and Fagan 1991; Wood *et al.*1993). An environmental monitoring program was developed to examine the post-operation performance of the outfalls. The objective was to i) assess the impact of the outfalls on the marine environment, and ii) to compare the operation of the outfalls against design assumptions and provide data for design of future outfalls. In view of the high level of natural variability of the physical environment, it was necessary to use a suite of initial dilution and circulation models to provide a framework for interpretation of field data. The model JETLAG was applied to many field situations and have been verified from the field data obtained by the large scale experiments - ".. with JETLAG proving to be the most versatile and useful model in most instances" (Gordon and Fagan 1991). It was also concluded that "the minimum dilution estimates obtained from the field work were typically twice those obtained from the plume model. This appears to be due to the additional dispersion resulting from the fluctuating ambient current which may not have been included in the laboratory experiments used to calibrate these models". The verified model was used during the outfall commissioning phase and also in the post-commissioning phase to provide temporal and spatial information on water quality. This information was integrated with the regular monthly water quality measurements. Fig. 10.18 shows the computed plume for the Malabar outfall at different times of the year, for a typical ambient current of 0.1 m/s. The 3.6 km long outfall (diameter 3.5 m) consists of shore normal tunnel with a 775 m long diffuser section at the seaward end. A total sewage flow of $Q_o = 2.3$ m^3/s is discharged through 28 risers at a depth of 80 m. Each riser is fitted with a "gas-burner" type head, discharging 4-6 horizontal jets (typical D=0.1 m). Because of the relatively large depth (77.5 m), it is acceptable to model the discharges from each riser by an equivalent buoyant jet with the same volume, momentum and buoyancy flux. It is seen that in June, the water is unstratified, and the plume surfaces with a dilution which can exceed 1000. In February however, there is a linear temperature stratification in the surface layer, with a 4.8 ^{o}C differential over the upper 50 m. The effect of the temperature stratification always results in a trapped plume; it is also seen that in the presence of a thermocline (6 ^{o}C jump over 10 m), the sewage field is trapped at about the level of the thermocline.

3.2 FIELD VERIFICATION AT NORTH WEST NEW TERRITORIES OUTFALL, URMSTON ROAD, HONG KONG

The North West New Territories outfall in Hong Kong consists of a 2.6 km long submarine pipeline into the Urmston Road tidal channel. A 600 m diffuser section at the offshore end contains 30 risers at a depth of about 20 m. Each diffuser riser has two discharge ports (port diameter 0.3 m), discharging horizontally in opposite directions. Several post-operation monitoring surveys of the outfall were carried out using dye and radio-isotope tracers, CTD (conductivity-temperature-depth) and acoustic doppler current profiling (ADCP); details can be found in Horton *et al.*(1997). The measurements enable the determination of the initial dilution and trap height. Fig. 10.19 shows the comparison of measured average dilution against the predictions by JETLAG for a wet season survey of July 1995. In the field experiments, the density stratification was evident, with up to a 15 ppt and 4 ^oC difference in salinities and temperatures respectively through the water column. The riser flow was in the range of 0.05-0.12 m^3/s; $U_a \approx 0.3 - 0.8 m/s$. Initial dilutions varied between 50 to over 1000, with a trap centerline depth of 16-19 m. Higher dilutions were associated with faster currents and lower discharges. Given the variabilities in the field conditions and the complexity of the phenomenon the overall agreement of the prediction with data is quite satisfactory. All model dilution estimates were well within a factor of two of the measured dilutions as shown on the figure. The predicted plume trapping height (with estimated upper and lower plume boundaries) is compared with the measured vertical location of peak tracer concentration at the same horizontal position. The observed trapping level was well within the effluent jet/plume as predicted by JETLAG.

3.3 ENVIRONMENTAL IMPACT ASSESSMENT OF THE HONG KONG SSDS OCEAN OUTFALL

As part of the Strategic Sewage Disposal Scheme (SSDS) of Hong Kong, it was proposed to discharge sewage from a population of about 3 million (after advanced primary treatment) through a 15 km long tunnelled outfall into the southern coastal waters of Hong Kong. Preliminary studies have resulted in an initial diffuser design with 20 risers (spaced at 60 m) each with six to eight jets discharged horizontally like a rosette (Howard *et al.*1991). For outfall design and impact assessment,

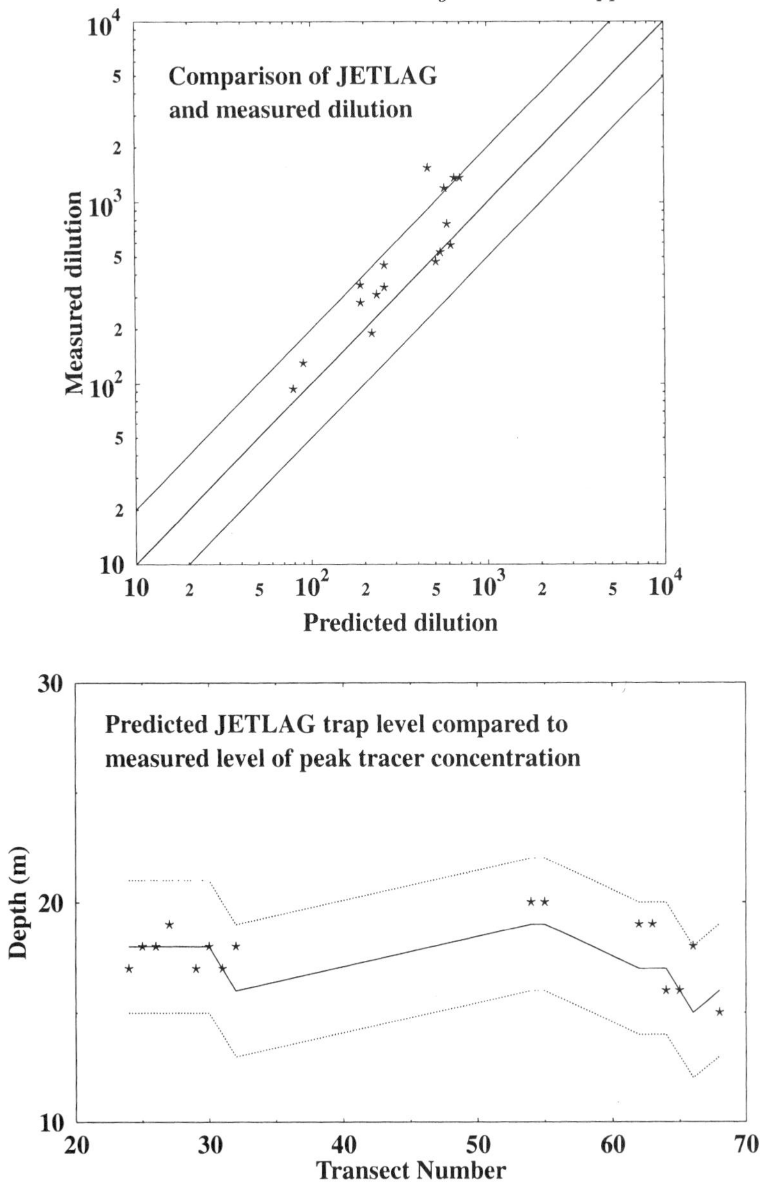

Figure 10.19. Comparison of model predictions of dilution and trap height with field data at North West New Territories sewage outfall, Urmston Road, Hong Kong

it was necessary to predict mixing under the entire range of ambient and discharge conditions. For example, the design 95 percentile current speed was 0.04 m/s, while velocities up to 0.6 m/s must also be considered. For this application, the limiting water quality objective that

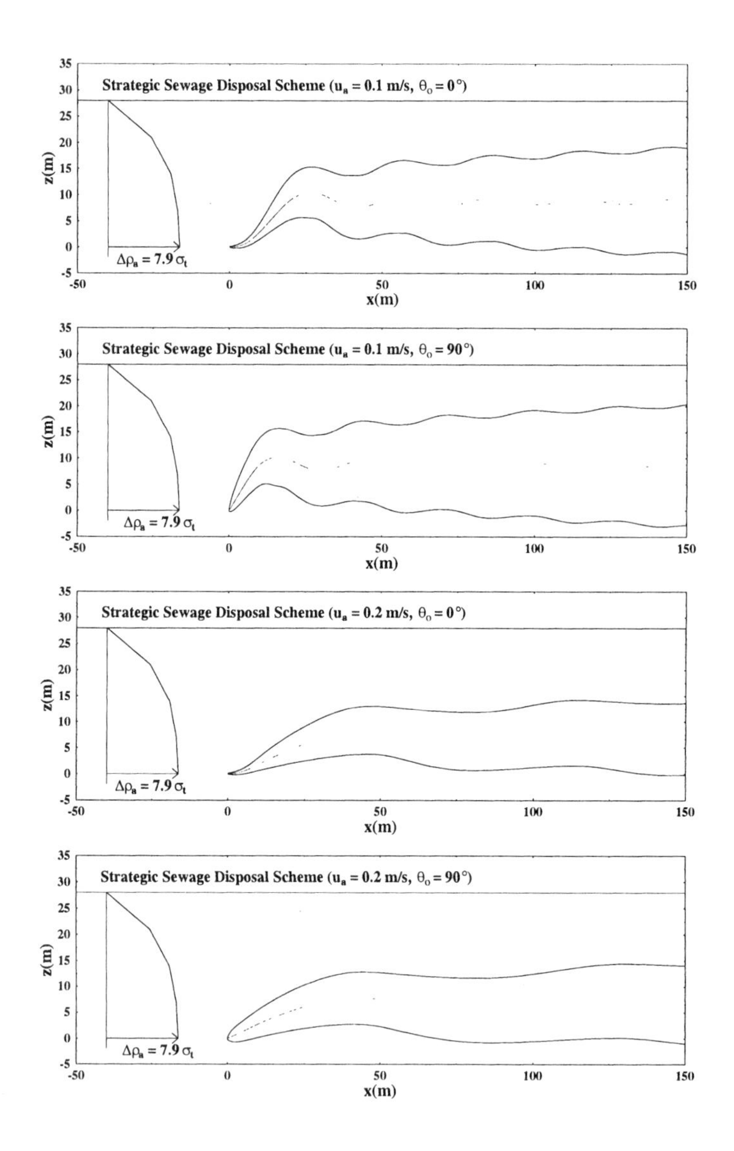

Figure 10.20. Computed plume trajectory of buoyant jets at different angles from the SSDS outfall

dictated the initial dilution requirement was the allowable un-ionised ammonia nitrogen concentration (0.021 mg/L). Fig. 10.20 shows an example of the predicted plume for a horizontal jet in coflow ($\phi_o = 0$, $\theta_o = 0$) and a horizontal jet in crossflow ($\phi_o = 0$, $\theta_o = 90^o$) for a summer condition, for $u_a = 0.1$ and 0.2 m/s. The riser and jet diameters are 3 m and 0.25 m respectively; the design flow is $Q_t = 23.1$ m^3/s. The jet densimetric Froude number is $Fr = 16.3$; the temperature and salinity vary respectively from 23 ^{o}C and 34.5 ppt at the sea bed to 28 ^{o}C and 25 ppt at the surface. It is seen that in all cases the plumes are trapped beneath the surface. For the stronger current, the increased dilution results in a greater submergence of the sewage field. Fig. 10.22 shows the computed buoyant sewage plumes issuing from the proposed Hong Kong Ocean outfall diffuser risers in the form of rosette jet groups. The complex turbulent buoyant shear flow, with jet configurations ranging from a coflow jet, oblique jet, to a counterflow buoyant jet, can clearly be seen. The computed plume trajectories enable a meaningful definition of the initial mixing zone.

3.4 VISJET - INTERACTIVE VIRTUAL REALITY MODEL

The Lagrangian model JETLAG has been further developed by incorporating three-dimensional computer graphics techniques. The new VISJET software package is a PC-based interactive model that gives virtual reality displays of the predicted plume. Figures 10.21 and 10.22 are samples of the VISJET graphic output for a multi-jet discharge from rosette diffuser. Views of the jets and the diffuser beneath the ocean surface can be obtained interactively; the user can manipulate the computer interface to examine the flow closeup from different orientations relative to the directions of the jets and the crossflow. The buoyant jets discharged in multi-directions interact with the crossflow to form a complex three-dimensional pattern. The predicted concentration and dilution are obtained directly simply by pointing the cursor at the location where quantitative information is desired. When the receiving water is density-stratified, VISJET depicts the predicted trapping level of the plume and spreading of the plume fluid in a horizontal layer. The interrogative computer graphics also provides the determination of the degree of merging between the bent-over jets from a multi-jet group in any downstream cross-section plane. The composite dilution of the jet group can hence be determined.

The VISJET software and application examples are included in the CD attached with this monograph. VISJET can also be downloaded from http://www.aoe-water.hku.hk/visjet .

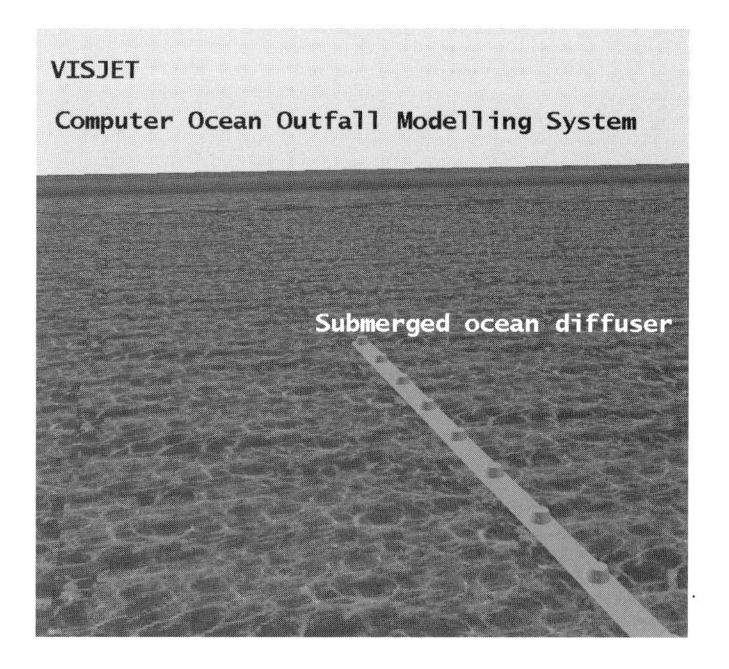

Figure 10.21. Submerged diffuser below the ocean surface (top). The closeup view of the same diffuser (bottom) is obtained interactively using VISJET's three-dimensional graphic rendering routines. The graphics also shows the air and water interface at the far end of the image on the top of the figure.

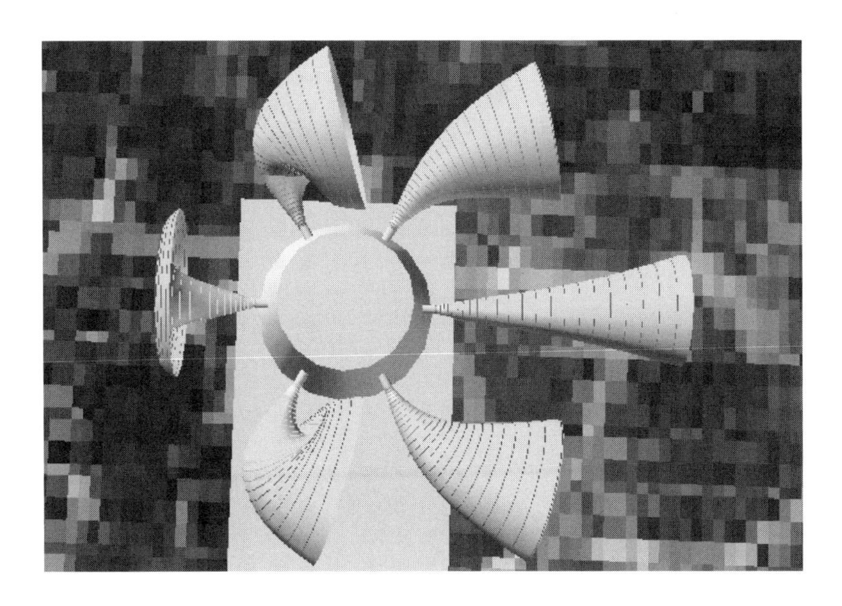

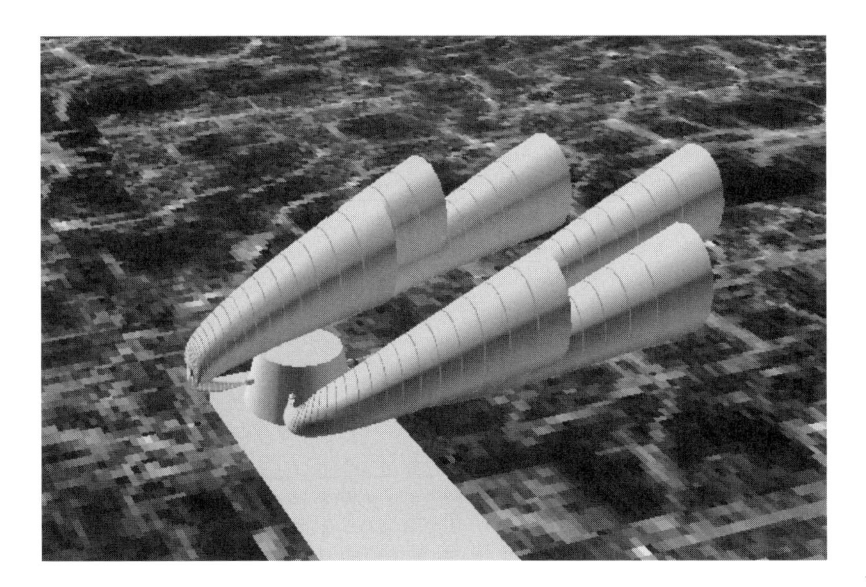

Figure 10.22. Merging of buoyant jets above a rosette-shaped ocean outfall riser. Top and side views of the jets are obtained interactively using the VISJET graphic rendering routines. A sequence of cross-sectional slices along the path of each buoyant jet is shown.

PROBLEMS

10.1 One approach to correlate the initial dilution in moving water, S_M, is to first compute the stillwater dilution S_s, using e.g. the Cederwall Equation, and then to correlate the ratio of the moving water to stillwater dilution (with dimensionless parameters such as the ambient to jet velocity ratio, $\frac{u_a}{u_j}$, and depth to jet diameter ratio, z/D). The following are three such semi-empirical equations which are based on field or laboratory data: i) the Agg and Wakeford Equations based on sampling of tracer concentration at sewage boils; ii) the Hydraulics Research Station (HRS) equation based on heated jet experiments in the laboratory; and iii) the Bennett equation based on dye tracing studies at the Hastings Outfall in the UK:

$$\frac{S_M}{S_s} = 12.794 \, (\frac{u_a}{u_j})^{0.938} \qquad \text{(Agg \& Wakeford 1972)}$$

$$\frac{S_M}{S_s} = 1 + 6(\frac{z}{D}) \, (\frac{u_a}{u_j})^{1.5} \qquad \text{(HRS 1977)}$$

$$\frac{S_M}{S_s} = 1.296 \, \frac{u_a^{0.682} z^{1.183}}{Q_o^{0.955}} \qquad \text{(Bennett 1983)}$$

Using the above equations, compute the initial dilution in moving water for two cases: i) a single port outfall with diameter $D = 0.5$ m, discharge jet velocity $u_j = 0.6$ m/s, depth $H = 10$ m, and ambient velocity $u_a = 0.1$ m/s; ii) a jet discharging from a multiport diffuser with widely spaced jets, with $D = 0.1$ m, $u_j = 1$ m/s, $H = 20$ m, and $u_a = 0.1$ m/s. Assume the same initial relative density difference for both cases, $\frac{\Delta\rho_o}{\rho_a} = 0.026$. Comment on your results.

10.2 A horizontal round buoyant jet with initial jet velocity $u_o = 1$ m/s, diameter $D = 0.15$ m, and relative density difference $\Delta\rho_o/\rho_a = 0.025$ discharges in depth of $H = 10$ m. Compute the initial dilution a) in still water; and b) in an unstratified crossflow of i) 0.02 m/s, ii) 0.1 m/s, and iii) 0.5 m/s. Comment on your results.

10.3 *Outfall Mixing Anomaly in Weak Current*

An ocean outfall discharges a total flow of $Q_t = 23.1 \; m^3/s$ into coastal water of 22 m depth. There are 21 widely spaced risers on the diffuser section. There are six horizontal jets on each riser (D=0.25 m). Initial mixing calculations for two discharge flows have been performed for a weak current $u_a = 0.04 \; m/s$ and a design stratification condition. The results show surprisingly that the initial dilutions for the two drastically different discharge flows are very similar.

Discharge flow (m^3/s)	23.1	10.4
Predicted neutral buoyancy level z_t (m)	7.7	8.4
Predicted average dilution S	31.3	27.8

a) Is the above result reasonable? Assuming a stagnant ambient condition and $\Delta\rho_o/\rho = 0.025$, and adopting the Cederwall equation or otherwise, compute the jet densimetric Froude number and the average dilution for the two discharge flow conditions at $z = 7.7$ m and $z = 8.4$ m respectively. Comment on the dependence of predicted dilution on the discharge flow.

b) Use the VISJET model to obtain the plume trajectory and average dilution for the following jet discharges: $\theta_o = 30, 90, 120^o$. What is the maximum lateral penetration distance of the perpendicular discharge ($\theta_o = 90^o$)? If the riser spacing is 60 m, will there be interference between the jets from adjacent risers at the surface or trap level. The effluent salinity and temperature are 2 ppt and 20°C respectively. Assume $u_a = 0.04$ m/s, and the following ambient profile:

Depth below surface (m)	Salinity (ppt)	Temperature (°C)
0.0	25.0	28.0
5.5	29.5	27.0
11.0	33.0	25.0
16.5	34.0	23.0
22.0	34.5	23.0

10.4 *Buoyant plume in stratified crossflow*
It is often puzzling that the predicted initial dilution of a round buoyant jet in a stratified crossflow appears to be rather insensitive to the discharge flow. This apparent insensitivity is examined as follows. Consider a buoyancy-dominated jet with specific buoyancy flux $F_o = Q_o g'_o$, discharging into a stratified crossflow U_a. Assume the receiving water is linearly stratified, with the stratification parameter $\epsilon = -\frac{g}{\rho_d}\frac{d\rho_a}{dz}$. If the mixing of the buoyant plume in crossflow is assumed to be analogous to that of an advected line thermal of initial buoyancy F_o/U_a, then the maximum height of rise z_{max} can be assumed to depend on the line thermal buoyancy F_o/U_a and the stratification parameter ϵ.

a) Show that the maximum height of rise is a function of the jet buoyancy flux, ambient current, and ambient stratification parameter:

$$z_{max} \sim \left(\frac{F_o}{U_a \epsilon}\right)^{1/3}$$

b) Assuming the dilution equations for a weak and strong current, i.e. the BDNF and BDFF equations respectively (Eq. 10.5 and Eq. 10.6), show that the dependence of the dilution at the trap level for a given discharge flow of Q_o is:

$$S_t = S(z_{max}) \sim \frac{g_o'^{8/9}}{(\epsilon U_a)^{5/9} Q_o^{1/9}}$$

for the buoyancy-dominated near field (BDNF). It is seen that the dilution at the trap level is not as sensitive to current or port discharge as the unstratified situation ($S \sim F_o^{1/3} H^{5/3}/Q_o \sim (g'_o)^{1/3} H^{5/3}/Q_o^{2/3}$). Similarly, show that the initial dilution at trap level for the buoyancy-dominated far field (BDFF) is given by:

$$S \sim \frac{U_a^{1/3} g_o'^{2/3}}{\epsilon^{2/3} Q_o^{1/3}}$$

10.5 a) For the dense jet in Prob. 3.7, use VISJET to determine the jet trajectory, width, terminal height of rise, and dilution for $\phi_o = 30, 45, 60^o$. Assume the jet is located at 1 m from the sea bottom.

b) For $\phi_o = 60^o$, use VISJET to determine the trajectory, maximum height of rise, and dilution of the dense jet in a crossflow $U_a = 0.1$ m/s.

10.6 Wastewater from a small town next to a popular beach (frequented by wind surfers in the dry season) is discharged in the form of a single horizontal round jet in coastal water of 12 m depth. The idea is to discharge the jet at high velocity from a rock cliff face away from the beach area as much as possible. Located at 1 m from the bottom, the jet diameter is $D = 0.036$ m and the discharge flow is $Q_o = 0.02$ m^3/s (see Prob. 5.8). Based on field measurements, the average ambient conditions in the dry season are as follows:

Depth (m)	S (ppt)	T(^{o}C)	U_a (m/s)
1	30.1	21.2	0.1
3	30.1	21.2	0.08
5	30.1	21.2	0.06
7	30.2	21.1	0.05
9	30.5	20.9	0.04
11	30.6	20.7	0.03

Use VISJET to predict the trajectory of this horizontal buoyant jet in a crossflow, with a 3D trajectory. Determine the offshore penetration distance and whether the jet is well-trapped beneath the free surface. Assume effluent characteristics of S=2 ppt and T=20°C.

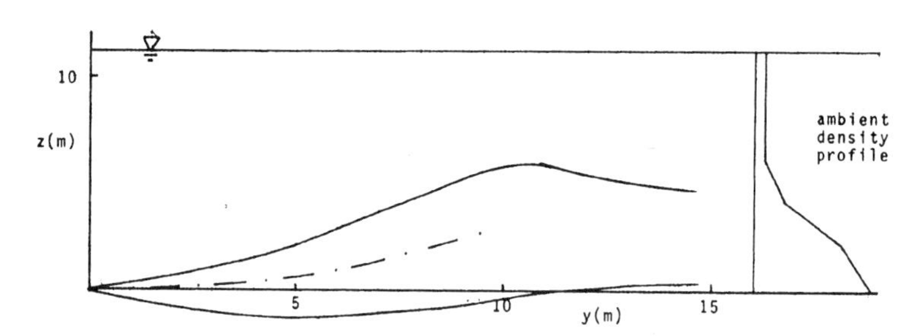

10.7 <u>*Lagrangian model formulation*</u>

The Lagrangian jet model (JETLAG) formulation in Eq. 10.7 to 10.27 (and its correspondence to the Eulerian formulation) can be best illustrated using a vertical buoyant jet in uniform stagnant ambient fluid. Using top-hat profiles, the Eulerian equations for conservation of mass and momentum are, respectively:

$$\frac{d}{dz}\int \rho w \, dA = \frac{d}{dz}(\rho A W) = \rho_a Q_e = \rho_a 2\pi B(\alpha W)$$
$$\frac{d}{dz}\int \rho w^2 dA = \frac{d}{dz}(\rho A W^2) = \Delta\rho g A$$

where B, W, $\Delta\rho = \rho_a - \rho$, and $A = \pi B^2$ are respectively the plume radius, average velocity and density deficit, and plume area respectively. Q_e is the entrainment flow per unit length along the plume axis, and α=entrainment coefficient.

Noting that $W = dz/dt$, we can cast the above equations in terms of *time*:

$$\begin{aligned}\frac{d}{dt}(\rho AW) &= \rho_a 2\pi B(\alpha W)W \\ \frac{d}{dt}(\rho AW^2) &= \Delta\rho g AW\end{aligned}$$

Suppose we multiply each of the above equations by an arbitrary (constant) time interval parameter Δt^*, we would have

$$\begin{aligned}\frac{d}{dt}[\rho A(W\Delta t^*)] &= \rho_a 2\pi B(\alpha W)(W\Delta t^*) \\ \frac{d}{dt}[\rho AW(W\Delta t^*)] &= \Delta\rho g A(W\Delta t^*)\end{aligned}$$

In the above equations, if we interpret $h = W\Delta t^*$ as a thickness of a plume element, then the equations can be discretized as:

$$\begin{aligned}\frac{\Delta M_k}{\Delta t} = \frac{\Delta}{\Delta t}(\rho_k A_k h_k) &= \rho_a \alpha W_k 2\pi B_k h_k \\ \frac{\Delta M_k W_k}{\Delta t} = \frac{\Delta}{\Delta t}(\rho_k A_k h_k W_k) &= \Delta\rho_k g A_k h_k\end{aligned}$$

where $M_k = \rho_k A_k h_k$ is the plume element mass. This is effectively our Lagrangian model using just the shear entrainment (Eq. 10.32). Note that the time step Δt is a free parameter (dictated by numerical stability and accuracy considerations). This is chosen to achieve a prescribed percentage increase in mass of plume element (typically one percent). With this framework, $h_k \propto W_k$, the volume of the plume element represents the volume flux. The initial choice of plume element thickness, $h_o \sim D$, is however immaterial and will not affect the solution of the plume sectional properties.

In this Lagrangian framework, we are tracking the changes of a plume element mass that is released from the source over a time interval (Δt^*). The consecutive plume elements are assumed to be independent and do not mix with each other. Note that there is no need for the Boussinesq approximation, $\Delta\rho/\rho \ll 1$ (see also Chapter3). For example, for the same absolute initial density difference $|\Delta\rho_o|$, small differences in the behaviour of positively or negatively buoyant jets can be simulated. Since added mass is not explicitly included in the formulation, it can be shown the bent-over jet or plume width predicted by JETLAG is related to the top-hat width B by $B' = \sqrt{1+k_n}B \approx \sqrt{2}B$ (cf Eqs. 7.22 and 8.27, and Prob. 8.5).

10.8 *Radial entrainment velocity distribution*

Consider a vertical jet or plume in stagnant fluid. The shear entrainment is the radial entrainment flow caused by the velocity gradient between the jet and the surroudings. Since the axial velocity $w_m(s = z)$ is known (Table 2.3), the entrainment flow per unit axial length into the jet/plume at any level s is known, $\frac{dQ}{ds} = 2\pi r V_r(r)$, where $V_r(r)$ is the radial inflow velocity at any radial position r. By symmetry, the entrainment flow field in the normal cross-section of the jet/plume is given by a 2D sink flow of strength $m = 2\pi r V_r(r) = 2\pi\alpha\Delta U_g b_g$.

a) By axisymmetry, show from the continuity equation the following expression for the entrainment velocity at any radius $r = r_1$:

$$\frac{d}{ds}\left(\int_o^{r_1} w(s,r)2\pi r dr\right) = 2\pi r_1 V_r(r_1)$$

b) Assuming Gaussian profile for $w(s,r)$, show that the radial entrainment velocity is given by:

$$\frac{d}{ds}\{\pi \Delta U_g b^2(1 - e^{-(r/b)^2})\} = 2\pi r V_r(r)$$

Hence show that $V_r(r)$ can be obtained from the local centerline velocity excess ΔU_g and Gaussian half-width $b = b_g$ as:

$$\frac{V_r(r)}{\alpha \Delta U_g} = \frac{(1 - e^{-(r/b)^2}) - A(r/b)^2 e^{-(r/b)^2}}{r/b}$$

where $A = \frac{1}{\alpha}\frac{db}{ds}$. The attached figure shows the radial variation of the entrainment velocity.

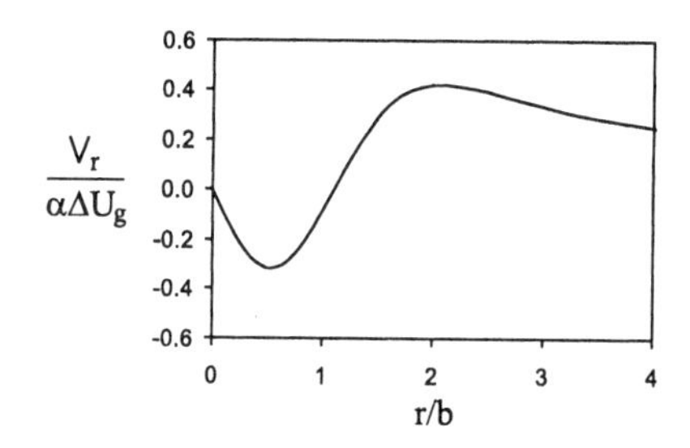

c) If $\alpha = 0.056$, and $db/ds \approx 0.109$, $A = 1.94$, show that the radial entrainment velocity has a maximum value of 0.421 at $r/b_g = 2.071$.

10.9 *External flow of a bent-over jet in crossflow*

Consider a slightly bent-over vertical buoyant jet in a weak horizontal crossflow. The jet axis is inclined at angle ψ_k to horizontal. Experiments show that the flow outside of the jet is the sum of the shear entrainment and the crossflow velocity $U_a \sin\psi_k$ (Fig. 10.11 and Fig. 10.23). In the cross-section perpendicular to the jet axis, outside of the jet, it can be shown from continuity considerations that $\frac{\partial r V_r}{\partial r} + \frac{\partial V_\theta}{\partial \theta} = 0$, where V_r, V_θ are the radial and tangential velocities. Hence a 2D stream function $\Psi(x,y) = \Psi(r,\theta)$ can be defined with

$$V_r = \frac{1}{r}\frac{\partial \Psi}{\partial \theta}; V_\theta = -\frac{\partial \Psi}{\partial r}$$

a) By super-imposing the radial entrainment velocity (Prob. 10.8) and the crossflow velocity, show that the 2D stream function is given by:

$$\Psi(r,\theta) = -\alpha \Delta U_g b f(r)(\theta - \theta_1) + U_a \sin\psi_k (y - y_1)$$

where $f(r) = (1 - e^{-(r/b)^2}) - A(r/b)^2 e^{-(r/b)^2}$, and $A = \frac{1}{\alpha}\frac{db}{ds}$. $(\theta_1 = \varphi_k, y_1 = r\sin\theta_1)$ are the angular and y-coordinate of the stagnation point.

For the limiting streamline that passes the stagnation point on the plume boundary, $r = r^* \approx 2b_g = \sqrt{2}b_k$, show that $(r \to \infty, \theta \to \pi)$

$$\frac{y_\infty - y_1}{b} = \frac{\alpha\Delta U_g}{U_a \sin\psi_k}(\pi - \theta_1)$$

where y_∞= half-width of streamtube far upstream.

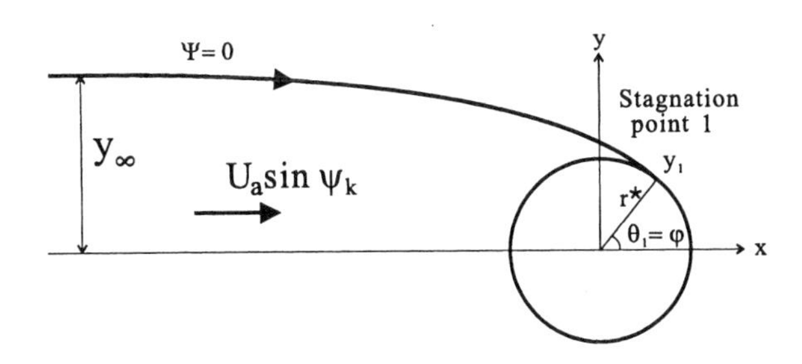

Figure 10.23. Entrainment flow pattern in cross-section of jet in crossflow (normal to jet axis)

b) Since the entrainment flow into the plume element is $q_e = 2U_a \sin\psi_k y_\infty$, show that

$$q_e = 2(\alpha\Delta U_g b(\pi - \theta_1) + U_a \sin\psi_k\ y_1)$$

where $\theta_1 = \psi_k$, and $r_1 = r^*$. Hence the relation:

$$\frac{q_e}{2\pi\alpha\Delta U_g b} = \frac{\pi - \varphi_k}{\pi} + \frac{2U_a \sin\psi_k\ r^*}{2\pi\alpha\Delta U_g b}\sin\varphi_k$$

As $2\pi\alpha\Delta U_g b_g$ and $2U_a \sin\psi_k\ r^*$ are respectively the shear and vortex entrainment (Fig. 10.11), this leads to the general entrainment computation (Eq. 10.36):

$$E = E_s\frac{\pi - \varphi_k}{\pi} + E_f \sin\varphi_k$$

where the 'separation angle' φ_k can be dynamically computed (Eq. 10.37).

Appendix A
Density of Seawater

Sea water density ρ is a function of salinity (S in ppt) and temperature (T in ^{o}C) and is often expressed in terms of the sigma unit σ_t, which is the density of water above the reference value of 1000 kg/m^3:

$$\rho\ (\mathrm{kg/m^3}) = 1000 + \sigma_t \tag{A.1}$$

The density of sea water under standard atmospheric pressure can be determined by one of the following two formulae:

1) The FOFONOFF'S (Fofonoff, 1962) formula for σ_t is calculated as follows:

$$\sigma_t = (\sigma_o + 0.1324) \times [1 - A + B(\sigma_o - 0.1324)] - \Sigma(T) \tag{A.2}$$

where the coefficients σ_o, A, B, Σ are functions of either S or T:

$$\sigma_o(S) = [(6.8 \times 10^{-6}S - 4.82 \times 10^{-4})S + .8149]S - 0.093$$

$$A(T) = .001T[(.0010843T - .09818)T + 4.7867]$$

$$B(T) = 1. \times 10^{-6}T[(.01667\ T - .8164)T + 18.03]$$

$$\Sigma(T) = (T - 3.98)^2 \frac{(T + 283.)}{503.57(T + 67.26)}$$

By the nature of the empirical equation derived from field data, Eq. A.2 does not reduce to the pure water formula (σ_t (freshwater) $= \Sigma(T)$) for $S = 0$. For determination of freshwater density the pure water density formula $\Sigma(T)$ can be used.

2) The UNESCO formula (Milero and Poisson, 1981) is based on density

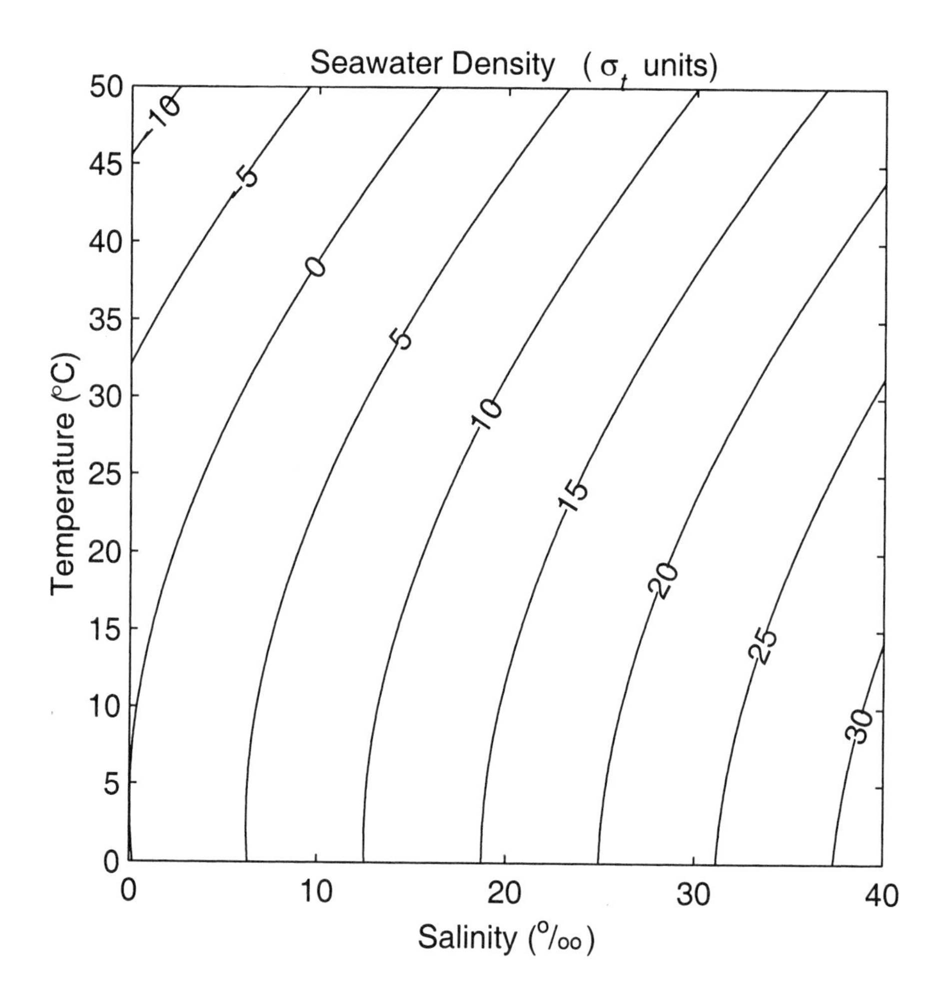

Figure A.1. Density of sea water (in σ_t units) as a function of salinity (part per thousand) and temperature (^oC) obtained using the FOFONOFF formula

measurements from 0 to 40°C and S=0.5 - 43 ppt. Density is calculated in two steps. First, the density of fresh water:

$$\rho_w = +999.842594 + 6.793952 \times 10^{-2}T - 9.095290 \times 10^{-3}T^2 + 1.001685 \times 10^{-4}T^3 - 1.120083 \times 10^{-6}T^4 + 6.536332 \times 10^{-9}T^5 \quad (A.3)$$

Then, the addition due to salinity:

$$\rho = \rho_w + AS + BS^{\frac{3}{2}} + CS^2 \quad (A.4)$$

where

$$A = +8.24493 \times 10^{-1} - 4.0899 \times 10^{-3}T + 7.6438 \times 10^{-5}T^2$$
$$-8.2467 \times 10^{-7}T^3 + 5.3875 \times 10^{-9}T^4$$

$$B = -5.72466 \times 10^{-3} + 1.0227 \times 10^{-4}T - 1.6546 \times 10^{-6}T^2$$

$$C = +4.8314 \times 10^{-4}$$

Finally the sigma unit,

$$\sigma_t = \rho - 1000. \qquad (A.5)$$

Eq. A.4, the one-atmosphere international equation of state of seawater, has been suggested for use by the UNESCO joint panel on oceanographic tables and standards (UNESCO 1981). The two formulae (FOFONOFF versus UNESCO) give results that differ by less than 0.1 σ_t in the range of T=0-40 ^oC and S=0-40 ppt, well within the accuracy required for typical engineering calculations. For density determination at near zero temperatures, the direction of buoyancy may be crucial. The table below compares the density computed by the two formulae at temperature $T = 0\,^oC$ and $4\,^oC$.

Temperature	FOFONOFF	UNESCO
σ_t at $0\,^oC$	σ_t = - 0.092594	σ_t = - 0.157406
σ_t at $4\,^oC$	σ_t = +0.038701	σ_t = - 0.025042

Fig.A.1 shows the density contours in σ_t units as a function S (in ppt) and T (in ^oC). For a given temperature, the density varies approximately linearly with salinity. On the other hand, for a given salinity, the density-temperature variation can be quite nonlinear for temperature changes greater than a few degrees. Figures A.2 shows the nonlinear density-temperature variation for salinity S = 0, 10, 20, 30 and 40 ppt. On average, the relative density above fresh water in sigma units, $(\rho - \rho_w)$, is equal to about 80% of the salinity.

The UNESCO formula shows the separate effect of salinity and temperature more clearly. Figure A.3 gives the relative density-to-salinity ratio, $(\rho - \rho_w)/S$, as a function of temperature for $S = 1$ and $S = 35$ ppt computed by the UNESCO equation. The ratio decreases slightly with increase of temperature. The linearized form of Equation A.4 is $\rho = \rho_w + AS$; for $T = 0 - 30^oC$, this linear approximation gives densities that differ from the UNESCO value by less than 0.2 σ_t units for

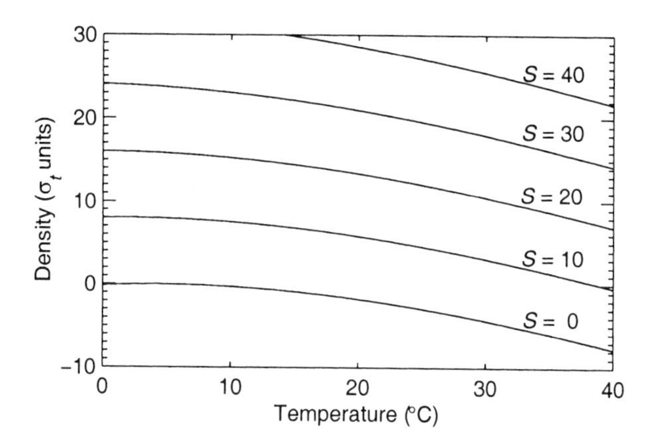

Figure A.2. σ_t as a function of temperature using the UNESCO formula.

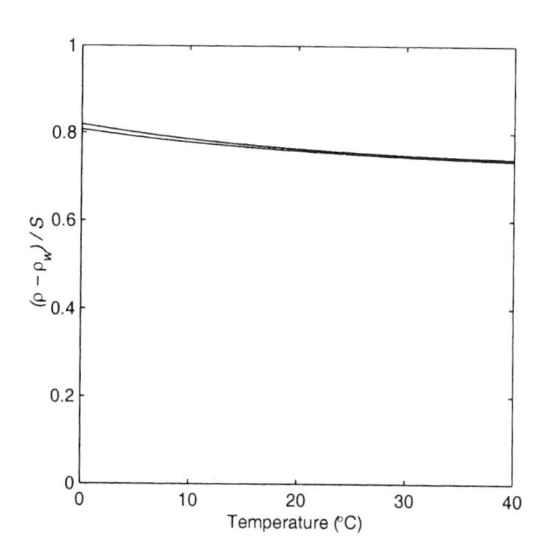

Figure A.3. Density-to-salinity ratio, $(\rho - \rho_w)/S$, as a function of temperature (T in ^{o}C), for $S = 1$ and $S = 35$ (upper and lower curve respectively)

$S = 0 - 10$ ppt. The error increases to around 0.8 σ_t when S increases to 40 ppt.

EXAMPLE A.2 <u>*Nonlinear $\rho - T$ relation*</u>
A buoyant jet is produced in the laboratory by discharging 40 $^o C$ warm water at a rate of 3 cm^3/s into a tank of water at 20$^o C$. (a) Find the buoyancy flux of the discharge at the source. (b) Estimate the buoyancy flux after the jet is diluted 100 times by entrainment of the ambient colder water into the jet.

SOLUTION: The density of the $T_o = 40^oC$ water at the source is $\rho_o = 992.2204$ kg/m^3. The density of the surrounding $T_a = 20^oC$ water in the tank is $\rho_a = 998.2063$ kg/m^3. The specific buoyancy flux at the source

$$F_o = g[\frac{\rho_a - \rho}{\rho}]Q = 981 \times \frac{998.2063 - 992.2204}{992.2204} \times 3 = \underline{17.8}\ \text{cm}^4/\text{s}^3$$

Entrainment of ambient fluid into the jet reduces the temperature difference. The reduction is determined by the heat conservation equation:

$$(T - T_a)Q = (T_o - T_a)Q_o$$

that gives the average temperature

$$T = T_a + (T_o - T_a)\frac{Q_o}{Q} = 20 + (40 - 20) \times \frac{1}{100} = 20.2\,^oC$$

and the corresponding average density $\rho = 998.1649$ kg/m^3. Therefore, the specific buoyancy flux becomes

$$F = g[\frac{\rho_a - \rho}{\rho}](100Q_o) = 981 \times \frac{998.2063 - 998.1649}{998.1649} \times (100 \times 3)$$

$$= \underline{12.2}\ \text{cm}^4/\text{s}^3$$

after the jet is diluted by 100 times. The heat flux is conserved in the buoyant jet but the buoyancy flux is reduced by as much as 31 %.

A closely related example is the problem of negative buoyancy at 4 ^{o}C water. Thermal effluent from power plants into the Great Lakes has been observed to sink in the winter as heated water in the thermal plume is diluted by the freezing (0 ^{o}C) water from the lake. Pockets of water in a thermal effluent, with a source temperature of say 12 ^{o}C, will be reduced to 4 ^{o}C in region of the plume where a dilution of 3 is attained. Such pockets of water at 4 ^{o}C temperature is heavier than its surrounding water and have been observed to sink and impact on the winter ecosystem of the lake (Griffiths, 1980).

Appendix B
Notation

α	entrainment coefficient
β	spreading rate
γ	intermittency factor, thermal expansion coefficient
Γ	mass flux
ϵ	dissipation rate of turbulent kinetic energy
θ	horizontal angle for buoyant jet or vertical angle for momentum jet
ϕ	vertical angle for buoyant jet
λ	ratio of concentration to velocity width
ν	kinematic viscosity
ρ	fluid density
σ_t	sigma unit
τ_{ij}	stress tensor
Ψ_m^*	normalized stream function
ψ_k	angle between jet axial vector and crossflow direction
A	cross-sectional area, projection area
b	half-width of jet/plume
B	half-width of top-hat profile
c, C	concentration
c_p	specific heat at constant pressure
d_o	dimension for line source, slot width
D	dimension for point source, initial jet diameter
E	kinetic energy flux
F	buoyancy flux, specific buoyancy flux
Fr	source densimetric Froude number
$\mathrm{Fr_L}$	local densimetric Froude number
Fr_p	plume Froude number

g	acceleration due to gravity
g'	reduced gravity, buoyancy force per unit mass of fluid
H	depth, heat flux of source
$\hat{\mathbf{k}}$	unit vector in the opposite direction of the gravity
k, k_n	turbulent kinetic energy, added mass coefficient
K	jet to ambient velocity ratio
l_Q	source geometry length scale
l_s	momentum length scale
L_b	crossflow buoyancy length scale
L_d	diffuser length
L_m^*	excess momentum length scale
L_{mv}	crossflow momentum length scale
M	momentum flux, specific momentum flux
N	Brunt-Väisälä frequency
p	pressure
Q	volume flux, volume
s, r	streamwise and radial co-ordinate
R	radius, gas constant
Re	Reynolds number
S	dilution, salinity
Sc_t	turbulent Schmidt number
t	time
T	temperature, time scale
u, v, w	velocity
U_i	velocity
U, W	velocity of top-hat profile
V	volume
x, y, z	Cartesian co-ordinate

Subscript/Superscripts

a	ambient
c	concentration, centerline
e	excess, entrainment, end of potential core
G, g	Gaussian profile
h	horizontal
j	jet
l	line source
m	centerline maximum
o	source
p	plume
s	shear, stagnant ambient
t	thermal, turbulent

v	vertical, velocity
$\bar{\ }$	time average of turbulent fluctuation
$\tilde{\ }$	associated with dominant eddies
*	dimensionless
$'$	turbulent fluctuation

References

Abraham, G. (1963a). "Jet diffusion in stagnant ambient fluid," Delft Hydr. Lab. Pub. No. 29.

Abraham, G. (1963b). "Horizontal jets in stagnant fluid of other density," *Proc. ASCE, J. Hydr. Div.,* Vol. 91, HY1, pp. 63-68.

Abraham, G. (1970). "The flow of round buoyant jets issuing vertically into ambient fluid flowing in a horizontal direction,", *Proc. 5th Int. Water Poll. Res. Conf.,* San Francisco, pp. III-15/1-7.

Agg, A.R. and Wakeford, A.C. (1972). "Field studies of jet dilution of sewage at sea outfalls," *J. Inst. of Public Health Engineers,* Vol.71, pp.126-149.

Albertson, M.L., Dai, Y.B., Jensen, R.A., and Rouse, H. (1950). "Diffusion of submerged jets," *Trans. ASCE,* Vol. 115, pp. 639-644.

Altai, W. and Chu, V. H. (2001). "Large Eddy Simulation by Lagrangian Block Method," *Computational Fluid Dynamics 2000*, N. Satofuka Ed., Springer-Verlag, New York, pp.467-472.

Andreopoulos, J. & Rodi, W. (1984). "Experimental investigation of jets in a crossflow." *J. Fluid Mech.*, **138**, pp.93-127.

Antonia, R.A. and Bilger R.W. (1973). "An experimental investigation of an axisymmetric jet in a co-flowing air stream," *J. Fluid Mech.,* Vol. 61, pp. 805-822.

Antonia, R.A. and Bilger, R.W. (1974). "The prediction of the axisymmetric turbulent jet issuing into a co-flowing stream," *Aeronautical Quarterly,* Vol. 251, pp. 69-80.

Anwar, H.O. (1969). "Behaviour of buoyant jet in calm fluid," *J. Hydr. Div., ASCE,* Vol. 95, HY4, Proc. paper 6688, pp. 1289-1303.

Ayoub, G.M. (1971). "Dispersion of buoyant jets in a flowing ambient fluid," *Ph.D. thesis,* Imperial College, the University of London.

Baines, W.D. (1975). "Entrainment by a plume or jet at a density interface," *J. Fluid Mech.*, Vol. 68, pp. 309-320.

Baines, W.D. (1983). "A technique for direct measurement of volume flux of a plume," *J. Fluid Mech.,* Vol. 132, pp. 247-256.

Baines, W.D. and Chu, V.H. (1996). "Jets and Plumes" in *Environmental Hydraulics,* Chapter 2, V. P. Singh and I. Hager Editors, Kulwar Academic Publishers, pp. 7-61.

Baines, W.D. and Murphy, T.M. (1986). "The temperature distribution in an enclosed space produced by a forced plume," *Proc. 8th Int. Heat Transfer Conf.,* San Francisco, Hemisphere, pp. 1507-1512.

Baines, W.D. and Turner, J.S. (1969). "Turbulent buoyant convection from a source in a confined region," *J. Fluid Mech.,* Vol. 37, pp. 51-81.

Baines, W.D., Corriveau, A.F. and Reedman, T.J. (1993). "Turbulent fountains in a closed chamber," *J. Fluid Mech.,* Vol. 255, pp. 621-646.

Baines, W.D., Turner, J.S. and Campbell, I.H. (1990). "Turbulent fountains in an open chamber," *J. Fluid Mech.,* Vol. 212, pp. 557-592.

Belatos, S. and Rajaratnam, N. (1973). "Circular turbulent jet in an opposing infinite stream," *Proc. 1st Canadian Hydrotechnical Conf., Edmonton, Canada,* pp. 220-237.

Bennett, N.J. (1981). "Initial dilution: a practical study on the Hastings long sea outfall", *Proc. Inst. of Civil Engineers, Part 1,* Vol. 70, pp. 113-122.

Bennett, N.J. (1983). "Design of sea outfalls - the lower limit concept of initial dilution", *Proc. Inst. of Civil Engineers, Part 2,* Vol. 75, pp. 113-121.

Birch, A.D., Brown, D.R., Dodson, M.R. and Thomas, J.R. (1978). "The turbulent concentration field of a methane jet," *J. Fluid Mech.,* Vol. 88, pp. 431-449.

Bradbury, L.J.S. (1965). "The structure of a self-preserving turbulent plane jet," *J. Fluid Mech.,* Vol. 23, pp. 31-64.

Briggs, G.A. (1965). "A plume rise model compared with observations," *J. Air Poll. Control Assoc.,* Vol. 15, No. 9, pp. 433-438.

Briggs, G.A. (1969). *Plume Rise,* U.S. Atomic Energy Commission Critical Review, 81 pp.

Brooks, N.H. (1972). "Dispersion in hydrologic and coastal environments," *Report No. KH-R-29, W. M. Keck Lab. of Hydraulics and Water Resources,* California Institute of Technology, Pasadena, California.

Cathers, B. and Peirson, W.L. (1991). "Verification of plume models applied to deep-water outfalls", in *Proc. of the Int. Symp. on Envir. Hydr.*, Hong Kong, Lee, J.H.W. and Cheung, Y.K. (ed.), December 1991, Vol.1, Balkema, pp.261-266.

Cederwall K. (1968). "Hydraulics of marine wastewater disposal", *Hydraulics Div., Chalmers Institute of Technology,* Sweden, Report No.42.

Cederwall, K. (1971). "Buoyant slot jets into stagnant or flowing environments," Report No. KH-R-25, W. M. Kech Laboratory, California Institute of Technology, Pasedena, California.

Chan, D. T. L., Lin J. T. and Kennedy, J. F. (1976) "Entrainment and drag forces of deflected jets," *J. Hydr. Div., ASCE,* Vol. 102, HY5, pp.615-635.

Chassaing, P., George, J., Claria, A. and Sananes (1974). "Physical characteristics of subsonic jets in a cross-stream," *J. Fluid Mech.,* Vol. 62, part 1, pp. 41-64.

Chen, G.Q. and Lee, J.H.W. (2002). "Advected turbulent line thermal driven by concentration difference", *Communications in Nonlinear Science and Numerical Simulation,* Vol. 7, pp.175-195.

Cheung, Valiant (1991). "Mixing of a round buoyant jet in a current" Ph.D. thesis, Dept of Civil and Structural Engineering, University of Hong Kong, 202 pp.

Cheung, V. and Lee, J.H.W. (1991). "Heated jets with three-dimensional trajectories", in *Proc. of the Int. Symp. on Envir. Hydr.*, Hong Kong, Lee, J.H.W. and Cheung, Y.K. (ed.), December 1991, Vol.1, Balkema, pp.71-75.

Cheung, V. and Lee, J.H.W. (1996). Discussion of "Improved prediction of bending plumes". *J. Hydr. Res.*, Vol.34, No.2, pp.260-262.

Cheung, V. and Lee, J.H.W. (1999). Discussion of "Simulation of oil spills from underwater accidents I: model development ", J. Hydr. Res., Vol.37, pp.425-429.

Chien, C.J. and Schetz, J.A. (1975). "Numerical solution of the three-dimensional Navier-Stokes equations with application to channel flows and a buoyant jet in a cross-flow." *Trans. ASME E: J. Appl. Mech.*, Vol. 42, pp. 575-579.

Chu, P.C.K. (1996). "Mixing of Turbulent Advected Line Puffs," Ph.D. thesis, Dept of Civil and Structural Engineering, University of Hong Kong, 222 pp.

Chu, P.C.K. and Lee, J.H.W. (1996). "Vorticity dynamics of an inverted jet in crossflow." *Proc. 2nd Int. Conf. on Hydrodynamics, December 1996, Hong Kong,* Vol. 2, pp. 743-748.

Chu, P.C.K., Lee, J.H.W., and Chu, V.H. (1999). "Spreading of a turbulent round jet in coflow", *J. Hydr. Engrg., ASCE,* Vol.125, No.2, pp.193-204.

Chu, V. H. (1975). "Three-dimensional forced plumes in a laminar crossflow", *Proc. 15th IAHR Congress,* Int. Assoc. for Hydr. Res., Delft, The Netherlands, Vol.4, pp.31-39.

Chu, V. H. (1977). "A line-impulse model for buoyant jet in a cross-flow," in *Heat Transfer and Turbulent Buoyant Convection,* B. D. Spalding and N. H. Afgan editors, Hemisphere Publishing Co., Washington D.C., pp. 325-337.

Chu, V.H. (1979). "L. N. Fan's data on buoyant jets in crossflow," *J. Hydr. Div., ASCE,* Vol. 105, No. HY5, pp. 612-617.

Chu, V.H. (1985). "Oblique turbulent jets in a crossflow," *J. Engrg. Mech.,* Vol. 111, No. 11, pp. 1343-1359.

Chu, V.H. (1994). "Lagrangian scalings of jets and plumes with dominant eddies," in *Recent Research Advances in the Fluid Mechanics of Turbulent Jets and Plumes,* NATO ASI Series E: Applied Sciences, Vol. 255, Davies P. A. and Valente Neves, M. J. Editors, Kluwer Academic Publishers, Dordrecht, pp. 45-72.

Chu, V.H. and Altai, W. (2001). "Simulation of shallow transverse shear flow by generalized second moment method," *J. Hydr. Res.*, Vol. 39, No.6, pp. 575-582.

Chu, V. H. and Altai, W. (2001). "Lagrangian Block Method," *Computational Fluid Dynamics 2000*, N. Satofuka Ed., Springer-Verlag, New York, pp.771-772.

Chu, V.H. and Baddour, R.E. (1984). "Turbulent stratified shear flows," *J. Fluid Mech.*, Vol. 132, pp. 353-378.

Chu, V. H. and Baines, W. D. (1989). "Entrainment by a buoyant jet between confined walls," *J. Hydr. Engrg., ASCE,* Vol. 115, No. 4, pp. 475-492.

Chu, V. H. and Baines, W. D. (1991). "Starting jets between confined walls," *Proc. of the 1st Int. Symp. on Envir. Hydr.*, Vol. 1, pp. 53-58.

Chu, V.H. and Goldberg, M.B. (1974). "Buoyant forced plumes in crossflow", *J. Hydr. Div., ASCE,* Vol.100, HY9, pp.1203-1214.

Chu, V.H. and Lee, J.H.W. (1996). "A general integral formulation of turbulent buoyant jets in crossflow", *J. Hydr. Engrg., ASCE,* Vol. 122, No.1, pp. 27-34.

Chu, V.H., Senior, C. and List, J. (1981). "Transition from a turbulent jet into a turbulent plume," *ASME Publication 81-FE-29,* Boulder, Colorado, 1981, 8 pp.

Corrsin, S. and Kistler, A.L. (1954). "The free stream boundaries of Turbulent Flows," *NACA Technical Note 3133.*

Csanady, G.T. (1965). "The buoyant motion within a hot gas plume in a horizontal wind". *J. Fluid Mech.,* Vol. 22, pp. 225-239.

Davidson, M.J., Gaskin, S., and Wood, I.R. (2002). "A study of a buoyant axisymmetric jet in a small coflow". *J. Hydr. Res.*, Vol. 40, No.4, pp. 477-489.

Davidson, M.J., Knudsen, M. and Wood, I.R. (1991). "The behaviour of a single, horizontally discharged, buoyant flow in a non-turbulent coflowing ambient fluid", *J. Hydr. Res.,* Vol.29, No.4, pp.545-566.

Davidson, M.J. and Pun, K.L. (1999). "Weakly advected jets in crossflow", *J. Hydr. Res., ASCE,* Vol. 125, No.1, pp. 47-58.

Dehghani, M.R. (1996). "Line thermals and buoyant jets in cross flow," Ph. D. thesis, Department of Civil Engineering, McGill Universty, 181 pp.

Demuren, A.O. (1983). "Numerical calculations of steady three-dimensional turbulent jets in cross flow." *Comp. Meth. Appl. Mech. & Engng.*, Vol. 37, No.3, pp. 309-328.

Dimotakis, P.E., Miake-Lye, R.C., and Papantoniou, D. A. (1983). "Structure and dynamics of round turbulent jets," *Phys. Fluid,* Vol. 26, no. 11, pp. 3185-3192.

Escudier, M.P. and Maxworthy, T. (1973). "On the motion of turbulent thermals," *J. Fluid Mech.,* Vol. 61, pp. 541-552.

Everitt, K.W. and Robins, A.G. (1978). "The development and structure of turbulent plane jets," *J. Fluid Mech.,* Vol. 88, pp. 563-583.

Fage, A. and Falkner, V.M. (1935). " Note on experiments on the temperature and velocity in the wake of a heated cylindrical obstacle," *Proc. Roy. Soc. London,* A135, 702-705.

Fan, L.N. (1967). "Turbulent buoyant jets into stratified or flowing ambient fluids," *Report No. KH-R-15,* W. M. Keck Laboratory of Hydraulics and Water Resources, California Institute of Technology, Pasadena, Calif.

Fan, L.N. and Brooks, N.H. (1969). "Numerical solutions of turbulent buoyant jet problems," Report No. KH-R-18, W. M. Keck Laboratory, California Institute of Technology, Pasadena.

Fischer, H.B., *et al.*(1979). *Mixing in Inland and Coastal Waters,* Academic Press, San Diego, California.

Fofonoff, N.P. (1962). "Physical properties of sea-water", in *The Sea* (Ed. M.N. Hill), Vol.1, Interscience, New York, pp. 3-30.

Fox, D.G. (1970). "Forced plume in a stably stratified fluid", *J. Geophys. Res.*, Vol. 75, pp.6818-6835.

Fric, T.F. and Roshko, A. (1994). "Vortical structure in the wake of a transverse jet", *J. Fluid Mech.*, Vol. 279, pp. 1-47.

Frick, W.E. (1984). "Non-empirical closure of the plume equations," *Atmos. Envir.*, Vol. 18, No. 4, pp. 653-662.

Gartshore, I. S. (1966). "An experimental examination of the large-eddy equilibrium hypothesis," *J. Fluid Mech.,* Vol. 24, pp. 89-98.

Gartshore, I. S. (1967). "Two-dimensional turbulent wakes," *J. Fluid Mech.,* Vol. 30, pp. 547-560.

Gaskin, S.J. (1995). "Single buoyant jet in a crossflow and the advected line thermal", Ph.D. thesis, University of Canterbury, Christchurch, New Zealand.

Gaskin, S.J., Papps, D.A. and Wood, I.R. (1995). "The axisymmetric equations for a buoyant jet in a crossflow", *Proc.12th Australasian Fluid Mech. Conf.*, Sydney, Dec.1995.

Gaskin, S.J. and Wood, I.R. (2001). "Flow structure and mixing dynamics of the advected line thermal", *J. Hydr. Res.*, Vol. 39, No.5, pp. 459-468.

George, W. K., Alphert, R. L., and Tamanini, F. (1977). "Turbulent measurements in an axisymmetrical buoyant plume," *Int. J. Heat and Mass Transfer,* Vol. 20, pp. 1145-1154.

Germeles, A. E. (1975). "Forced plumes and mixing of liquids in tanks," *J. Fluid Mech.,* Vol. 71, pp. 601-625.

Glezer, A. and Coles D. (1990). "An experimental study of a turbulent vortex ring," *J. Fluid Mech.,* Vol. 211, pp. 243-283.

Gordon, A.D. and Fagan, P.W. (1991). "Ocean outfall performance monitoring", in *Proc. of the Int. Symp. on Envir. Hydr.*, Hong Kong, Lee, J.H.W. and Cheung, Y.K. (ed.), December 1991, Vol.1, Balkema, pp.243-248.

Griffiths, J.S. (1980). "Potential effects of unstable thermal discharges on incubation of round whitefish eggs," Ont. Hydro Res. Div. Rep. No. 80-140-K.

Gutmark, E. and Wygnanski, I. (1976). "The planar turbulent jet," *J. Fluid Mech.*, Vol. 73, pp. 465-495.

Harleman, D.R.F. (1982). "Hydrothermal analysis of lakes and reservoirs", *J. Hydr. Div., ASCE*, Vol.108, HY3, pp.302-325.

Heskestad, G. (1984). "Engineering relations for fire plumes", *Fire Safety Journal*, Vol.7, pp. 25-32.

Hewett, T.A., Fay, J.A., and Hoult, D.P. (1971). "Laboratory experiments of smoke-stack plumes in a stable atmosphere," *Atmos. Envir.*, Vol.5, pp.767-789.

Hinze, J. O. (1959). *Turbulence - an Introduction to its Mechanism and Theory*, McGraw-Hill, 586 pp.

Hinze, J.O. and van der Hegge Zijnen, B.G. (1949). "Transfer of heat and matter in turbulent mixing zone of an axially symmetric jet," *Appl. Sci. Res.*, Vol., A1, pp. 435-461.

Hirst, E. (1972). "Buoyant jets with three-dimensional trajectories", *J. Hydr. Div., ASCE*, Vol.98, HY11, pp.1999-2014.

Hodgson, J.E. and Rajaratnam, N. (1992). "An experimental study of jet dilution in crossflows", *Canadian J. Civil Engrg.*, Vol.19, No.5, pp.733-743.

Hodgson, J.E. Moaward, A.K., and Rajaratnam, N. (1999). "Concentration field of multiple circular turbulent jets", *J. Hydr. Res.*, Vol.37, No.2, pp. 249-256.

Horton, P.R., Lee, J.H.W. and Wilson J.R. (1997). "Near-field JETLAG modelling of the Northwest New Territories Sewage Outfall, Urmston Road, Hong Kong", *Proc. 13th Australasian Coastal and Ocean Engrg. Conf.* (Pacific Coasts and Ports '97), Christchurch, New Zealand, Sept. 97, Vol.2, pp. 561-566.

Hoult, D.P., Fay, J.A., and Forney, L.J. (1969). "A theory of plume rise compared with field observations", *J. Air Poll. Control Assoc.*, Vol. 19, pp. 585-590.

Hoult, D.P. and Weil, J.C. (1972). "Turbulent plume in a laminar crossflow", *Atmos. Envir.*, Vol. 6. pp. 513-531.

Howard, R.A., Charlton, J.A., and Webb, M.B. (1991). "The design of the Stage I outfall for the Hong Kong Strategic Sewage Disposal Scheme", in *Proc. of the Int. Symp. on Envir. Hydr.*, Hong Kong, Lee, J.H.W. and Cheung, Y.K. (ed.), December 1991, Vol.1, Balkema, pp.335-339.

Hydraulics Research Station. (1977). "Horizontal outfalls in flowing water,", *Hydraulics Research Station Report EX763*, Wallingford, United Kingdom.

Imberger, J. and Patterson, J. C. (1990). "Physical Limnology," in *Advances in Applied Mechanics*, edited by J. W. Hutchinson and T. Y. Wu, Academic Press, Vol. 27, pp. 303-475.

Iribarne, J.V. and Godson, W.L. (1981). *Atmospheric Thermodynamics,* R. Reidel Publishing Co., Dordrecht, 259 pp.

Jenkins, P. E. and Goldschmidt, V. W. (1973). "Mean temperature and velocity in a plane turbulent jet," *J. Fluid Engrg., ASME,* Vol. 95, pp. 581-584.

Jirka, G.H., (1982). "Turbulent buoyant jets in shallow fluid layers", In *Turbulent buoyant jets and plumes*, W. Rodi (ed.), *HMT, The Science and Applications of Heat and Mass Transfer*, Vol. 6, Pergamon, pp. 69-119.

Jirka, G.H., Abraham, G. and Harleman, D.R.F. (1975). "An assessment of techniques for hydrothermal prediction", M.I.T. Parsons Lab. for Water Resources & Hydrodyn., Report 203.

Jirka, G.H. and Doneker, R.L. (1991). "Hydrodynamic classification of submerged single-port discharges", *J. Hydr. Engrg., ASCE*, Vol. 117, No.9, pp. 1095-1112.

Jirka, G.H. and Harleman, D.R.F. (1973). "The mechanics of submerged multiport diffusers for buoyant discharges in shallow water", Tech.Rep. 169, R. Parson Lab. for Water Resources and Hydrodynamics, MIT.

Jirka, G.H. and Harleman, D.R.F. (1979). "Stability and mixing of a vertical plane buoyant jet in confined depth", *J. Fluid Mech.*, Vol. 94 (2), pp. 275-304.

Kamotani, Y. & Greber, I. (1972). "Experiments on a turbulent jet in a cross-flow." *AIAA Journal*, Vol. 10, pp. 1425-1429.

Keffer, J.F. & Baines, W.D. (1963). "The round turbulent jet in a cross wind." *J. Fluid Mech.*, Vol. 15, pp. 481-496.

Killworth, P.D. and Turner, J.S. (1982). "Plumes with time varying buoyancy in a confined region," *Gephys. Astrophys. Fluid Dynamics,* Vol. 20, pp. 265-291.

Knudsen, M. (1988). "Buoyant horizontal jets in an ambient flow," Ph. D. thesis, University of Canterbury, Christchurch, New Zealand.

Knudsen, M. and Wood, I.R. (1985). "An axisymmetric jet in a moving fluid," *Proc. 9th Australasian Fluid Mechanics Conf.,* Auckland, New Zealand, pp. 484-487.

Koh, R.C.Y. and Brooks, N.H. (1975). "Fluid mechanics of waste disposal in the ocean", *Ann. Rev. Fluid Mech.*, Vol. 7, pp. 187-211.

Kotsovinos, N.E. (1975). "A study of entrainment and turbulence in a plane buoyant jet," *Report No. KH-R-32,* W. M. Keck Laboratory of Hydraulics and Water Resources, California Institute of Technology, Pasadena, Calif.

Kotsovinos, N.E. (1976). "A note on the spreading rate and virtual origin of a plane turbulent jet," *J. Fluid Mech.,* Vol. 77, pp. 305-311.

Kotsovinos, N.E. (1977). "Plane turbulent buoyant jet. Part 2. Turbulence structure," *J. Fluid Mech.,* Vol. 81, pp. 55-62.

Kotsovinos, N.E. and List, E.J. (1977). "Plane turbulent buoyant jet. Part I. Integral properties.," *J. Fluid Mech.,* Vol. 81, pp. 25-44.

Lam, K.M. and Chan, H.C. (1997). "Round jet in ambient counterflowing stream", *J. Hydr. Engrg.,* Vol.123, pp. 895-903.
(Discussions: Vol.125, 1999, pp. 428-432)

Lamb, H. (1932). *Hydrodynamics*, Cambridge University Press, U.K.

Launder, B. E. and Spalding, D.B. (1974). "The numerical computation of turbulent flows," *Comp. Meth. in Appl. Mech. Engrg.,* Vol.3, pp.269-289.

Launder, B. E. (1978). "Heat and Mass Transport," in Chapter 6, *Turbulence,* P. Bradshaw editor, Springer-Verlag Series on Topics in Applied Physics, Vol. 12, pp. 228-331.

Lawrence, G.A., Burke, J. M., Murphy, T.P. and Prepas, E.E. (1997). "Exchange of water and oxygen between the two basin of Amisk Lake," *Can. J. Fish. Aquat. Sci.*, Vol. 54, pp. 2121-2132.

Lee, J.H.W. (1989). "Note on Ayoub's data of horizontal round buoyant jet in current," *J. Hydr. Engrg., ASCE,* Vol. 115, No. 7, pp. 969-975.

Lee, J.H.W. (1993). Discussion of "An experimental study of jet dilution in crossflows", *Canadian J. Civil Engrg.*, Vol. 20, pp.1073-1076.

Lee, J.H.W. and Jirka, G.H. (1981). "Vertical round buoyant jet in shallow water", *J. Hydr. Div., ASCE,* Vol. 107, HY12, pp.1651-1675.

Lee, J.H.W. and Cheung, V.W.L. (1986). "An inclined plane buoyant jet in stratified fluid", *J. Hydr. Engrg., ASCE,* Vol. 112, pp.580-589.

Lee, J.H.W. and Neville-Jones, P. (1987a). "Initial dilution of horizontal jet in cross-flow," *J. Hydr. Engrg.,* Vol. 113, No. 5, pp. 615-629.

Lee, J.H.W. and Neville-Jones, P. (1987b). "Sea outfall design - prediction of initial dilution," *Proc. Inst. of Civil Engineers,* Vol. 82, No. 1, pp. 981-994.

Lee, J.H.W., Cheung, Y.K., and Cheung, V. (1987c). "Mathematical modelling of a round buoyant jet in a current: an assessment", *Proc. Int. Symp. on River Poll. Control and Management,* Shanghai, China, Oct. 1987, pp. S62-S82.

Lee, J.H.W. and Cheung, V. (1990). "Generalized Lagrangian model for buoyant jets in current," *J. Envir. Engrg.,* ASCE, Vol. 116, No. 6, pp. 1085-1106.

Lee, J.H.W. and Cheung, V. (1991). "Mixing of buoyancy-dominated jets in a weak current," *Proc. Inst. of Civil Engineers,* Vol. 91, no. 2, pp. 113-129.

Lee, J.H.W. and Wong, C.F. (1993). "Experiments on advected line momentum puffs", *Proc. 25th IAHR Congress,* Tokyo, Vol.5, pp. 250-257.

Lee, J.H.W. and Chu, P.C.K. (1995). "Application of video image processing in the study of environmental flows", *Proc. 10th ASCE Engrg. Mech. Conf.,* Boulder, Colorado, pp.1014-1017.

Lee, J.H.W., Rodi, W. and Wong, C.F. (1996). "Turbulent line momentum puffs", *J. Engrg. Mech.,* ASCE, Vol.122, No.1, pp. 19-29.

Lee, J.H.W. and Chen, G.Q. (1998). "A numerical study of turbulent line puffs via the renormalization group (RNG) k-ϵ model,", *Int. J. Numer. Meth. Fluids,* Vol. 26, pp.217-234.

Lee, J.H.W., Li, L., and Cheung, V. (1999). "A semi-analytical self-similar solution of a bent-over jet in crossflow", *J. Engrg. Mech., ASCE,* Vol.125, pp.733-746.

Lee, J.H.W., Cheung, V., Wang, W.P., and Cheung, S.K.B. (2000). "Lagrangian modeling and visualization of rosette outfall plumes", *Proc. Hydroinformatics 2000, University of Iowa, July 23-27, 2000,* (CDROM)

Lee, J.H.W., Chen, G.Q., and Kuang, C.P. (2002a). "Mixing of a turbulent jet in crossflow - the advected line puff", in *Environmental Fluid Mechanics: theories and applications* (Ed. Hayley H. Shen *et al*), ASCE, Chapter 3, pp. 47-83.

Lee, J.H.W., Kuang, C.P., and Chen, G.Q. (2002b). "The structure of a turbulent jet in crossflow - effect of jet-to-crossflow velocity", *China Ocean Engineering,* Vol. 16, pp. 1-20.

Lee, J.H.W. and Chen, G.Q. (2002c). "Numerical experiment on two-dimensional line thermal", *China Ocean Engineering,* Vol. 16, pp. 453-467.

Lilly, D.K. (1964). "Numerical solution for the shape preserving 2-D thermal convection element", *J. Atm. Sci.* Vol.21, pp. 83-98.

Linden, P.F. (1999). "The fluid mechanics of natural ventilation," *Annu. Rev. Fluid Mech.*, Vol. 31, pp. 201-238.

List, E.J. (1979). "Turbulent jets and plumes," in *Mixing in Inland and Coastal Waters,* Academic Press, Chapter 9, pp. 315-389.

List, E.J. (1982). "Turbulent jets and plumes," *Ann. Rev. Fluid Mech.* Vol.14, pp. 189-212.

List, E.J. and Imberger, J. (1973). "Turbulent entrainment in buoyant jets and plumes," *J. Hydr. Div., ASCE,* Vol.99, HY9, pp.1461-1474.

Maczynski, J.F.J. (1962). "A round jet in an ambient co-axial stream", *J. Fluid Mech.,* Vol.13, pp. 597-608.

McClimans, T. and Eidnes, G. (2000). "Forcing nutrients to the upper layer of a fjord by a buoyant plume", Proc. 5th Int. Symp. on Stratified Flows, University of British Columbia, (ed. G. Lawrence), July 2000, Vol.1, pp.199-204.

McQuivey, R.S., Keefer, T.N. and Shirazi M.A. (1971). *Basic Data Report on the Turbulent Spread of Heat and Matter,* U. S. Department of the Interior Geological Survey and U. S. Environmental Protection Agency, Fort Collins, Colorado.

Mih, W.C. and Hoopes, J.A. (1972). "Mean and turbulent velocities for plane jet", *J. Hydr. Div., ASCE,* Vol. 98, HY7, pp.1275-1294.

Miller, D. and Comings, E. (1957). "Static pressure distribution in the free turbulent jet", *J. Fluid Mech.*, Vol. 3, pp.1-16.

Millero, F.J. (1978). "Freezing point of sea water", In "Eighth Report of the Joint Panel on Oceanographic Tables and Standards," *UNESCO Tech. Pap. Mar. Sci.*, No. 28, Annex 6, UNESCO, Paris.

Millero, F.J. and Poisson, A. (1981). "International one-atmosphere equation of state for seawater," *Deep Sea Res.*, Vol. 28A, pp. 625-629.

Morton, B.R., Taylor, G.I., and Turner, J.S. (1956). "Turbulent gravitational convection from maintained and instantaneous sources," *Proc. Royal Soc., London,* Vol. A234, pp. 1-23.

Moussa, Z.M., Trischka, J.W. and Eskinazi, S. (1977). "The near field in the mixing of a round jet with a cross-stream.", *J. Fluid Mech.*, Vol. 80, pp. 49-80.

Muellenhoff, W.P., Soldate, A.M., Baumgartner, D.J., Schuldt, M.D., Davis, L.R., and Frick, W.E. (1985). Initial mixing characteristics of municipal ocean discharges. *Report EPA-600/3-85-073*, United States Environmental Protection Agency, Newport, Oregon.

Newman, B.G. (1967). "Turbulent jets and wakes in a pressure gradient," *Fluid Mechanics of Internal Flow,* (ed. G. Sovran), Elsevier, pp.176-209.

Nickels, T.B. and Perry, A.E. (1996). "An experimental and theoretical study of the turbulent coflowing jet," *J. Fluid Mech.,* Vol. 309, pp. 157-182.

Papantoniou, D. A. (1986). "Observations in turbulent buoyant jets by use of Laser-induced fluorescence," *Ph.D. thesis,* California Institute of Technology, Pasadena, California.

Papanicolaou, P. and List, E.J. (1987). "Statistical and spectral properties of tracer concentration in round buoyant jets," *Int. J. Heat Mass Transfer,* Vol. 30, No. 10, pp. 2059-2071.

Papanicolaou, P. and List, E.J. (1988). "Investigations of round vertical turbulent buoyant jets," *J. Fluid Mech.,* Vol. 195, pp. 341-391.

Papantoniou, D. and List, E.J. (1989). "Large-scale structure in the far field of buoyant jets," *J. Fluid Mech.,* Vol. 209, pp. 151-190.

Patankar, S.V. (1980). *Numerical Heat Transfer and Fluid Flow*, Hemisphere Publishing Co., New York.

Patankar, S.V., Basu, D.K. and Alpay, S.A. (1977). "Prediction of the three-dimensional velocity field of a deflected turbulent jet.", *Trans. ASME I: J. Fluids Engrg.*, Vol. 99, pp. 758-762.

Patel, R.P. (1971). "Turbulent jets and wall jets in uniform streaming flow," *Aeronautical Quarterly,* Vol. XXII, pp. 311-326.

Phillips, O.M. (1966). *The Dynamics of the Upper Oceans*, Cambridge University Press, 261 pp.

Pratte, B.D. and Baines, W.D. (1967). "Profiles of the round turbulent jet in a cross flow," *J. Hydr. Div., ASCE,* Vol. 92, HY6, pp. 53-64; see also 1968 correction, Vol. 93, No. HY3, pp. 815-816.

Pratte, B.D. and Keffer, J.F. (1972). "The swirling turbulent jet" *J. Basic Engrg., Trans. ASME,* Vol.94, No.4, pp. 739-748.

Proni, J.R., Huang, H., and Dammann, W.P. (1994). "Initial dilution of Southeast Florida ocean outfalls", *J. Hydr. Engrg., ASCE,* Vol. 120, No. 12, pp. 1409-1425.

Pun, K.L. and Davidson, M.J. (1999). "On the behaviour of advected plumes and thermals", *J. Hydr. Res.*, Vol. 37, No.4, pp. 519-540.

Rajaratnam, N. (1976). *Turbulent jets.*, Elsevier, Amsterdam, Netherlands.

Richards, J.M. (1961). "Experiments on the penetration of an interface by buoyant thermals," *J. Fluid Mech.,* Vol. 11, pp. 369-384.

Richards, J.M. (1963). "Experiments on the motion of isolated cylindrical thermals through unstratified surroundings," *Int. J. Air Water Poll.,* Vol. 7, pp. 17-34.

Richards, J.M. (1965). "Puff motion in unstratified surroundings," *J. Fluid Mech.,* Vol. 21, pp. 97-106.

Ricou, F.P. and Spalding, D.B. (1961). "Measurements of entrainment by axisymmetrical turbulent jets,", *J. Fluid Mech.,* Vol. 11, pp. 21-32.

Roberts, P.J.W. (1979). "Line plume and ocean outfall dispersion", *J. Hydr. Div., ASCE,* Vol.105, HY4, pp.313-331 (discussion by G.H.Jirka, HY12, pp.1573-1575).

Roberts, P.J.W. (1980). "Ocean outfall dilution: effect of currents", *J. Hydr. Div., ASCE,* Vol.106, HY5, pp.769-782.

Roberts, P.J.W., Snyder, W.H. and Baumgartner, D.J. (1989). "Ocean outfalls. I: submerged wastefield formation", *J. Hydr. Engrg., ASCE,* Vol.115, No. 1, pp.1-25

Rodi, W. (1980). *Turbulence models and their application in hydraulics,* Int. Assoc. for Res., Delft, Netherlands.

Roshko, A. (1976). "Structure of turbulent shear flows: a new look," *AIAA Journal,* Vol. 14, pp. 1349-1357; Vol. 15, 1977, p. 768.

Rouse, H., Baines, W.D., and Humphreys, H.W. (1953). "Free convection over parallel sources of heat," *Proc. of the Physical Soc. B,* Vol. 56, pp. 393-399.

Rouse, H., Yih, C.-S., and Humphreys, H.W. (1952). "Gravity convection from a boundary source," *Tellus,* Vol. 4, pp.201-210.

Schatzmann, M. (1979). "An integral model of plume rise," *Atmos. Envir.,* Vol. 13, pp. 721-731.

Schatzmann, M. (1981). "Mathematical modeling of submerged discharges into coastal waters," *Proc. 19th IAHR Congress, New Delhi,* Vol. 3, pp. 239-246.

Schladow, S.G. and Fisher, I.H. (1995). "The physical response of temperate lakes to artificial destratification," *Limnol. Oceanogr.,* Vol. 40, pp. 359-373.

Scorer, R. S. (1957). "Experiments on convection of isolated mass of buoyant fluid," *J. Fluid Mech.,* Vol. 2, pp. 583-594.

Scorer, R.S. (1959). "The behaviour of chimney plumes," *Int. J. Air Poll.,* Vol.1, pp.198-220.

Scorer, R.S. (1978). *Environmental Aerodynamics.* Ellis Horwood.

Sharp, J.J. and Vyas, B.D. (1977). "The buoyant wall jet", *Proc. Institution of Civil Engineers,* Part 2, Vol.63, pp. 593-611.

Shaughnessy, E.J. and Morton, J.B. (1977). "Laser light-scattering measurements of particle concentration in a turbulent jet," *J. Fluid Mech.,* Vol. 80, pp. 120-148.

Shirazi, M.A., McQuivey, R.S., and Keefer, T.N. (1974). "Heated water jet in coflowing turbulent stream", *J. Hydr. Div., ASCE,* Vol. 100, HY7, pp. 919-934.

Sobey, R.J., Johnston, A.J. and Keane, R.D. (1988). "Horizontal round buoyant jet in shallow water", *J. Hydr. Engrg., ASCE,* Vol. 114, No. 8, pp. 910-929.

Sykes, R.I., Lewellen, W.S. and Parker, S.F. (1986). "On the vorticity dynamics of a turbulent jet in a crossflow." *J. Fluid Mech.*, Vol. 168, pp. 393-413.

Syrbin, A. N. and Lyakhovskiy, D. N. (1936). "Aerodynamics of an elementary flame," Soobshch, Tsentr. Nauchn-Issled, Kotloturoinnyi Inst.

Taylor, G.I. (1932). "The transport of vorticity and heat through fluids in turbulent motion," *Proc. Roy. Soc. London,* Vol. A135, pp. 685-702.

Thomas, R.M. (1973). "Conditional sampling and other measurements in a plane turbulent wake," *J. Fluid Mech.,* Vol. 57, pp. 549-581.

Townsend, A.A. (1976). *The Structure of Turbulent Shear Flow,* second edition, Cambridge University Press, London.

Tsang, G. (1971). "Laboratory study of line thermals." *Atmos. Envir.*, Vol. 5, pp. 445-471.

Turner, J.S. (1960). "A comparison between buoyant vortex rings and vortex pairs," *J. Fluid Mech.* Vol. 7, pp. 419-432.

Turner, J.S. (1966). "Jets and plumes with negative and reversing buoyancy," *J. Fluid Mech.,* Vol. 26, pp. 779-792.

Turner, J.S. (1973). *Buoyancy Effects in Fluids,* Cambridge University Press, 367 pp.

Turner, J.S. (1986). "Turbulent entrainment: the development of the entrainment assumption, and its application to geophysical flows," *J. Fluid Mech,* Vol. 173, pp. 431-471.

UNESCO (1981). Background papers and supporting data on the International Equation of State of Seawater 1980. *UNESCO Tech. Pap. Mar. Sci.*, No.38, UNESCO, Paris.

Ven der Hegge Zijnen, B.G. (1957). "Measurements of the distribution of heat and matter in a plane turbulent jet of air," *Appl. Sci. Res.,* Vol. 7A, pp. 277-292.

Wallace, R.B. and Wright, S.J. (1984). "Spreading layer of two-dimensional buoyant jet," *J. Hydr. Engrg.*, Vol. 110, No. 6, pp. 813-828.

Wang, H.J. and Davidson, M.J. (2001). "A profile tracking system for investigating the behaviour of discharges in moving environments", *Experiments in Fluids,* Vol.31, No.5, pp. 533-541.

Wilkinson, D.L. (1988). "Avoidance of seawater intrusion into ports of ocean outfalls", *J. Hydr. Engrg., ASCE,* Vol. 114, No. 2, pp. 218-228.

Winiarski, L.D. and Frick, W.E. (1976). "Cooling tower plume model", *Report EPA-600/3-76-100,* U.S. Envir. Protection Agency, Corvallis, Oregon.

Wong, C.F. (1991). "Advected line thermals and puffs," M. Phil. thesis, Department of Civil and Structural Engineering, University of Hong Kong, Hong Kong.

Wong C.F. and Lee, J.H.W. (1991). "Experiments on advected line thermals", *Proc. Int. Symp. on Envir. Hydr.*, Hong Kong, Vol. 1, pp. 153-157.

Wong, D.R. and Wright, S.J. (1988). "Submerged turbulent jets in stagnant linearly stratified fluids," *J. Hydr. Res.,* Vol. 26, No. 1, pp. 199-223.

Wood, I.R. (1993). "Asymptotic solutions and behaviour of outfall plumes," *J. Hydr. Engrg., ASCE,* Vol. 119, pp. 555-580.

Wood, I.R., Bell, R.G. and Wilkinson, D.L. (1993). *Ocean Disposal of Wastewater,* World Scientific, Singapore.

Worster, M.G. and Leith, A.M. (1985). "Laminar free convection in confined regions," *J. Fluid Mech.,* Vol. 156, pp. 301-319.

Wright, S.J. (1977a). "Mean behavior of buoyant jet in a crossflow." *J. Hydr. Div., ASCE,* Vol.103, HY5, pp. 499-513.

Wright, S.J. (1977b). "Effects of ambient crossflow and density stratification on the characteristic behaviour of round turbulent buoyant jets," *Report No. KH-R-36, W. M. Keck Lab. of Hydraulics and Water Resources,* California Institute of Technology, Pasadena, California.

Wright, S.J. and Wallace, R.B. (1979). "Two-dimensional buoyant jets in stratified fluid," *J. Hydr. Div., ASCE,* Vol. 105, HY11, pp. 1393-1406.

Wright, S.J., Wong, D.R., Zimmerman, K.E., and Wallace, R.B. (1982). "Outfall diffuser behavior in stratified ambient fluid," *J. Hydr. Div.*, ASCE, Vol. 108, HY4, pp. 483-501.

Wright, S.J. (1984). "Buoyant jets in density-stratified crossflow," *J. Hydr. Engrg., ASCE,* Vol. 110, No. 5, pp. 643-656.

Wright, S.J. and Wallace, R.B. (1991). "Surface dilution of round submerged buoyant jets," *J. Hydr. Res.*, Vol. 29, No.1, pp. 67-89.

Wright, S.J. (1994). "The effect of ambient turbulence on jet mixing," in *Recent Research Advances in the Fluid Mechanics of Turbulent Jets and Plumes, NATO*

ASI Series E: Applied Sciences, Vol. 255, (Davies P.A. and Valente Neves, M.J., eds.), Kluwer Academic Publishers, Dordrecht, pp. 13-27.

Wygnanski, I.J. and Fiedler, H. (1969). "Some measurement in the self-preserving jet," *J. Fluid Mech.,* Vol. 38, No.3, pp. 577-612.

Wygnanski, I.J. and Fiedler, H. (1970). "The two-dimensional mixing region," *J. Fluid Mech.,* Vol. 41, pp. 327-361.

Yih, C.S. (1981). "Similarity solutions for turbulent jets and plumes," *ASCE J. Engrg. Mech.* Vol.107, pp. 455-478.

Yoda, M., Hesselink, L., and Mungal, M.G. (1992). "The evolution and nature of large-scale structures in the turbulent jet," *Phy. Fluids A,* Vol. 4, pp. 803-811.

Zhang, H. and Baddour, R. E. (1997). "Maximum vertical penetration of plane turbulent negatively buoyant jets," *J. Engrg. Mech.*, Vol. 123, No. 10, pp. 973-977.

Zhang, H. and Baddour, R.E. (1998). "Maximum penetration of vertical round dense jets at small and large Froude number," *J. Hydr. Engrg.*, Vol. 124, No. 5, pp. 550-552.

Zic, K. and Stefan, H.G. (1988). "Lake aerator effect on temperature stratification analyzed by MINLAKE model," *Lake and Reservoir Management, Vol. 4, pp. 73-83.*

Index